Protein Reviews

Series Editor:

M. Zouhair Atassi
Baylor College of Medicine
Houston, Texas

For other titles published in this series, go to
www.springer.com/series/6876

Fenyong Liu · Sidney Altman
Editors

Ribonuclease P

Springer

Editors
Fenyong Liu
University of California
Berkeley, CA
USA
liu_fy@uclink4.berkeley.edu

Sidney Altman
Yale University
New Haven, CT
USA
sidney.altman@yale.edu

ISBN 978-1-4419-1141-4 e-ISBN 978-1-4419-1142-1
DOI 10.1007/978-1-4419-1142-1
Springer New York Dordrecht Heidelberg London

Library of Congress Control Number: 2009933259

Printed on acid-free paper

Springer is part of Springer Science+Business Media (www.springer.com)

Preface

The Discovery of Ribonucleasc P and Enzymatic Activity of Its RNA Subunit

Sydney Brenner and Francis H. C. Crick had a specific project in mind when they offered Sidney Altman a position in their group in 1969 to conduct postdoctoral research at the Medical Research Council Laboratory of Molecular Biology (LMB) in Cambridge, England. At the time, an intense international competition was ongoing in as many as a dozen labs to determine the three-dimensional structure of tRNA. At the LMB, Aaron Klug was attacking the structure by crystallographic analysis with Brian F. C. Clark providing large amounts of purified phenylalanine tRNA. (Eventually, Aaron announced his empirically determined 3-D structure of yeast phenylalanine tRNA, a structure that is generally common to tRNAs, due in part to several conserved, novel three-way nucleotide interactions.) Concurrently, Michael Levitt, a Ph. D. student of Francis, was visually scrutinizing the cloverleaf secondary structure of the 14 tRNA sequences known at the time. Levitt was searching for nucleotide covariation in different parts of the molecules that were conserved in the 14 sequences known at the time. He identified a possible covariation of an apparent Watson-Crick pairing type between the residues at position 15 from the 5' end of the tRNA and residue 48. This association implied these parts of the tRNA, namely the D loop containing residue 15 and the 5' end of the T stem-adjoining residue 48, folded on one another in a tertiary structure shared by different tRNAs. In presenting this finding to Francis, Michael was concerned because Holley's alanine tRNA sequence did not fit this covariation. Without skipping a beat, Francis quipped, 'Well, of course, Bob Holley's [Nobel-winning] sequence is wrong!' a retort subsequently confirmed by Guy Dirheimer and associates.

Sydney and Francis, always reaching beyond the forefront of experimental methods and systems, wanted Sid to attack the 3-D structure of tRNA using nuclear magnetic resonance methods, which had not yet been applied to an RNA macromolecule. It had not escaped their notice that Sid had taken an undergraduate physics degree at the Massachusetts Institute of Technology, and had written his senior thesis in nuclear physics. However, when Sid arrived at the LMB and met Sydney and Francis, they told him explicitly that tRNA structure analysis by NMR was no longer necessary, and they suggested him to go away and think of a project on his own. Sid, politely scared out of his mind, went home and thought about a new project for 2 weeks and then met again with his mentors. This change in research direction was the first committed step toward Sid's discovery of precursor tRNA, ribonuclease P, its RNA-protein composition, the enzymatic activity of the RNA, and ultimately sharing the 1989 Nobel Prize in Chemistry with Thomas R. Cech, who independently discovered non-enzymatic, catalytic RNA self-cleavage.

Sid's initial work at the LMB involved obtaining a tyrosine tRNA mutant that produced little mature tRNA, but when RNA was isolated by a new, rapid method, the mutant accumulated a metabolically unstable larger RNA, as indicated by its slower mobility during denaturing polyacrylamide gel electrophoresis. Next, Sid and John D. Smith demonstrated that the larger RNA was the mutant precursor tRNA by base sequence determination and by its site-specific cleavage at the 5' end of mature tRNA by a nuclease (later named ribonuclease P) present in cell-free extracts. Others readily accepted these straightforward observations; in fact, within the week I had confirmed the site-specific cleavage by RNase P of T4 phage precursor tRNAs. In retrospect, the existence of a precursor tRNA was implied when Robert W. Holley determined the first tRNA sequence, for it contained a monophosphate at its 5' end, whereas primary transcripts contain triphosphate residues.

Despite the novelty of Sid's research, the Anna Fuller Fund fellowship supporting his work at the LMB was about to expire, just as he was embarking on the purification of RNase P activity. Recognizing the potential novelty and importance of Sid's work (and his diligent work ethic), Sydney Brenner orchestrated a fresh fellowship for Sid to continue his work at the LMB. For the purification, Sid joined forces with Hugh D. Robertson whose expertise included nuclease purification. Sid's last paper from the LMB describing the purification and properties of RNase P was published in The *Journal of Biological Chemistry*. The Discussion of this paper contained a provocative, terse sentence. While the statement was not especially noteworthy, and perhaps underappreciated by Sid, but not by his perspicacious coauthor, Hugh, its significance would eventually generate much ridicule and consternation, not only between Sid and his future students, but also vehement opposition from a number of senior scientists in other laboratories involved in characterizing the enzyme and among many biochemists in general. The pertinent sentence reads, "In light of these [purification] properties, it is possible that the active form of RNase P, which must have a strong negative charge, could be associated with some nucleic acid." This sentence set the stage for a paradigm shift in defining the allowable chemical composition of enzymes.

Within 6 months, the enzyme had been substantially purified, but not freed of its nucleic acid component. Sid departed the LMB to accept a faculty position at Yale University and establish his own laboratory. One of his many gifted graduate students, Benjamin C. Stark, undertook further purification of RNase P to achieve homogeneity. Despite committed and rigorous work, Ben was unable to remove RNA from the active enzyme. Frustration and disappointment mounted at many levels within and outside the lab. Sid's major extramural research grant was in serious jeopardy, as was his promotion to Professor. Furthermore, Ben became concerned about the content and sufficiency of his work for the Ph.D. degree.

Ben was nevertheless convinced that one or more RNAs of a specific molecular weight(s) were part of the enzyme. He repeatedly lamented to fellow graduate student, Sheldon I. Feinstein, that Sid was publically critical of his heretical proposal regarding the RNA. After hearing enough of Ben's grieving, Sheldon told Ben to request a meeting of his Ph. D. committee to convince them that he should devise experiments to test his radical idea. At the meeting, Sid and Donald M. Crothers asked Ben the pivotal question: how would he go about proving his proposal? Ben had an idea, but it was not a very good one. He would determine if pre-treating RNase P preparations with a ribonuclease degraded RNase P activity. The flaw with this approach was to assay the treated RNase P that would require a tRNA precursor substrate, and that an inevitable carry over of a miniscule amount of the degrading ribonuclease would abolish the precursor substrate. Peter M. M. Rae, another committee member, suggested that Ben pre-treat RNase P with a calcium-dependent microccocal nuclease, dialyze out the calcium from the digestion mix, and then assay RNase P. The latter enzyme requires magnesium but not calcium for its activity. Indeed, microccocal nuclease treatment abolished RNase P activity, demonstrating that the RNA component of the enzyme is essential for the reaction. These findings were published in the *Proceedings of the National Academy of Sciences, U.S.A.* in 1978, with Ben Stark as the lead author. Subsequently, Sid and lab members designed a series of biochemical experiments to further characterize the RNA component of the enzyme.

One of several possibilities for the function of the RNA was to align the enzyme on the substrate tRNA precursor by interactions with nucleotides common to most tRNAs, which insured site-specific cleavage. (However, "The nucleolytic activity of the enzyme [was] reserved for the protein moiety.") Perhaps, the tRNA segment of the precursor was folded into the conserved structure of mature tRNA, which the enzyme recognized. Searching the precursor-specific sequences in other tRNA precursors did not reveal an obvious consensus sequence for traditional base pairing. Another possibility, loosely derived from P22 phage biology, was the RNA served as a scaffolding structure, in this case centered on RNA, to configure the protein in its catalytically active conformation.

Another of Sid's respected students, Ryszard Kole, was the lead author of the next *PNAS* paper, published in 1979, describing the dissociation of highly purified RNase P into component RNA and protein fractions. While each subunit lacked significant enzyme activity when assayed in the typical RNase P buffer containing a low Mg^{2+} concentration (10 mM), mixing equal molar proportions of the two reconstituted activity.

Subsequent definitive experiments, published in *Cell* in 1983, demonstrated that the RNase P RNA component of both *Escherichia coli* and *Bacillus subtilis* contains the enzymatic subunit of RNase P, and that activity was not detected in the protein subunit. A senior associate in Sid's lab, Dr. Cecilia Guerrier-Takada made this observation and was the lead author on the subsequent *Cell* paper. The activity of the RNA was maximal in a buffer that contained high Mg^{2+} concentration (50 mM), unlike the normal 10 mM concentration in the buffer typically employed for assay of the intact enzyme. Serendipity prevailed in these experiments as the RNA-only experiment was intended as a negative control. When this control manifested cleavage, Cecilia suspected a simple accident had been made in the experiment, possibly by adding the protein component to the RNA-only sample. However, repeating the RNA-only experiment gave the same remarkable result! Collaborator Norman R. Pace, working at the University of Colorado Medical Center in Denver, had provided partially purified *B. subtilis* RNase P subunits for these experiments.

This experiment had been designed to test the activity of interspecies reconstituted RNase P, not RNA-only cleavage. Nevertheless, Sid offered to include Norm and his graduate student as coauthors of the Guerrier-Takada et al. (1983) *Cell* paper.

During this time, diverse and important factors contributed to Sid's ability to maintain his work ethic and sanity. They included the strong support he received from his students, Yale faculty colleagues, particularly renowned biochemists, and other associates. The most enduring support, however, was the love of his family.

Following the difficult if not epic beginnings of research on ribonuclease P, the chapters that follow in this book describe the rich repertoire of basic and applied work to dissect, define, and exploit possible biomedical applications of the enzyme.

William H. McClain
Madison, Wisconsin

The author's post-doctoral tenure with Sydney Brenner and Francis Crick coincided with those of Sidney Altman and Hugh Robertson.

Sidney Altman, joint recipient of the Nobel Prize for Chemistry, receiving his prize on December 10, 1989, from King Carl XVI Gustaf of Sweden.

I dedicate this preface to the memory of Hugh Robertson and the staff of the LMB whose unfailing commitment to training younger scientists, both directly and by example, have significantly contributed to the exceedingly productive careers of many influential international scientists, and finally to the copious RNA scientists around the Globe whose generally congenial collaboration and friendly competition have thrust RNA research into the vanguard of contemporary sciences.

Contents

Contributors

Sidney Altman
Department of Molecular, Cellular and Developmental Biology,
Yale University, 266 Whitney Avenue New Haven, CT 06520, USA
sidney.altman@yale.edu

James W. Brown
Department of Microbiology, North Carolina State University,
Campus Box 7615, Raleigh, NC 27695, USA
james_brown@ncsu.edu

Wen-Yi Chen
Department of Biochemistry The Molecular Cellular and Developmental
Biology Program, The Center for RNA Biology, The Ohio State University,
Columbus, OH 43210, USA

I-Ming Cho
Department of Molecular Genetics The Center for RNA Biology,
The Ohio State University, Columbus, OH 43210, USA

Yana Dehtiar
Department of Molecular Biology, The Hebrew University-Hadassah
Medical School, Jerusalem 91120, Israel

J. Christopher Ellis
Laboratory of Signal Transduction, National Institute of Environmental
Health Sciences Research Triangle Park, NC 27709, USA,
Department of Microbiology, North Carolina State University,
Campus Box 7615, Raleigh, NC 27695, USA

David R. Engelke
Department of Biological Chemistry, 1150 W. Medical Center Drive,
University of Michigan, Ann Arbor, MI 48109-0606, USA
engelke@umich.edu

Carol A. Fierke
Department of Chemistry, University of Michigan, 930 N. University Avenue, Ann Arbor, MI 48109, USA
fierke@umich.edu

Venkat Gopalan
Department of Biochemistry, The Molecular Cellular and Developmental Biology Program, The Center for RNA Biology, The Ohio State University, 484 West 12th Avenue, Columbus, OH 43210, USA

Michael Harris
Department of Biochemistry, RNA Center, Case Western Reserve University School of Medicine, Cleveland, OH 44106, USA
michael.e.harris@case.edu

Roland K. Hartmann
Institut für Pharmazeutische Chemie, Philipps-Universität Marburg, Marbacher Weg 6, D-35037 Marburg, Germany
roland.hartmann@staff.uni-marburg.de

John Hsieh
Department of Chemistry, University of Michigan, 930 N. University Avenue, Ann Arbor, MI 48109, USA

Nayef Jarrous
Department of Molecular Biology, The Hebrew University-Hadassah Medical School, Jerusalem 91120, Israel
jarrous@md.huji.ac.il

Fabrice Jossinet
Architecture et Réactivité de l'ARN, Université de Strasbourg, IBMC, CNRS, 15 rue René Descartes, Strasbourg 67084, France

Leif A. Kirsebom
Department of Cell and Molecular Biology, Biomedical Center, Box 596, Uppsala University, Uppsala, Sweden
Leif.Kirsebom@icm.uu.se

Kristin S. Koutmou
Department of Chemistry, University of Michigan, 930 N. University Avenue, Ann Arbor, MI 48109, USA

Gerhard Klebe
Institut für Pharmazeutische Chemie, Philipps-Universität Marburg, D-35037 Marburg, Germany
klebe@staff.uni-marburg.de

Lien B Lai
Department of Biochemistry, The Center for RNA Biology, The Ohio State University, Columbus, OH 43210, USA

Fenyong Liu
Program in Comparative Biochemistry, Division of Infectious Diseases,
School of Public Health, University of California, Room 326,
16 Barker Hall, Berkeley, CA 94720, USA
liu_fy@berkeley.edu

Michael C. Marvin
Department of Biological Chemistry, 1150 W. Medical Center Drive,
University of Michigan, Ann Arbor, MI 48109-0606, USA

Benoît Masquida
Architecture et Réactivité de l'ARN, Université de Strasbourg, IBMC, CNRS,
15 rue René Descartes, Strasbourg 67084, France
b.masquida@ibmc.u-strasbg.fr

William H. McClain
Department of Bacteriology, University of Wisconsin-Madison, Madison,
WI 53706, USA

Alfonso Mondragón
Department of Biochemistry, Molecular Biology, and Cell Biology,
Northwestern University, 2205 Tech Drive, Evanston, IL 60208, USA
a-mondragon@northwestern.edu

Tao Pan
Department of Biochemistry and Molecular Biology, University of Chicago,
929 East 57th street, Chicago, IL 60637, USA
taopan@uchicago.edu

Patrick Pfeffer
Institut für Pharmazeutische Chemie, Philipps-Universität Marburg,
D-35037 Marburg, Germany
pfefferp@staff.uni-marburg.de

Robert Reiner
Department of Molecular Biology, The Hebrew University-Hadassah Medical
School, Jerusalem 91120, Israel

Klaus Reuter
Institut für Pharmazeutische Chemie, Philipps-Universität Marburg,
D-35037 Marburg, Germany
reuterk@staff.uni-marburg.de

Agustín Vioque
Instituto de Bioquímica Vegetal y Fotosíntesis, Universidad de Sevilla and CSIC,
Américo Vespucio 49, 41092 Sevilla, Spain
vioque@us.es

Scott C. Walker
Department of Biological Chemistry, 1150 W. Medical Center Drive,
University of Michigan, Ann Arbor, MI 48109-0606, USA

Eric Westhof
Architecture et Réactivité de l'ARN, Université de Strasbourg, IBMC, CNRS, 15 rue René Descartes, Strasbourg 67084, France

Dagmar K. Willkomm
Institut für Pharmazeutische Chemie, Philipps-Universität Marburg, D-35037 Marburg, Germany
willkomm@staff.uni-marburg.de

Lindsay Yandek
Department of Biochemistry, RNA Center, Case Western Reserve University School of Medicine, Cleveland, OH 44106, USA
michael.e.harris@case.edu

Chapter 1
History of RNase P and Overview of Its Catalytic Activity

Sidney Altman

1.1 RNase P and Life

The presence of temperature sensitive lethal mutations of *E. coli* in genes for the protein and RNA components of RNase P easily led to the conclusion that this ribonucleoprotein (RNP) enzyme was essential for viability (Schedl and Primakoff 1973; Ozeki et al. 1975). Although subsequently more mutants in RNase P function were available only in *Bacillus subtilis* (Waugh and Pace 1990) and in *Saccharomyces cerevisiae* (Stolc and Altman 1997), enzymatic studies of many more species, bacterial, archaeal and eukaryotic, confirmed the notion that the enzyme seemed ubiquitous and essential for growth. While there is some discrepancy about the need for an RNA component in RNase P from certain mitochondria (Puranam and Attardi 2001; Rossmanith and Karwan 1998) and chloroplasts (Wang et al. 1988), this problem will be resolved as the biochemistry of these problems becomes clearer (Holzmann et al. 2008).

Any statement that RNase P is not essential for life is presently incorrect (Rendan et al. 2008). The enzyme might be absent in absolute symbionts, but those particular species cannot grow on their own in any case. They, as with phage and viruses, have to be in an appropriate host that does require RNase P for growth. Statements about archaea show clearly that the RNase P is present in those organisms (Kouzuma et al. 2003; Andrews et al. 2001).

1.2 The Initial Substrate

All the early work on RNase P relied on tRNA precursors (ptRNAs) as substrates (Altman 1971; Altman and Smith 1971; Guthrie et al. 1973) In retrospect, this seemed obvious as the first substrate for activity was a probe that consisted of a radioactive pure, tRNA precursor to $tRNA^{Tyr}$ (pTyr) in *E. coli* (Altman 1971).

S. Altman (✉)
Department of Molecular, Cellular and Developmental Biology, Yale University, New Haven, CT 06520, USA
sidney.altman@yale.edu

F. Liu and S. Altman (eds.), *Ribonuclease P*, Protein Reviews 10,
DOI 10.1007/978-1-4419-1142-1_1, © Springer Science+Business Media, LLC 2010

As worked progressed, some other substrates were found in *E. coli* that were long hairpins with a single strand attached at the 5′ end of the hairpin (e.g., the precursor to 4.5 S RNA; Bothwell et al. 1976) or tRNA-like substrates (tmRNA; Komine et al. 1999) or plant viral RNAs (Guerrier-Takada et al. 1988). The enzyme cleaved these substrates at the junction of a hairpin and a single stranded region to generate a 5′ phosphate at the terminus of the hairpin. Detailed studies of the essential substrate requirements indicated that the minimal substrate was one extra nucleotide at the 5′ terminus attached to about half a turn of an RNA double stranded helix (Forster and Altman 1990; Peck-Miller and Altman 1991). While the rate of cleavage is low on such a substrate, the general picture of a minimal substrate for RNase P is correct. Recently, other substrates have been identified that are within the larger RNAs (Li and Altman 2003). One of these is located in the region between the lacY and lacA genes in the lac operon of *E. coli* (Li and Altman 2004).

1.3 The Purification of RNase P

The identification of a radioactive, pure precursor $tRNA^{Tyr}$ from *E. coli* led immediately to the conclusion that there must be an enzyme that was responsible for removing the long extra sequence at the 5′ end of this molecule. At the 3′ end of the precursor tRNA, only a few nucleotides were present and it was assumed that an exonuclease ultimately would be responsible for the removal of the few nucleotides. Only later it was found that there were long sequences at the 3′ ends of precursor tRNAs and new endonucleases were involved in those cases, too (Bikoff et al. 1975; Zuo and Deutscher 2002;). In any case, a search was immediately started for an enzyme that would remove the extra 5′ sequences from ptRNAs.

An assay of a crude extract of *E. coli* with the $ptRNA^{Tyr}$ as a probe revealed that a nuclease with the apparent correct activity was present although, in this case, contaminating exonucleases degraded the long 5′ segment that was removed as a consequence of the cleavage event. Further partial purification confirmed the desired properties of the enzyme (Robertson et al. 1972). An endonuclease that made one cut in the ptRNA and produced a 5′ phosphate on the tRNA product was demonstrated after the first purification step from the crude extract. (Contaminating exonucleases were removed in one step on a DEAE–Sephadex column.) The complete purification of RNase P was carried out by Ben Stark, a graduate student at the Yale University (Stark et al. 1978).

1.4 Diversion in Mammalian Cells

Early in the 1970s, Sherman Weissman encouraged our laboratory to assay HeLa cell extracts for RNase P activity. This assay, using the bacterial $ptRNA^{Tyr}$ as a substrate, was successful (Bothwell and Altman 1975). Other mammalian cells were also assayed, all giving the same result. The success in using the bacterial ptRNA was significant in that pTyr was a reliable substrate for many reactions, and has always been successful regardless of the species from which a cell extract was assayed. Yeast and human ptRNAs were subsequently used for assays as eukaryotic substrates.

The purification of human RNase P was commenced by Ray Koski, a graduate student, and completed many years later, first by Marcjanna Bartkiewicz (Bartkiewicz et al. 1989) and then by Paul Eder and Nayef Jarrous, postdoctoral fellows (Eder et al. 1997; Jarrous and Altman 2001). Koski partially purified the enzyme from RNase P through the DEAE–Sephadex and glycerol gradient steps (Koski et al. 1976).

1.5 The Requirement for an RNA Component

Ben Stark started purifiying RNase P from *E. coli* from 5 g of cells, an inadequate amount. Soon, he could identify the activity after a few steps of purification but found no Coomassie brilliant blue staining bands on an acrylamide gel. However, he then stained these gels with methylene blue, in the belief that it would stain the protein bands better than Coomassie. He did see two bands, both of which were RNA. Stark concluded the enzyme consisted of RNA, an idea that I tried to disabuse him of. He then went on to purify the enzyme from 500 g of cells and, indeed, could now see a prominent protein band after Coomassie staining. He continued to do an additional staining with methylene blue and an RNA band was still visible. At that point, it became an exercise in the "Avery" experiment (Avery et al. 1944). Could one treat the enzyme with various enzymes and show that the obliteration of RNA destroyed the activity of the enzyme? In fact, at the suggestion of Peter Rae, a colleague at that time, this experiment did work with micrococcal nuclease, which required Ca^{2+} for action. After use, the Ca^{2+} could be removed with EGTA (it had already been determined that RNase P required Mg^{2+} for activity). This experiment worked, as did another in which RNase A was covalently linked to the beads so that it could be removed easily when necessary. Ultimately, after many obstacles of editorial ignorance and the malicious cupidity of other scientists, the paper describing the requirement for an RNA component was published (Stark et al. 1978). Subsequently, more experiments on the nature of the RNA component, the separation of the protein and the RNA subunits and their reconstitution into an active enzyme, were performed over the next few years.

1.6 Separation of Two Components

The separation of the two components of RNase P was a straightforward problem in biochemistry once one knew what the general chemical nature of the subunits were. It was relatively easy to separate the RNA component on gels after it had been identified. One obviously then knew that the protein must be positively charged. (This hybrid complex of RNA and protein made Stark's initial attempts at purification somewhat frustrating.) Several methods were incapable of getting a clear purification until a CM–Sephadex column was used. After that, the purification was simple. Today, both components have been cloned and can be overexpressed in *E. coli*. The RNA is generally made by transcription in vitro. The protein is over-expressed in *E. coli* from a plasmid that contains its genes and can then be purified. The protein can also be tagged with a histidine hexamer to facilitate the immobilized metal affinity chromatography.

1.7 Reconstitution

Once the two subunits from *E. coli* had been separated, reconstitution proceeded easily (Kole and Altman 1981). In fact, today when pure holoenzyme is required from *E. coli*, the separated (see above) subunits, after purification, are mixed together to form the holoenzyme. To prove that, we actually had reconstituted the actual enzyme from *E. coli* subunits prepared from temperature sensitive strains were used and the purified, reconstituted enzyme was shown (Kole and Altman 1981; Baer et al. 1989), as appropriate, that it was temperature sensitive. Subsequently, the reconstitution assay was also used to test the viability of fragments of M1 RNA in terms of their ability to make active enzyme (Guerrier-Takada and Altman 1992). Several pieces made from fragments of the enzyme were shown to be active when mixed together in this assay. In fact, the demarcation of a fragment of the RNA subunit that could interact with ptRNAs was made in this fashion. The conventional tRNA binding site, including the TUC loop, is located on one part of the RNA subunit. Another fragment that could only interact with the hairpin-like substrates was also identified. The catalytic center of the enzyme was located in this way on the RNA subunit along with the aid of an extensive program of mutagenesis.

1.8 Small Ribosome?

About 1980, the question of the function of the RNA component was being debated. The notion of RNase P as a small ribosome was put forward as an idea (Kole and Altman 1981) that others might consider. The ribosome was familiar. It also was a very complex organization of RNAs and proteins. At that time, it was thought that the RNA was a scaffold for hanging the enzymatically active proteins in the right configurations, whichever ones they were. The same could be said of RNase P, although there was only one protein in the bacteria it was unclear what formed the details of the active site of the enzyme. Years later, and certainly not due to any predictions, many RNA–protein complexes were identified and their functions and structures were studied in detail.

The ribosome, after crystallization (Steitz and Moore 2003), was described as an RNA enzyme although this was stated without experimental proof: in fact, there were no proteins or parts of proteins within about 15 Å of the peptidyl transferase site.

1.9 Catalytic Properties of RNA

At a meeting at Cold Spring Harbor laboratory it was decided that hybrid RNase P enzymes made of the *E. coli* and the *B. subtilis* subunits would be assayed to see if the function of the subunits was at least universal to some degree. N. Pace supplied our laboratory with the RNA and protein subunits of RNase P from *B. subtilis*. The experiments were undertaken by C. Guerrier-Takada and immediately showed some surprising results. (One also has to recall that the *E. coli* subunits were much more highly purified than the ones from *B. subtilis* and that Guerrier-Takada was using a natural ptRNA substrate, whereas the *B. subtilis* substrate was an artificial, less efficient one (Guerrier-Takada et al. 1983)). In a control experiment, in which the RNA subunit was tested alone, Guerrier-Takada showed that M1 RNA, assayed at a higher than normal Mg^{2+} concentration for the holoenzyme, had catalytic activity. The higher Mg^{2+} (100 mM) concentration was used because Pace's group had shown it was necessary for the activity of the *B. subtilis* holoenzyme (Gardiner et al. 1985). This experiment was repeated and the same result was soon shown for the *B. subtilis* RNA subunit. All the experiments described in the 1983 paper were repeated quickly: it was a matter of simply doing the experiments that one knew outlined the characteristics of a real enzyme. About 3 weeks later a manuscript was finished. The authorship was decided in a telephone call by Altman to Pace, who was stupefied by our result. No other authors claimed the first discovery of the catalytic activity of M1 RNA or *B. subtilis* RNase P RNA before this paper was written by S. A. with the collaboration of C. Guerrier-Takada and published. Guerrier-Takada's results were absolutely clear and totally original. It is certain that Cecilia Guerrier-Takada first discovered the catalytic properties of the RNase P RNA. No direction from anyone had any influence over the Altman lab at that time.

Several advances have been made in the past decade on the prevalence of catalytic activity in the RNA component of RNase P. First, the enzyme has been characterized in several archaea and the RNA has shown to be catalytic (Pannucci et al. 1999), or catalytic in the presence of at least two of the four protein subunits of the complex (Kouzuma et al. 2003; Boomershine et al. 2003; Tsai et al. 2006). Second, two eukaryotic RNA subunits, from human cells and Giardia, have also shown to be catalytic at a very much lower level than the M1 RNA from *E. coli* (Kikovska et al. 2007). These results will be discussed further in detail by Kirsebom but surely more data can be expected on these latter findings before they are universally accepted.

While what has been reported above describes the appearance of the catalytic activity of the RNA subunit of the enzyme, one also should remark on the mechanism of action of the catalysis. Allen and colleagues (Haydock and Allen 1985; Guerrier-Takada et al. 1986) made the first theoretical statement in which they postulated an S_N2 mechanism of interaction in which Mg^{2+} ions played a major role in promoting bond cleavage of a target substrate's phosphodiester chain. While this mechanism might be right, there are several variations regarding the number and positions of the divalent cation that are important in the reaction and that are described below in the chapter by Harris. Other experiments on the nature and number of the metal ions and monovalent ions have been carried out (Guerrier-Takada et al. 1986; Kazakov and Altman 1991; Persson et al. 2003; Zahler et al. 2005) and have contributed some information but the catalytic mechanism is still somewhat cloaked in mystery.

1.10 Structure of RNase P

Subsequent to the sequence determination of the RNA subunit of *E. coli* RNase P and then other RNA subunits of the enzyme in different species, a method developed by Woese was used to accurately predict the secondary structure of the enzyme in prokaryotes (James et al. 1988; Darr et al. 1992). This method was also used for eukaryotic RNA subunits of RNase P when further sequences were determined. The sequence identification also allowed chemical and cross-linking studies to be carried out that elucidated some aspects of the three-dimensional structure of *E. coli* RNase P (Westhof et al. 1996; Chen et al. 1998). In fact, the computer assisted model, using both theoretical and experimental details as information for the computer programs, made by Westhof and collaborators was remarkably similar to the crystallographic model produced by Mondragon (Krasilnikov et al. 2003) and subsequently Pace (Kazantsev et al. 2005). The exact nature of the catalytic site, which was not defined well by the computer assisted model, was also poorly defined by crystallography. The specific location of the nucleotides in what was thought to be the active site were not located properly. The question arises as to whether or not there is one, or more than one, catalytic structure of the enzyme in the free form. Previous experiments had indicated that there in fact was more than one structure of RNase P, possibly because it is an enzyme. There had to be one structure of the enzyme alone and at least one other with the substrate/product.

While there is as yet no defined structure of the holoenzyme or the holoenzyme plus the substrate, some work has been done on the structure of the protein subunit of bacterial RNase P (Stams et al. 1998; Spitzfaden et al. 2000; Kazantsev et al. 2003). This small protein has been described in structural detail and chemical cross-linking of this protein with the RNA subunit has been carried out to determine where the protein sits on the RNA (Tsai et al. 2003). These data have been useful but further clarification and usefulness of the results in defining the active site of an RNA enzyme awaits the proof of the structure through crystallography.

1.11 More Substrates

The advent of the transcription of many different RNA molecules presented the opportunity to test some of these molecules as substrates for RNase P. In particular, the finding that several different molecules, not just ptRNAs or other small molecules stable RNAs, were substrates as defined by their appearance under conditions when no RNase P was made (Yang and Altman 2007; Samanta et al. 2006; Chamberlain et al. 1996; Coughlin et al. 2008), was particularly interesting. Indeed, operon RNAs in *E. coli* were checked in this fashion and some proved to have cleavage sites for the RNase P between ORF regions. The prevalence of these sites may be important for the degradation or regulation of the operons but as yet we have no experimental evidence concerning the relation of RNase P to their function. Riboswitch RNAs were also tested, although no one has yet shown that they appeared under RNase P-deficient conditions, and a few were also found to be substrates but not under any conditions were their canonical structures were present (Seif and Altman 2008). Aside from the lac operon, where there is good evidence that this operon molecule is degraded in a polar fashion, the riboswitches may be susceptible to RNase P when in a folded conformation that does not include the full-length molecule. The structural variants might occur naturally in solution (Greenleaf et al. 2008) or are remnants from the transcription process in which the final, canonical form of the RNA was not assumed properly.

In eukaryotes, tRNA precursors are substrates for RNase P (Burdon 1975; Bothwell and Altman 1975). There are also data that indicate that one of the precursor rRNAs is also a substrate in *S. cerevisiae* (Chamberlain et al. 1996; Stolc and Altman 1997) and possibly in human cells. A ncRNA, made in yeast under conditions where RNase P is deficient which is also a substrate (Yang and Altman 2007; Samanta et al. 2006). The relation of this last substrate to cell function is not at all clear, but many such RNA molecules are present only in the RNase P deficient cells in yeast. Their function in cell physiology is not understood.

1.12 EGS Scheme

The origin of the EGS scheme has been outlined in a paper in which we determined the minimal nature of the substrate for RNase P. Forster showed that a small substrate that was quite efficient in terms of its ability to be cleaved by the enzyme consisted of two RNA molecules, one of which was the target and the other was the EGS (Forster and Altman 1990). Over the next few years we developed this concept and showed its viability in terms of usage in *E. coli* and in mammalian cells in tissue culture (Goplan et al. 2002).

The first EGS we used was a simple, linear RNA complementary to the target RNA of about 15 nt and which ended in the ubiquitous (A/G) CCA found in all tRNAs (Guerrier-Takada et al. 1995). That scheme has worked well in prokaryotes, in fact, even better than a "3/4" EGS (that contains three quarters of a tRNA molecule but lacks the 5′ sequence from nt 1 through the D loop) that includes most of a tRNA sequence. In mammalian cells, the 3/4 EGS works well (Liu and Altman 1995) but the M1-EGS, in which the usual stem EGS is covalently linked to the 3′ end of an M1 RNA molecule, works even better (a non-functional spacer sequence of about 20 nt exists between the 3′ end of M1 RNA and the beginning of the EGS sequence). We do not understand yet why the M1-EGS works so well in mammalian cells but probably part of the reason is that the catalytic site of M1 RNA is very close to the target in the complex with the M1-EGS and this structure binds to some host (RNase P?) proteins. (M1-EGS does not work as well as the stem EGS in bacteria, possibly because of competition with the host M1 RNA.) Of particular value is the fact that the T loop of the 3/4 EGS functions as a recognition signal for the host RNase P in the mammalian cells. Changing that signal is a method of testing whether or not the cleavage process we observe is mediated by the RNase P function. In general, the remaining antisense effect (the effect of hydrogen bonding of the EGS to the target RNA with no subsequent cleavage of the complex by RNase P) is about 10% of the EGS-mediated RNase P effect (Plehn-Dujowich and Altman 1998). In bacteria, the EGS effect is negated when *E. coli* temperature sensitive cells in RNase P function are shifted to the restrictive temperature (Guerrier-Takada et al. 1995).

We emphasize that there is no other scheme that works as well as the EGS technology in prokaryotes in terms of inhibiting gene expression. In eukaryotes, the EGS method works at least as well as siRNA in terms of the absolute numbers it achieves in inhibiting the expression of any particular target (Zhang and Altman 2004). If an siRNA site is chosen first in a target RNA (Shan G, personal communication), an EGS then directed against this site functions even more efficiently than the siRNA.

1.13 More EGS Experiments

A summary of published EGS results in both prokaryotes and eukaryotes is presented in Tables 1.1 and 1.2. These data indicate clearly that this technology is very successful in cells of both types. Furthermore, the effect of multiple EGSs against the same target is additive, rather than multiplicative (see the results with the inhibition of drug resistance in *E. coli*; Guerrier-Takada et al. 1995). This last result has recently been repeated with EGSs against a *Y. pestis* mRNA (Ko et al. 2008). When two targets are attacked in a virus, as shown with the results on flu virus production, the resultant effect yields essentially a 100% reduction in particle production (Plehn-Dujowich and Altman 1998). Another equally important result is that for stem EGSs, three mismatches in H-bonding of the target with the EGS still yield an effective EGS agent (provided the three mismatches are not contiguous; McKinney et al. 2001). This result is in sharp contrast to data on drug sensitive targets, in which a one nucleotide mutation can change sensitivity to resistance.

Table 1.1 Summary of EGS experiments with bacteria

Organism	Target	Inhibition (%)
E. coli	β-Galactosidase	~60
	Alkaline phosphatase	~60
	Protein subunit, RNase P	>60
	Gyrase	>60
	Phage mRNA	~70
	Drug resistance	~100
	Virulent (non-*E. coli*) mRNAs	>50
S. typhimurium	Type III injection	~80

One or two EGS used against each target

Table 1.2 Summary of EGS experiments with mammalian cells

Organism	Target	Inhibition (%)
Mouse/canine/	RNA polymerase (early)	~85
Influenza virus	Nucleocapsid (late)	~85
	Both	~95
Human	Protein subunits (RNase P)	60–85
	Lamin A/C	60–85
Herpes virus	Thymidine kinase	~85
HCMV[a]	Protease function	~80
	NMDA receptor	>60
HIV[b]		>60 (99.9% killing)

One or two EGS against each target
[a]Kim et al. 2004; [b]Kraus et al. 2002

As a comparison with the siRNA technology, when a target is attacked either with EGSs or with siRNAs, the absolute amount of gene inhibition of expression is approximately equal, as shown in the experiments with a target that is an mRNA of one of the protein subunits of human RNase P (Zhang and Altman 2004) as alluded to above. RNAi also does not work in certain systems, zebrafish (Pei et al. 2008) and maize (Rangarajan et al. 2004), where the EGS method does.

Hnatysyn and colleagues (Kraus et al. 2002) have shown that five clades of HIV are inhibited effectively in gene production, and viral production, in tissue culture with the EGS technology. Liu and colleagues (Bai et al. 2008) have also demonstrated that the production of CMV in mice is also blocked by administration of plasmid encoded EGSs by intravenous administration.

1.14 Human RNase P

The first full scale purification of human RNase P from HeLa cells was undertaken by M. Bartkiewicz (Bartkiewicz et al. 1989) and succeeded, albeit on a microscale, of purifying the activity several hundred-fold. The final step of purification was a Cs_2SO_4 gradient.

These experiments indicated that there was one, distinct and complete RNA (H1 RNA) associated with the enzymatic complex. It was sequenced and has remained a standard for several years. It was also found that both the DEAE–Sephadex and first glycerol gradient fractions were relatively stable when stored on ice for several months. Subsequently, P. Eder started more ambitious preparations from 30 L of cells. Eder (Eder et al. 1997), particularly, and Jarrous (Jarrous and Altman 2001) did manage to purifiy the enzyme to about 30% purity from an estimate of purity of RNase P bands on a gel and showed, at least nominally, that there were at least ten protein subunits, aside from the RNA subunit. This monumental task also allowed the preparations of rabbit polyclonal antibodies, which were extremely useful, to several of the proteins.

An analysis of proteins that interacted with the proteins of human RNase P, was carried out by Jiang using the yeast two hybrid (Jiang and Altman 2001) and three hybrid (Jiang et al. 2001) assays and did show that there were other proteins that had some interaction with the RNase P protein subunits. No further work has been undertaken yet on these other proteins. Jiang also indicated that some of the P subunits had enzymatic activity of their own as individual proteins (Jiang and Altman 2002) but no correlation of these activities from the total complex was observable.

1.15 Reconstitution and Regulation of Human RNase P

Some time ago the result was published that the activity of human RNase P could be reconstituted in vitro with the RNA subunit, H1 RNA, and a few of the protein subunits (Mann et al. 2003). This interesting result should be repeated in the other laboratories.

Over the past few years, there has been an increasing evidence that the synthesis of the subunits of RNase P is under regulation in human cells and that the enzyme, itself, is involved in the aspects of transcription with one or more RNA polymerases. Kovrigina et al. (2003) showed that inhibition of the expression of a protein subunit of RNase P, also result in a decreased expression of some of the other subunits, but about half were unaffected in this regard. Further data indicated that the other proteins in the nucleus were also inhibited, but the nature of this kind of regulation seems to involve translation. Lamin A/C protein is inhibited by translation after Rpp38 is decreased in expression (Shan G, personal communication). However, Jarrous has shown that RNase P does have a physical interaction with pol III (Reiner et al. 2006) and possibly with pol I, and the enzyme or parts of it seem to affect the ability of the polymerases in transcription experiments. The previous data on the inhibition of subunit expression pointed to a particular and specific depression of pol II activity, but that remains a speculation at the moment. What is clear is that this complex enzyme, RNase P, does not exist in isolation inside the cell nucleus or nuceleoli where it is normally located (Jarrous et al. 1999), and has no interaction of one kind or another with any other processes inside the cell.

1.16 An Overview of RNase P Research and our Current World

There are several outstanding problems related to the RNase P that are worthy and valuable in terms of research in molecular biology. The nature of the catalytic reaction, in all its fine detail, still awaits a complete description. Such work will ultimately result from biochemical experiments and from structural studies. With the latter, a clear picture of the catalytic site in the presence of a substrate will be necessary in order to pinpoint where the different nucleotides are situated in the structure. To be complete, this picture ultimately should involve both the RNA subunit of a procaryote, the protein subunit, and the substrate. The work of Harris' group and that of Kirsebom's, both of which have produced good, indirect pictures of different parts of the catalytic process, needs to be finished.

The picture of the evolution of RNase P, or the interrelation between the structure in prokaryotes, archaea and eukaryotes is a tempting problem. How did proteins gradually overtake the function of the RNA in providing a complete description of how catalysis worked—or did the proteins, with other functions, successively impinge on the RNA function and honed the effect of the complete process into what we see today? Equally important is the nature of different substrates in the three classes of organisms. What is the extent of substrate range? Do the different substrates identified now in *E. coli* function presumably in some capacity have physiological factors inside the cell? How do the plethora of ncRNAs in yeast act as substrates for RNase P?

The regulation of RNase P synthesis and the effect of this complex on the other transcription and translation reactions in mammalian cells is critical to understand the role of the enzyme in higher cell types. It has been known for some time that the RNase P in *E. coli* is made at a constitutive level and does not change appreciably unless, for example, protein synthesis is stopped (Hansen et al. 1985). We know little about the RNase P in mammalian cells except that there are regulatory features that we do not understand.

Occasionally, we receive extensive charts of the protein enzymatic pathways that have been elucidated to date. These charts represent years of person-hours at the bench and the best of biochemical expertise and the innovation for the past century or so. How soon will it be before we receive charts of the interconnectedness of RNase P, the various kinds of regulatory RNA, miRNA, shRNA, etc., and their effect on the response of cells to events that control their phenotypes and their lifetimes? Perhaps, when such information will be provided, we can change the informal name for today's world as we see it. The current world can no longer be named as the "protein" world but rather as an "RNA–protein" world.

Acknowledgments The author thanks profusely all the members of his own laboratory over the years who have contributed their experimental efforts and perceptive discussions that formulated the results from Yale University. The author also thanks Rose Stone for sterling service as an administrative assistant over a period of many years.

References

Altman S (1971) Isolation of tyrosine tRNA precursor molecules. Nat New Biol 229:19–21

Altman S, Smith JD (1971) Tyrosine tRNA precursor molecule polynucleotide sequence. Nat New Biol 233:35–39

Andrews AJ, Hall TA, Brown JW (2001) Characterization of RNase P holoenzyme from *Methanococcus jannaschi* and *Methanothermobacter thermoautotrophicus*. Biol Chem 382:1171–1177

Avery OT, MacLeod CM, McCarty M (1944) Studies of the chemical nature of the substance inducing transformation of pneumonococal types. J Exp Med 79:137–158

Baer MF, Wesolowski D, Altman S (1989) Characterization in vitro of the defect in a temperature sensitive mutant of the protein subunit of ribonuclease P from *Escherichia coli*. J Bacteriol 171:6862–6866

Bai Y, Trang P, Li H, Kim K, Zhou T, Liu F (2008) Effective inhibition in animals of viral pathogenesis by ribozyme derived from RNase P catalytic RNA. Proc Natl Acad Sci U S A 105: 10919–10924

Bartkiewicz M, Gold H, Altman S (1989) Identification and characterization of an RNA molecule that copurifies with RNase P activity from HeLa cells. Genes Dev 3:488–499

Bikoff EK, LaRue BF, Gefter ML (1975) In vitro synthesis of transfer RNA. J Biol Chem 250:6248–6255

Boomershine WP, McElroy CA, Tsai H, Wilson RC, Gopalan V, Foster MP (2003) Structure of Mth11/Mth Rpp 29, an essential protein subunit of archaeal and eukaryotic RNase P. Proc Natl Acad Sci U S A 100:15398–15403

Bothwell ALM, Altman S (1975) Partial purification and properties of an endoribonclease indicated from human ICIS cells. J Biol Chem 250:1451–1459

Bothwell ALM, Garber R, Altman S (1976) Isolation and nucleotide sequence of precursor molecules to *E. coli* 4.5 S RNA. J Biol Chem 251:7709–7716

Burdon RH (1975) Processing of tRNA precursors in higher organisms. Brookhaven Symp Biol 26:138–153

Chamberlain JR, Pagan-Ramos E, Kindelberger DW, Engelke DR (1996) An RNase P RNA subunit mutation affects ribosomal RNA processing. Nucleic Acids Res 24:3158–3166

Chen JL, Nolan JM, Harris ME, Pace NR (1998) Comparative photocross-linking analysis of the tertiary structures of *Escherichia coli* and *Bacillus subtilis* RNase P RNASs. EMBO J 17:1515–1525

Coughlin DJ, Pleiss JA, Walker SC, Whitworth GB, Engelke DR (2008) Proc Natl Acad Sci U S A 105:12218–12223

Darr SC, Brown JW, Pace NR (1992) The varieties of ribonclease P. Trends Biochem Sci 17:178–182

Eder PS, Kekuda R, Stolc V, Altman S (1997) Characterization of two scleroderma autoimmune antigens that copurify with human ribonuclease P. Proc Natl Acad Sci U S A 94:1101–1106

Forster AC, Altman S (1990) External guide sequences for an RNA enzyme. Science 249:783–786

Gardiner KJ, Marsh TL, Pace NR (1985) Ion dependence of the *B. subtilis* RNase P reaction. J Biol Chem 260:5415–5419

Gopalan V, Vioque A, Altman S (2002) RNase P: variations and uses. J Biol Chem 277: 6759–6762

Greenleaf WF, Frieda KL, Foster DA, Woodside MT, Block SM (2008) Direct observation of hierarchical folding in single riboswitch optamers. Science 319:630–633

Guerrier-Takada C, Altman S (1992) Reconstitution of enzymatic activity from fragments of M1 RNA. Proc Natl Acad Sci U S A 89:1266–1270

Guerrier-Takada C, Gardiner K, Marsh T, Pace NR, Altman S (1983) The RNA moiety of ribonuclease P is the catalytic subunit of the enzyme. Cell 35:849–857

Guerrier-Takada C, Haydock K, Allen L, Altman S (1986) Metal ion requirements and other aspects of the reaction catalyzed by M1 RNA, the RNA subunit of ribonuclease P from *Escherichia coli*. Biochemistry 25:1509–1515

Guerrier-Takada C, van Belkum A, Pleij CWA, Altman S (1988) Novel reactions of RNase P with tRNA-like structures in turnip yellow mosaic virus RNA. Cell 53:267–272

Guerrier-Takada C, Li Y, Altman S (1995) Artificial regulation of gene expression in *Escherichia coli* by RNase P. Proc Natl Acad Sci U S A 92:11115–11119

Guerrier-Takada C, Salavati R, Altman S (1997) Phenotypic conversion of drug-resistant bacteria to drug sensitivity. Proc Natl Acad Sci U S A 94:8468–8472

Guthrie C, Seidman JG, Altman S, McClain WH, Smith JD, Barrell BG (1973) Identification of transfer RNA precursor molecules made by bacteriophage T4. Nat New Biol 246:6–11

Hansen FG, Hansen EB, Atlung T (1985) Physical mapping and nucleotide sequence of the rnpA gene that encodes the protein component of ribonuclease P in *Escherichia coli*. Gene 38:85–93

Haydock K, Allen LC (1985) Molecular mechanism of catalysis by RNA. Prog Clin Biol Res 172A:87–98

Holzmann J, Frank P, Loffler E, Bennett KL, Gerner C, Rossmanith W (2008) RNase P without RNA: identification and functional reconstitution of the human mitochondrial tRNA processing enzyme. Cell 135:462–474

James BD, Olsen GJ, Liu JS, Pace NR (1988) The secondary structure of ribonuclease P RNA, the catalytic element of a ribonucleoprotein enzyme. Cell 52:19–26

Jarrous N, Altman S (2001) Human ribonuclease P. Methods Enzymol 342:93–100

Jarrous N, Wolenski JS, Wesolowski D, Lee C, Altman S (1999) Localization in the nucleolus and coiled bodies of protein subunits of the ribonucleoprotein ribonuclease P. J Cell Biol 146:559–571

Jiang T, Altman S (2001) Protein-protein interactions with subunits of human nuclear RNase P. Proc Natl Acad Sci U S A 98:920–925

Jiang T, Altman S (2002) A protein subunit of human RNase P, Rpp 14, and its interacting partner, OIP2, have 3′→5′ exoribonuclease activity. Proc Natl Acad Sci U S A 99:5295–5300

Jiang T, Guerrier-Takada C, Altman S (2001) Protein–RNA interactions in the subunits of human nuclear RNase P. RNA 7:937–941

Kazakov S, Altman S (1991) Site-specific cleavage by metal ion cofactors and inhibitors of M1 RNA, the catalytic subunit of RNase P from *E. coli*. Proc Natl Acad Sci U S A 88:9193–9197

Kazantsev AV, Krivenko AA, Harrington DJ, Carter RJ, Holbrook SR, Adams PD, Pace NR (2003) High-resolution structure of RNase P protein from *Thermotoga maritima*. Proc Natl Acad Sci U S A 100:7497–7502

Kazantsev AV, Krivenko AA, Harrington DJ, Holbrook SR, Adams PD, Pace NR (2005) Crystal structure of a bacterial ribonuclease P RNA. Proc Natl Acad Sci U S A 102:13392–13397

Kikovska E, Svard SG, Kirsebom LA (2007) Eukaryotic RNase P mediates cleavage in the absence of protein. Proc Natl Acad Sci U S A 104:2062–2067

Kim K, Umamoto S, Trang P, Hai R, Liu F (2004) Intracellular expression of engineered RNase P ribozymes effectively blocks gene expression and replication of human cytomegalovirus. RNA 10:438–447

Ko JH, Izadjoo M, Altman S (2008) Inhibition of expression of virulence genes of *Yersinia pestis* in *Escherichia coli* by external guide sequences and RNase P. RNA 14:1656–1662

Kole R, Altman S (1981) Properties of purified RNase P from *E. coli*. Biochemistry 20:1902–1906

Komine Y, Kitibataka M, Yokogawa T, Nishikawa K, Inokuchi H (1999) A tRNA-like structure is present in l0Sa RNA, a small stable RNA from *Escherichia coli*. Proc Natl Acad Sci U S A 91:9223–9227

Koski R, Bothwell ALM, Altman S (1976) Partial purification and characterization of a ribonucleasse P-like activity from human KB cells. Cell 9:101–116

Kouzuma Y, Mizoguchi M, Takagi H, Fukuhara H, Tsukamoto M, Numata T, Kimura M (2003) Reconstitution of archaeal ribonuclease P from RNA and four protein components. Biochem Biophys Res Commun 306:666–673

Kovrigina E, Wesolowski D, Altman S (2003) Coordinate inhibition of expression of several genes for protein subunits of human nuclear RNase P. Proc Natl Acad Sci U S A 100:1598–1602
Krasilnikov AS, Yang X, Pan T, Mondragon T (2003) Crystal structure of the specificity domain of ribonuclease P. Nature 421:760–764
Kraus G, Geffin R, Spruill G, Young AK, Seivright R, Cardona D, Burzawa J, Hnatyszyn HJ (2002) Cross-clade inhibition of HIV-1 replication and cytopathology using sequence-specific RNase P molecules. Proc Natl Acad Sci U S A 99:3406–3411
Li Y, Altman S (2003) A specific endoribonuclease RNase P, affects gene expression in polycistronic operon mRNAs. Proc Natl Acad Sci U S A 100:13213–13218
Li Y, Altman S (2004) Polarity effects in the lac operon of *E. coli*. J Mol Biol 339:31–39
Liu F, Altman S (1995) Inhibition of viral gene expression by the catalytic RNA subunit of RNase P from *Escherichia coli*. Genes Dev 9:471–480
Mann H, Ben-Asouli Y, Schien A, Moussa S, Jarrous N (2003) Eukaryotic RNase P: role of RNA and protein subunits of a primordial catalytic ribonucleoprotein in RNA-based catalysis. Mol Cell 12:925–935
McKinney J, Guerrier-Takada C, Wesolowski D, Altman S (2001) Inhibition of *Escherichia coli* viability by 221 external guide sequences complementary to two essential genes. Proc Natl Acad Sci U S A 98:6605–6610
Ozeki H, Sakano H, Youada S, Ikemura T, Shimura Y (1975) Temperature sensitive mutants of *Escherichia coli* defective in tRNA biosynthesis. Brookhaven Symp 26:89–105
Pannucci JA, Haas ES, Hall TA, Harris JK, Brown JW (1999) RNase P RNAs from some Archaea are catalytically active. Proc Nat Acad Sci U S A 96:7803–7808
Peck-Miller K, Altman S (1991) Kinetics of the processing of the precursor to 4.5 S RNA, a naturally occurring substrate for RNase P from *E. coli*. J Mol Biol 221:1–5
Pei D-S, Sun Y-H, Long Y, Zhu Z-Y (2008) Inhibition of no tail (ntl) gene expression in zebrafish by external guide sequence (EGS) technique. Mol Biol Rep 35:139–143
Persson T, Cuzic S, Hartmann RK (2003) Catalysis by RNase P RNA: unique features and unprecedented active site plasticity. J Biol Chem 278:43394–435401
Plehn-Dujowich D, Altman S (1998) Effective inhibition of influenza virus production in cultured cells by external guide sequences and ribonuclease P. Proc Natl Acad Sci U S A 95:7327–7332
Puranam RS, Attardi G (2001) The RNase P associated with HeLa cell mitochondria contains an essential RNA component identical in sequence to that of nuclear RNase P. Mol Cell Biol 21:548–561
Rangarajan S, Raj ML, Hernandez JM, Grotewold E, Gopalan V (2004) RNase P as a tool for disruption of gene expression in maize cells. Biochem J 380:611–616
Reiner R, BenAsouli Y, Krivetsky I, Jarrous N (2006) A role for the catalytic riboneucleoprotein in RNA polymerase III transcription. Genes Dev 20:1621–1635
Rendan L, Schroder I, Soll D (2008) Life without RNase P. Nature 453:120–123
Robertson HD, Altman S, Smith JD (1972) Purification and properties of a specific *Escherichia coli* ribonuclease which cleaves a tyrosine transfer ribonucleic acid precursor. J Biol Chem 247:5243–5251
Rossmanith W, Karwan RM (1998) Characterization of human mitochondrial RNase P: novel aspects in tRNA processing. Biochem Biophys Res Commun 247:234–241
Samanta MP, Tongprasit W, Sethi H, Chin CS, Stolc V (2006) Global identification of noncoding RNAs in *Sacchaaromyces cerevisiae* by modulating an essential RNA processing pathway. Proc Natl Acad Sci U S A 103:4192–4197
Schedl P, Primakoff P (1973) Mutants of *Escherichia coli* thermosensitive for the synthesis of transfer RNA. Proc Natl Acad Sci U S A 70:2091–2095
Seif E, Altman S (2008) RNase P cleaves the adenine riboswitch and stabilizes pbuE mRNA in *B. subtilis*. RNA 14:1237–1243
Shan G, personal communication
Spitzfaden C, Nicholson N, Jones JJ, Guth S, Lehr R, Prescott CD, Hegg LA, Eggleston DS (2000) The structure of ribonuclease P protein from *Staphylococcus aureus* reveals a unique binding site for single-stranded RNA. J Mol Biol 295(1):105–115

Stams TS, Niranjanakumari S, Fierke CA, Christianson DW (1998) Ribonuclease P protein structure: evolutionary origins in the translational apparatus. Science 280:752–755

Stark BC, Kole R, Bowman EJ, Altman S (1978) Ribonuclease P: an enzyme with an essential RNA component. Proc Natl Acad Sci U S A 75:3717–3721

Steitz TA, Moore PB (2003) RNA, the first macromolecular catalyst: the ribosome is a ribozyme. Trends Biochem Sci 28:411–418

Stolc V, Altman S (1997) Rpp 1, an essential protein subunit of nuclear RNase P required for processing of precursor tRNA and 35 S precursor rRNA in *Saccharomyces cerevisiae*. Genes Dev 11:2414–2425

Tsai HY, Masquida B, Biswas R, Westhof E, Gopalan V (2003) Molecular modeling of the three-dimensional structure of the bacterial RNase P holoenzyme. J Mol Biol 325:661–675

Tsai HY, Pulukkunat DK, Woznick WK, Gopalan V (2006) Functional reconstitution and characterization of *Pyrococcus furiosus* RNase P. Proc Natl Acad Sci U S A 103:16147–16152

Wang MJ, Davis NW, Gegenheimer P (1988) Novel mechanisms for maturation of chloroplast transfer RNA precursors. EMBO J 7:1567–1574

Waugh DS, Pace NR (1990) Complementation of an RNase P (rnpB) glue deletion in *Escherichia coli* by homologous genes from distantly related bacteria. J Bacteriol 172:6316–6322

Westhof E, Wesolowski D, Altman S (1996) Mapping in three dimensions of regions in a catalytic RNA protected from attack by an FE(II)-EDTA reagent. J Mol Biol 258:600–613

Yang L, Altman S (2007) A non-coding RNA in *Saccharomyces cerivisiae* is a RNase P substrate. RNA 13:1–9

Zahler NH, Sun L, Christian EL, Harris ME (2005) The pre-tRNA nucleotide base and 2′-hydroxyl at N-1 contribute to fidelity in tRNA processing by RNase P. J Mol Biol 345:969–985

Zhang H, Altman S (2004) Inhibition of the expression of the human RNase P protein subunits Rpp 21, Rpp25, Rpp29 by external guide sequences (EGSs) and siRNA. J Mol Biol 342:1077–1083

Zuo Y, Deutscher MP (2002) Mechanism and action of RNase T. J Biol Chem 277:50160–50164

Chapter 2
The Evolution of RNase P and Its RNA

J. Christopher Ellis and James W. Brown

2.1 Introduction

Ribonuclease P (RNase P) is a ribonuclease responsible for the 5′ maturation of transfer RNA (tRNA). The RNase P holoenzyme is most commonly comprised of a single RNA and one or several associated proteins. It is the RNA, not any protein, that is the catalytic subunit of the enzyme; RNase P is an RNA enzyme. The reaction it carries out is the hydrolysis of a phosphodiester bond in the pre-tRNA, generating two RNA products: the 5′-mature tRNA and the 5′ leader RNA fragment. The RNase P RNA subunit is present in all three Domains of life (Bacteria, Archaea, and Eukarya), and in at least some mitochondria and plastids. RNase P RNA, nuclear splicing RNAs (snRNAs), and the ribosomal RNA are the only catalytic RNAs described to date that can conduct multiple catalytic cycles; other "ribozymes" are self-reactive. The presence of RNase P RNA in all branches of living things and its ability perform catalytically have lead to the hypothesize that RNase P is a relic of the RNA world.

RNase P has been studied primarily in terms of the maturation of tRNA, but the holoenzyme has also been shown to cleave the other substrates and play other important roles within the cell. It is also involved in the maturation of a host of other noncoding RNA such as the 2 S, 4.5 S, tmRNAs, and snoRNAs, and in the leader sequences of mRNAs it has also been shown to recognize and cleave riboswitchs such as the coenzyme B_{12} riboswitch in *E. coli* and *Bacillus subtilis* and the adenine riboswitch of the adenine efflux pump transcript in *Bacillus subtilis*. Interestingly neither the coenzyme B_{12} riboswitch nor the adenine riboswitch contains a predicted structure that is known to be recognized as an RNase P RNA substrate (Hori et al. 2000a; Gimple and Schon 2001; Tous et al. 2001; Altman et al. 2005; Coughlin et al. 2008; Seif and Altman 2008).

Although the sequences of RNase P RNAs are highly variable and recognize many different substrates, a core of the sequence and secondary structure is conserved among examples in all living things (Fig. 2.1). Sequences and sequence length vary

J.C. Ellis and J.W. Brown (✉)
Department of Microbiology, North Carolina State University, Campus Box 7615, Raleigh, NC 27695, USA
e-mail: james_brown@ncsu.edu

F. Liu and S. Altman (eds.), *Ribonuclease P*, Protein Reviews 10,
DOI 10.1007/978-1-4419-1142-1_2, © Springer Science+Business Media, LLC 2010

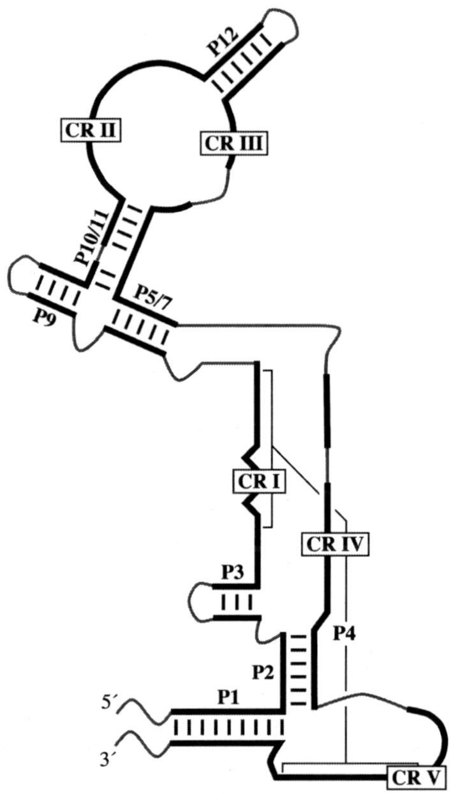

Fig. 2.1 Consensus secondary structure of bacterial, archaeal and eukaryotic nuclear RNase P RNAs. Elements of structure absent in only single, small evolutionary groups are included: P12 (absent in *Mycoplasma fermentans*), P2 and everything distal to P10/11 (reduced or absent in *Pyrobaculum*). *Grey lines* indicate connections that vary in structure between conserved elements. *Helices* are labeled P1-12 according to Haas and Brown (1998), and *conserved sequence regions* are labeled CR I-V according to Chen and Pace (1997)

considerably among RNase P RNAs, especially in eukaryotes, but in Bacteria and Archaea only five major distinct classes of structures have been described so far: "A" (ancestral) type, "B" (named for *Bacillus*, in which it was first discovered) type, "C" (Chloroflexi) type, "M" (Methanococci) type, and "P" (*Pyrobaculum*) type (Figs. 2.2 and 2.3). RNase P RNAs in eukaryotes (including organelles in addition to the nuclear enzymes) are more variable in both sequence and structure, and have not yet been divided into clear structure classes.

Molecular therapeutics is a field of particular current interest, in particular the engineering of RNA molecules for therapeutic use in humans and other animals. An important approach has been the engineering of External Guide Sequences (EGS) that create RNase P substrates in pathogenic RNAs, thereby directing specific destruction of these RNAs by the innate RNase P. This approach has been applied with some success in viral pathogenesis (Reyes-Darias et al. 2008). Another approach

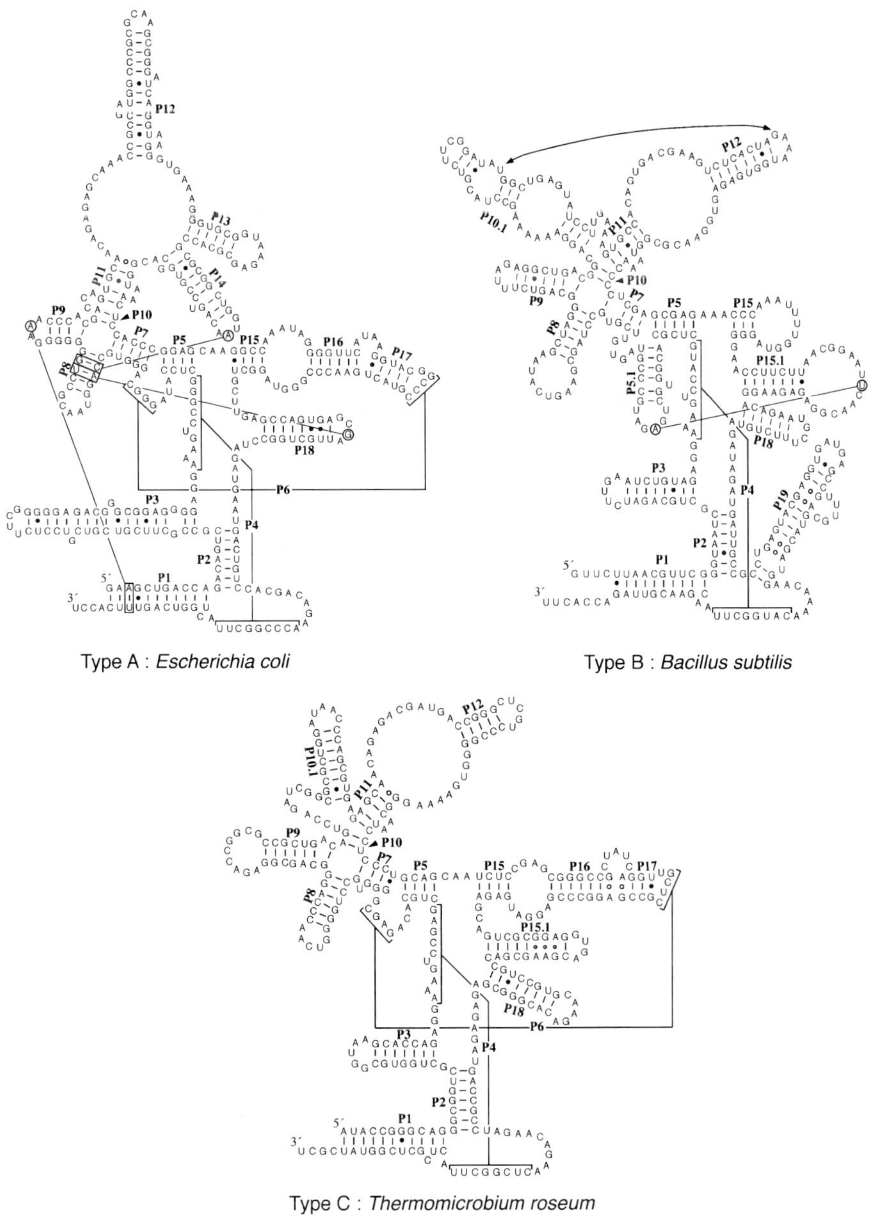

Fig. 2.2 Bacterial RNase P RNA structures classes. Representative RNase P RNA secondary structures of bacterial types A, B, and C. Helices are labeled P1-19 according to Haas and Brown (1998). Type A RNAs are found in most Bacteria, and sometimes lack P13 and P14 (in beta-proteobacteria) or P18 (Chlorobi). Type B RNAs are present in most Firmicutes, and in at least some mollicutes lack both P12 and P10.1. Type C RNAs are found in some Chloroflexi; structural variation in this group has not been investigated. The presence of P19 is variable in all types of RNase P RNAs. Many bacterial RNAs also contain additional peripheral helices in P12 or between P15 and P16, or between P16 and P17. The lengths of P3 and P12 are highly variable. Structures are from the RNase P Database (Brown 1999)

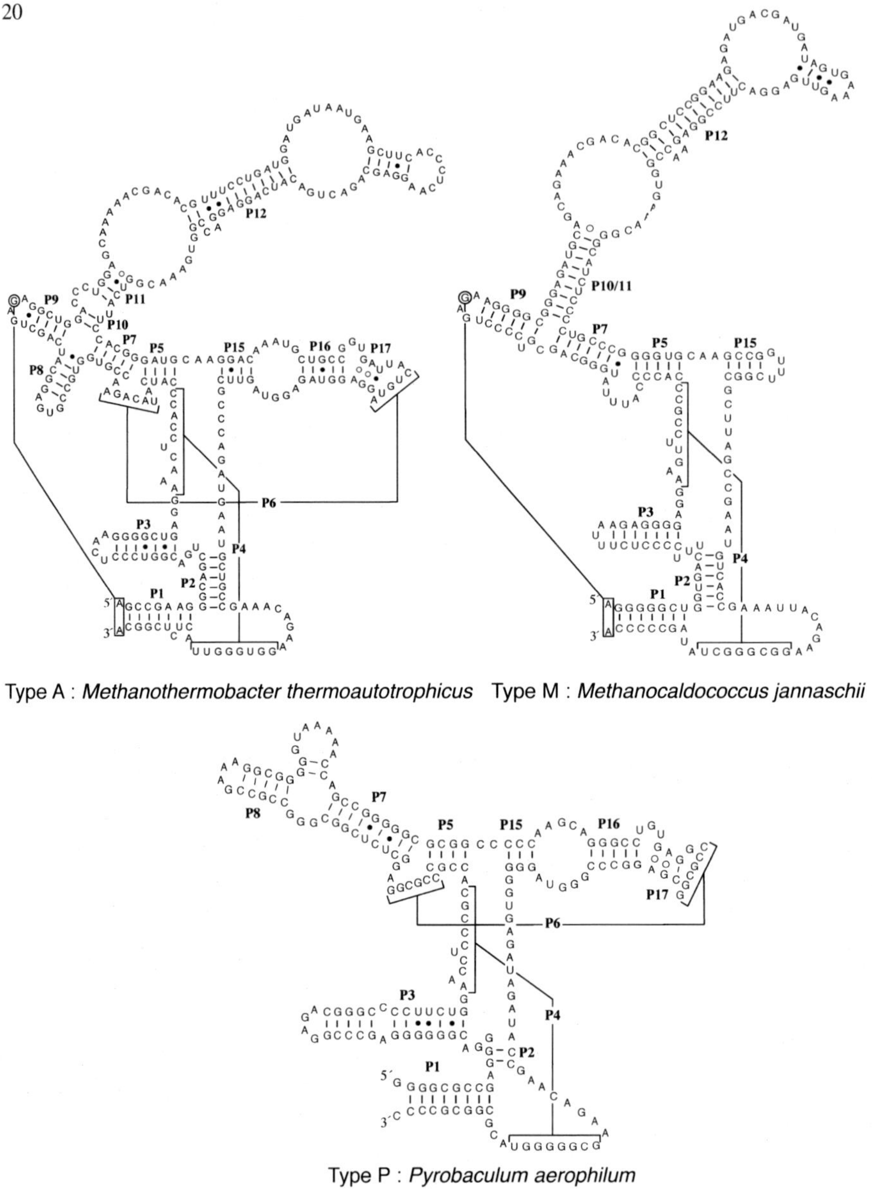

Fig. 2.3 Archaeal RNase P RNA structure classes. Representative RNase P RNA secondary structures of archaeal type A, M, and P. Helices are labeled P1-P17 according to Haas and Brown (1998). Type A RNAs are found in most Archaea, and most are, like their bacterial homologs, catalytically active in vitro in the absence of protein. P19 (see Fig. 2.2) is present in some archaeal type A RNAs. The structure of P12 is highly variable in these RNAs; the structures of P3 and P15/P16/P17/P6 also vary. Type M RNAs are found in Methanococci and *Archaeoglobus*; these RNAs are not catalytically proficient in the absence of protein, and vary little in structure. Type P RNAs have been seen only in species of the genus *Pyrobaculum*, and lack nearly all of the "S" domain seen in other bacterial and archaeal RNase P RNAs. In addition, conserved regions IIV and V (CR II and III are in the S-domain) are poorly conserved. The presence of the truncated "P2" is not confirmed by sequence covariation. Structures are from the RNase P Database (Brown 1999)

has been to engineer the RNase P RNA itself. One example is the recent engineering of an RNase P ribozyme to target the overlapping coding region of two capsid proteins in the murine cytomegalovirus (MCMV) (Bai et al. 2008). The engineered RNase P ribozyme was found to decrease viral growth by 2,000-fold (Bai et al. 2008).

Here we examine our current understanding of the evolutionary variation of the RNase P holoenzyme in the three domains of life (Bacteria, Archaea, and Eukaryotes), especially with respect to evolutionary diversity in RNA structure and protein composition.

2.2 Bacterial Ribonuclease P

In Bacteria, the RNase P holoenzyme is a ribonucleoprotein complex comprised of a single RNA (~400 nt) and a single small protein subunit (~14 kDa). The secondary structure of the RNA was determined by phylogenetic comparative analysis of several hundred sequences over the course of many years (James et al. 1988; Harris et al. 2001). In Bacteria, these secondary structures fall into two structure classes: A-type, the most common, and B-type, found in only in Gram-positive Bacteria (Fig. 2.2) (Brown et al. 1991, 1993, 1996; Haas et al. 1996b). In addition, some of the *Chloroflexi* (green non-sulfur Bacteria) RNase P RNAs share features intermediate between type A and B RNAs, and these have sometimes been referred to as type C (Haas and Brown 1998).

The tertiary structure of the RNase P RNA is formed by coaxially stacked helical domains. These domains are stabilized and joined together by long range docking interactions and result in a remarkably planar tertiary structure (Kazantsev et al. 2005; Torres-Larios et al. 2005). The most evolutionarily conserved sequences in the RNA are located near the substrate binding surface of the molecule, whereas the most variable sequences (variable in sequence, secondary, and tertiary structure) are located on the periphery of the RNA and are involved in stabilizing the core RNA structure (Kakuta et al. 2005; Kazantsev et al. 2005). Substrate recognition outside of the catalytic site is performed almost entirely by the P7/P8/P9/P10 cruciform and L15 (Christian and Harris 1999). The loop of P8 in the cruciform (L8) interacts with the catalytic center in P4, and this interaction is stabilized by tertiary contacts between P14 and P18 with L8 (Brown et al. 1996; Massire et al. 1998). P8 has been shown to be involved in the recognition of the substrate T-loop, but the details of this interaction are not known (Nolan et al. 1993). The other substrate recognition element L15 is the loop distal of P15; in type A RNAs, this is an internal loop between P15 and P16, in type B RNAs, it is the terminal loop of P15. Although the two helices proximal and distal to L15 are not well conserved in sequence, the sequence GGUA immediately preceding the 3′ strand of P15 is conserved in most bacteria. This sequence is directly involved in substrate recognition by basepairing to the 3′-NCAA tail of the pre-tRNA (Kirsebom and Svärd 1994). Some bacterial RNase P RNAs (notably those of many cyanobacteria and *Chlamydiae*) lack this conserved sequence and yet are functional in the absence of protein; recognition of the 3′-NCCA tail of pre-tRNA by these RNAs is uncharacterized.

2.2.1 Bacterial RNase P RNA Structure Classes

A-type RNase P RNAs are the most common form of the RNA in Bacteria (and Archaea) and *Escherichia coli* is the model for this structure class. The A-type RNase P RNA molecule have several RNA elements not found in other RNase P RNA types: P6, P13, P14, P16, and P17 (Haas and Brown 1998). Although these elements vary in sequence and in length and are not found in all RNase P RNAs, they play important roles in the tertiary structure of the RNA itself by creating long-range interactions that stabilize the functional core of the RNA. P16 and P17 are an extension of L15, and the loop of P17 basepairs with an asymmetric bulge between P5 and P7 to create helix P6 and a pseudoknot in the global RNA structure. Another pseudoknot is found in all known RNase P RNAs, created by P2 and P4, and some contain yet another pseudoknot created by the basepairing of P8 with the 3′-tail of the RNA. P13 and P14 stack coaxially; the loop of P14 forms a tertiary interaction with P8, whereas P13 interacts with the base of P12; the details of this interaction remain unclear (Brown et al. 1996).

B-type RNase P RNAs differ dramatically from A-type, but not in regions involved directly in substrate recognition (P8 and L15) or in the catalytic core (primarily P4 and surrounding joining regions). B-type RNase P RNAs are found in most Firmicutes (low G+C Gram-positive Bacteria) and the RNase P RNA of *Bacillus subtilis* is the model example. Type B RNAs lack both the P16/P17/P6 and P13/P14 structures, but these seem to be replaced by type B-specific elements P5.1/P15.1 and P10.1 (Haas et al. 1996b). The sequences and structures shared by both type A and type B RNase P RNAs define a "core" present in the RNase P RNAs that are catalytically proficient in vitro in the absence of protein. That this core contains all of the sequences and structures required for catalytic function has been confirmed by the characterization of "mini" and "micro" RNase P RNAs consisting only of this core (Waugh et al. 1989; Siegel et al. 1996). Although catalytically active, these core RNAs require substantial external stabilization; the implication of this is that although the core contains all of the elements of sequence and structure required for the substrate recognition and catalysis, the phylogenetically variable regions contribute substantially to the stabilization of this functional core (Kazantsev et al. 2005; Torres-Larios et al. 2005).

The evolutionary transition between type A and type B RNAs appears to have happened abruptly, with no evolutionary intermediates or alternative descendents described to date (Haas et al. 1996b). However, the RNase P RNAs found in the phylogenetically distant *Chloroflexi* (green non-sulfur Bacteria) are intermediate between types A and B. This seems to be an example of convergent evolution, rather than horizontal transfer, because although the secondary structures of these RNAs are related, their sequences are not specifically similar.

2.2.2 Dimerization Mediated by the RNA

The RNase P holoenzyme in Bacteria is a heterodimer of one molecule each of the RNA and protein subunits. The holoenzyme heterodimer of both *E. coli* (type A) and *Bacillus subtilis* (type B) can in turn dimerize in solution (Buck et al. 2005a). Buck et al. demonstrated that although both enzymes could form these dimers, the *E. coli* holoenzyme forms a heterogeneous mixture of holoenzyme monomers and dimers, and this mixtures shifts almost entirely to monomers in the presence of mature tRNA (Buck et al. 2005a). However, the *B. subtilis* holoenzyme exists almost exclusively as dimers, and does not shift to monomers in the presence of mature tRNA (Buck et al. 2005a). Buck et al. argue convincingly that it is the attributes of the RNA, not the protein, that are responsible for dimer formation in both *E. coli* and *B. subtilis*. Therefore, it is presumably the additional RNA elements (i.e., P5.1, P10.1, P15.1, P15.2) in Type B RNase P RNA that are associated with stable dimer formation in B-type RNAs, and the absence of these RNA elements in A-type RNase P RNA that disfavor dimer complex formation in the presence of substrate. The biological relevance (and even existence) of these dimers in vivo is unknown and, perhaps doubtful.

2.2.3 The Role of the Protein Subunit

With so much variation at the level of the ancient catalytic RNA, one might anticipate that the associated protein would be highly variable in sequence and structure, and that perhaps the proteins that associate with type A and B RNAs might be readily distinguishable. This is not the case. The protein is highly conserved not only in the organisms encoding the ancestral A-types RNA but also with the organisms encoding the type B RNase P RNAs. It is approximately 14 kDa and shares an unusual left-handed crossover connection and a large central cleft (Stams et al. 1998). This motif is shared by ribosomal protein S5 and ribosomal translocase elongation factor, which suggests these proteins evolved from a common ancestor of the ancient translational machinery (Stams et al. 1998).

Although the protein sequence and structure is highly conserved in Bacteria, the functional impact of the protein differs somewhat depending on the type of RNase P RNA to which it binds. In *B. subtilis,* the protein increases the holoenzymes substrate specificity (Crary et al. 1998; Niranjanakumari 1998), whereas the *E. coli* protein stabilizes the tertiary structure of the corresponding RNA (Guerrier-Takada et al. 1983; Westhof et al. 1996; Buck et al. 2005a). Furthermore, in *E. coli* the protein decreases the Mg^{2+} dependence of the RNA to fold into the native state from the intermediate state but the protein in *B. subtilis* does not change the RNase P RNA's Mg^{2+} dependence (Loria and Pan 1996; Buck et al. 2005a). Despite some functional differences that are dependent on RNase P RNA type it is clear that the protein plays an important role in both types of RNAs in substrate recognition and tertiary structure of the RNA, and the significance of these biochemical differences in vitro is unclear, given that these proteins have been shown to be interchangeable in vivo where-ever tested (Waugh and Pace 1990; Gösringer and Hartmann 2007).

2.2.4 RnpA is Part of a Conserved Genomic Arrangement

In bacteria, the gene that encodes the RNase P RNA protein subunit (*rnpA*) is located immediately downstream and in the same orientation as *rpmH*, which encodes for the ribosomal protein L34. This gene arrangement and their location near the origin of replication are highly conserved in bacteria with only a couple of known exceptions. The two genes in *E. coli* have been shown to be part of an operon with one minor and two major promoters upstream of *rpmH* and two transcriptional termination signals downstream of *rpmH* (upstream of *rnpA)* (Hansen et al. 1982, 1985; Panagiotidis et al. 1992). Most of the transcripts from the operon do not include the *rnpA* coding region and therefore the proteins are differentially expressed. Ribosomal protein L34 is abundant in cells but the RNase P protein is expressed at much lower levels, in part to the transcriptional disparity but also because the *rnpA* transcript is a poor match to the codon bias of *E. coli*, decreasing the translation efficiency further.

A handful variations of this genetic arrangement have been described, that of *Aquifex aeolicus* which we will be discussed below, and that of species of the genus *Thermus*. *Thermus* species are thermophilic, and their RNase P enzymes are composed of the usual type A RNA and *rnpA* protein. However, in all species of *Thermus* examined, *rnpA* and *rpmH* are completely overlapping (Fig. 2.4). The start codons of two genes are separated by only four base pairs and share the same orientation (Feltens et al. 2003). The first start codon initiates the *rnpA* coding sequence in the −1 register relative to *rpmH* and the second start codon initiates the coding sequence for *rpmH* (Feltens et al. 2003). This results in an *rnpA* protein that

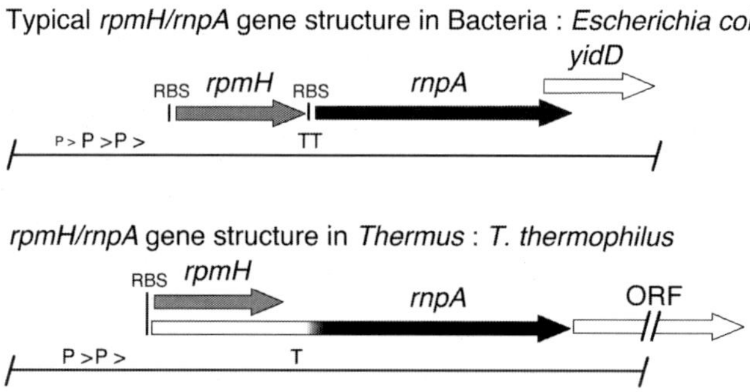

Fig. 2.4 RNase P protein gene (*rnpA*) arrangement in *Thermus*. The *rnpA* gene in Bacteria is generally encoded near the chromosomal origin of replication and immediately downstream of and in the same orientation as the gene for ribosomal protein L34 (*rpmH*) and often immediately upstream of *yidD* (glutathione-S-transferase). In species of the genus *Thermus*, the *rnpA* gene is extended upstream to completely overlap the *rpmH* gene, and these two open reading frames share the same ribosomal binding site (RBS) for translational initiation (Feltens et al. 2003). Promoters for the expression of this transcriptional unit are designated "P," transcriptional termination signals (which allow some transcriptional read-through) are designated "T." The homologous sequences in *rnpA* are shaded black

is substantially longer protein (163 amino acids) than the usual ~120 amino acids. In the lactic acid bacterium *Lactococcus lactis subsp. cremoris*, the rnpA gene is preceded by an ORF (LACR0128) rather than *rpmH*, and an *rmpH* homolog has not be identified in the genome (Wegmann et al. 2007).

2.2.5 And Then There was One: Aquifex and the Missing Link

Aquifex aeolicus is a deep branching hyperthermophilic bacterium. No candidate RNase P RNA- or protein-encoding genes have been identified in the genome of this specie, despite the fact that tRNA gene organization in this genome implies the need for this enzyme; tRNA genes are (as is usual in Bacteria) clustered and presumably co-transcribed, and some are located within the rRNA operon (Li and Altman 2004). Biochemical investigation of this surprising finding has been hampered by the fact that this organism is very difficult to cultivate, however some progress has been made. Substitution of Mg^{2+} with other divalent cations such as Co^{2+} completely inhibited the RNase P activity in extracts, consistent with what is seen in other bacteria (Kazakov and Altman 1991; Lombo and Kaberdin 2008; Marszalkowski et al. 2008). Furthermore, depletion of RNA from the cell extracts of *A. aeolicus* eliminates RNase P activity. Both of these are indirect evidence that an RNA is required for RNase P activity, but if this is the case, this RNA must not resemble any previously described RNase P RNA (Lombo and Kaberdin 2008).

Although no RNase P RNA gene has been found in the *A. aeolicus* genome, what about the conserved gene arrangement typically associated with *rnpA*? In relatives of *A. aeolicus* such as *Sulfurihydrogenibium azorense* and *Pseudoscourfieldia marina* (which contain more-or-less typical bacterial type A RNase P RNAs lacking P18), the usual bacterial arrangement and orientation of *rpmH* → *rnpA* is present (Ogasawara et al. 1985; Salazar et al. 1996). However, in *A. aeolicus, rnpA* is not found downstream of *rpmH*, nor anywhere else in the genome (Li and Altman 2004; Marszalkowski et al. 2008). How, then, is *A. aeolicus* processing pre-tRNAs in the apparent absence of an *rnpA* protein, which is essential in vitro in other bacteria? Is it possible that the novel variant of an RNase P RNA can process pre-tRNAs in the absence of the usually associated *rnpA* protein? Or has this species invented an entirely new RNase P enzyme of some type?

2.3 Archaeal Ribonuclease P

Like Bacteria, Archaea are single cell non-eukaryotic microorganisms. Archaea are divided into two major phyla: Crenarchaea and Euryarchaea. Although Archaea are distinct from both eukaryotes and bacteria, they are in many ways like eukaryotes, and in many other ways like bacteria. The RNase P holoenzyme in Archaea is an excellent evolutionary example of this. Most archaeal RNase P RNAs are of the type A found also in bacteria (Fig. 2.3), and at sufficiently high salt concentrations (4 M ammonium acetate and 300 mM $MgCl_2$) the RNase P RNAs of some of many of these Archaea are catalytically active in the absence of protein in vitro, and the activity of these RNAs is enhanced at more moderate ionic strength by the addition of the *Bacillus subtilis* (a bacterium) RNase P protein (Pannucci et al. 1999). However, the archaeal RNase P holoenzyme generally contain four proteins, none of which are clearly homologous to the bacterial *rnpA* RNase P protein, but these *are* clearly homologous to the core RNase P proteins of the eukaryotic nucleus (Hall and Brown 2004; Hartmann and Hartmann 2003). Because there is not any clear-cut homology between the bacterial RNase P protein on one hand and the archaeal and eukaryotic nuclear proteins on the other, it may be that the RNA subunit of RNase P evolved first, perhaps as part of the RNA World, and the proteins evolved independently following the divergence of the bacterial and eukaryotic/archaeal lineages.

2.3.1 Archaeal RNase P RNA Structure Classes

As in bacteria, most archaeal RNase P RNAs are A-type RNAs, largely resembling bacterial RNase P RNAs in terms of conserved sequences and structures. Archaeal type A RNase P RNAs do differ from those of most bacteria in the absence of P13/P14 (also absent in the type A RNAs of beta-proteobacteria) and P18 (also absent in the type A RNAs of Chlorobi and Aquificae). However, despite this structural similarity between the bacterial and archaeal RNase P RNA in general, the first available archaeal RNase P RNA sequences were not as bacterial-like as others, and could not be shown to be catalytically active in the absence of protein. It therefore became a thought that, like eukaryotic RNase P RNAs, the archaeal RNase P RNA was absolutely dependent on associated protein for catalytic activity. However, it was subsequently shown that many archaeal RNase P RNAs *are* catalytically active in the absence of protein, but require extreme ionic conditions (4 M ammonium acetate, 300 mM MgCl2) for this to be expressed (Pannucci et al. 1999). As in Bacteria, the protein subunits in the archaeal enzymes seem not to contribute directly toward catalysis, but at least predominantly toward the stabilization of the superstructure of the RNA subunit.

One of the subsets of archaeal RNase P RNAs that does not display catalytic activity even under high ionic conditions are those with a different structure class, the M-type RNase P RNAs (Fig. 2.3) To date, only five archaeal genera (*Archaeoglobus, Methanocaldococcus, Methanothermococcus,* and *Methanococcus*) are known to have M-type RNase P RNAs; this suggests that *Archaeoglobus* may be related to the Methanococci, despite uncertainly in the placement of this genus in phylogenetic trees based on rRNA sequences (Brown 1999). M-type RNAs are essentially similar to archaeal A-type RNAs but lack two RNA structural elements that are essential (at least in Bacteria) for substrate recognition: P8 and L15 (along with P16/P17/P6) (Fig. 2.3). P8 is part of the highly conserved cruciform consisting of P7, P8, P9, and P10 (Brown and Haas 1995; Harris et al. 2001). P8 is required for recognition of the T-loop of the substrate pre-tRNA. In addition, in Bacteria P8 stabilizes P18, P4, and P14 (via tertiary contacts) and P9 (via stacking). L15 recognizes and binds the 3′-NCCA tail of pre-tRNA, which is necessary for an efficient substrate cleavage (Svard and Kirsebom 1992; Svard et al. 1996). Type M RNase P RNAs, then, have specifically lost all of the elements of the RNA that are known to be directly involved in the substrate recognition outside of the catalytic center, and are remarkable like the minimal consensus RNase P RNA structure (Fig. 2.1).

A more extreme case of reduction in RNase P RNA is found in species of the genus *Pyrobaculum*. These "type P" RNase P RNAs were long unrecognized because they consist essentially of only the catalytic domain for the enzyme; P9 (or is it P8?) and everything distal to P10 is absent (Personal communication with Todd Lowe) However, this finding is quite recent and the possibility that a second RNA is associated with this catalytic core of the RNA, either in *trans* or covalently, has not been ruled out. The conserved sequences in these RNAs are not as good matches to the consensus as are those of other Archaea, and the biochemical properties of the enzyme are unknown.

2.3.2 *Pyrococcus horikoshii OT3 as a Model For RNase P in Archaea*

The structure of human RNase P proteins is of critical importance, but unfortunately human RNase P proteins, and more generally eukaryotic RNase P proteins, have been difficult to crystallize for structural analysis. However, the archaeal proteins are homologous to those of eukaryotes (including humans), and perhaps because they are thermophilic have proven to be more amenable to crystallographic analysis. The emerging model system is the RNase P of *Pyrococcus horikoshii*. The common four archaeal proteins have been experimentally confirmed to be a part of the RNase P holoenzyme and contribute to its function in this specie. All four proteins subunits of *P. horikoshii* (Ph1481p, Ph1601p, Ph1771p, and Ph1877p) share clear homology to four of the RNase P protein subunits identified in *Homo sapiens*: Pop5, Rpp21, Rpp29, and Rpp30 respectively. Only three of these subunits are necessary for the RNase P activity in vitro (Ph1481p, Ph1601p, and Ph1771p) but all four protein subunits are required to for robust activity (Kouzuma et al. 2003).

The *P. horikoshii* RNase P holoenzyme is composed of the RNase P RNA and 2 heterodimers that are formed by Ph1877p:Ph1481p and Ph1601p:Ph1771p protein interactions (Kawano et al. 2006; Honda et al. 2008). The two heterodimers interact weakly with each other and all four proteins interact with the RNase P RNA to form the complete holoenzyme (Kouzuma et al. 2003; Kifusa et al. 2005). The three-dimensional structures of all four of these proteins have been determined.

RNase P protein Ph1771p is a homolog of human RNase P protein Rpp29 and is composed of four α-helices and six antiparallel β-sheets and one β-strand near the C-terminus that protrudes away from the globular portion of the protein forming a β-barrel structure (Numata et al. 2004). The protruding β-strand (β7) forms a β-sheet with β4 which may be involved in protein–protein interaction with the other RNase P subunits (Numata et al. 2004). Two possible RNA binding sites were identified in Ph1771p. The first is a loop region connecting strands β2 and β3 which is composed of hydrophilic residues exposed to solvent and the other is composed of α-helices 1-4 and β-strand β6 forming a cluster of positively charged amino acids (Numata et al. 2004). Ph177p shares structural similarity to the other RNA binding proteins: *Staphylococcus aureus* translational regulator Hfq and *Haloarcula marismortui* ribosomal protein L21E (Numata et al. 2004).

RNase P protein Ph1877p is a homolog of human RNase P protein Rpp30 and is composed of ten α-helices and seven β-strands forming a TIM barrel (Takagi et al. 2004; Kawano et al. 2006). Ph1877p interacts with Ph1481p in the RNase P holoenzyme, which is consistent with the interactions determined in other archaeal RNase P protein–protein interactions (Hall and Brown 2004; Kifusa et al. 2005). The structure of the Ph1481p-Ph1877p dimmer has been determined in co-crystals.

RNase P protein Ph1601p is a homolog to human RNase P protein Rpp21 and is composed of an N-terminal domain of two α-helices, and the central and C-terminal regions form a zinc ribbon domain, giving the protein an L-shape (Kakuta et al. 2005). Several clusters of positively charged amino acids along one face of the L-arms, suggest an RNA binding role, whereas four Cys residues bind a zinc molecule stabilizing the N-terminus and C-terminus domains (Kakuta et al. 2005).

RNase P protein Ph1481p is a homolog to human RNase P protein Pop5 and is composed of five antiparallel β-sheets and five α-helices forming a α/β globular protein (Kawano et al. 2006). Whether or not this structure is meaningfully similar to the bacterial RNase P protein structures remains a matter of disagreement (Stams et al. 1998; Spitzfaden et al. 2000; Kazantsev et al. 2003; Takagi et al. 2004), but it is generally believed that any such similarity is most likely the result of convergence rather than homology. Certainly, the structure of Ph1481 is similar to the generic ribonucleoprotein (RNP) domains found in a number of RNA binding proteins, suggesting a role in the binding of the protein to RNase P RNA or pre-tRNA. Ph1877p and Ph1481p form a heterotetramer in solution, with a homodimer of Ph1481p in the center of two monomers of Ph1877p (Kawano et al. 2006). Whether this heterotetramer is associated with one or two molecules of the RNA subunit of RNase P has not been determined.

2.3.2.1 Is L7Ae a Fifth RNase P Protein in *Pyrococcus horikoshii* OT?

L7Ae protein was first described as a protein associated with the small ribosomal subunit. However, the crystal structure of the complete large ribosomal subunit (50 S) with the associated proteins reveled L7Ae was actually associated with the large subunit (Ban et al. 2000). More recently, L7Ae was found to be a component of C/D and H/ACA small nucleolar RNAs (sRNA) (Kuhn et al. 2002; Rozhdestvensky et al. 2003). L7Ae's association with sRNAs is at kink-turns (k-turns), found in both ribosomal RNA and sRNAs (Rozhdestvensky et al. 2003).

L7Ae protein in *Haloarcula marismortui* is similar in sequence to human RNase P protein subunit Rpp38 (Fukuhara et al. 2006). Furthermore, the *Pyrococcus horikoshii* RNase P reconstituted from the RNA and four known proteins has a lower than expected optimal reaction temperature of 55°C, much lower than the 70°C of enzyme purified from cell extracts (Kouzuma et al. 2003). Fukuhara et al. have examined the role L7Ae might play in the *Pyrococcus horikoshii* RNase P, and although they have no *direct* evidence that Ph1496p is an actual protein component of the enzyme in vivo, they have substantial circumstantial evidence that this is the case. Most notably, when L7Ae (Ph1496p) is added to the four known RNase P proteins (Ph1481p, Ph1601p, Ph1877p, and Ph1771p) and RNase P RNA in reconstitution experiments, optimal enzymatic activity is restored to the wild type temperature of 70 °C (Fukuhara et al. 2006). Although the binding region of L7Ae in the *P. horikoshii* RNase P RNA is not a generic K-turn, there is substantial variation in the structural motifs recognized by the K-turn binding proteins (Kuhn et al. 2002; Rozhdestvensky et al. 2003). Given that this region of the RNA is not found in Archaea outside of the Thermococci, and the fact that this protein has not been found in the RNase P of other Archaea, if L7Ae is indeed an RNase P subunit in *P. horikoshii,* this is probably an idiosyncrasy of this small phylogenetic group.

2.3.3 Two Flies in the Ointment: Nanoarchaeum and Pyrobaculum

No RNase P RNA or associated proteins were annotated in the genomic sequences of *Nanoarchaeum equitans* or *Pyrobaculum aerophilum*. Searches specifically for either the RNA or the proteins were likewise fruitless (Li and Altman 2004). What, then, is the nature of RNase P in these organisms?

Nanoarchaeum is a deep branching obligate symbiont of *Ignicoccus*. It has a very small genome (490,885 bp) and encodes some of its tRNAs in fragments, each half contains the appropriate flanking sequence and structures required for them to be joined by splicing in *trans* (Waters et al. 2003; Randau et al. 2005b; Di Giulio 2006). However, each tRNA gene (or 5′gene fragment) is preceded by an archaeal TATA-box promoter sequence 26 nucleotides upstream of location corresponding to the first nucleotide of the mature tRNA (Randau et al. 2008). This suggested that tRNAs in this organism might not be transcribed with a leader sequence, bypassing the necessity of RNase P for tRNA biosynthesis (Randau et al. 2008). This hypothesis has been confirmed by the identification of 5′-triphosphate in mature tRNAs (Randau et al. 2008). The implication is that RNase P has been dispensed with in this unusual organism; this is the only organism known to lack this important enzyme.

The genes encoding RNAs in *Pyrobaculum* are likewise usually immediately preceded by identifiable promoters, and so the presumption has been that this organism might also not require RNase P activity. However, promoter spacing is not a precise as in *Nanoarchaeum,* and would be predicted in most cases to result in pre-tRNAs with 1-3 nt leaders, which would require an RNase P activity or a suitable replacement. An unusual RNase P-like RNA has recently been identified in the genomes of several species of the genus *Pyrobaculum*, and this RNA is associated with RNase P activity. This RNA consists of only the catalytic domain of RNase P RNA (Fig. 2.3). No genes encoding proteins specifically related to the RNase P proteins of any kind have been identified in the genomes of *Pyrobaculum* species.

2.4 Eukaryotic Ribonuclease P

RNase P in eukaryotes is much more complex than in either Bacteria or Archaea. Eukaryotes contain a nuclear RNase P, that processes pre-tRNAs in the nucleolus for cytoplasmic translation, a mitochondrial RNase P for processing mitochondrially encoded tRNAs, and in plants and algae a plastid RNase P for plastid-encoded tRNAs. These enzymes are entirely distinct from each other. In addition, the nuclear RNase P is a member of an enzyme family with RNase MRP, which is involved in the rRNA processing also in the nucleolus.

Eukaryotic nuclear RNase P enzymes are more diverse in terms of RNA structure and protein composition than are those of Bacteria or Archaea, and so are less well understood. However, the secondary structures of RNase P RNAs in some phylogenetic groups have been determined in some detail. The structure of the core of the nuclear RNA is common to all and contains many of the core elements of the bacteria and archaeal RNAs. The protein composition of the nuclear enzyme is only well known in the human and yeast systems, in which the enzyme is composed of at least 9 proteins, some quite large. The nuclear RNase P RNA is not catalytically proficient in the absence of protein.

Organellar RNase P enzymes are even less well understood than are those of the nucleus. In some primitive mitochondria and plastids, the RNA subunits of RNase P (Fig. 2.5) are clearly related to those of the proteobacteria and cyanobacteria from which these organelles arose, although the proteins are not. In humans, and presumably metazoans in general, the traditional RNase P has apparently been dispensed with entirely, and replaced by a "Rube Goldberg" amalgamation of proteins that carry out the same function (Holzmann et al. 2008; Walker and Engelke 2008). The nature of chloroplast RNase P in green plants is unclear; there is evidence that this RNase P, like the protein only "RNase P" found in human mitochondria, might also be an all-protein enzyme, but the components of this enzyme have yet to be identified.

2.4.1 *Saccharomyces cerevisiae a Model for RNase P in Eukaryotes*

The best understood eukaryotic RNase P system is that of the yeast *Saccharomyces cerevisiae*. Much is also known about the human system, but our insight is clearer

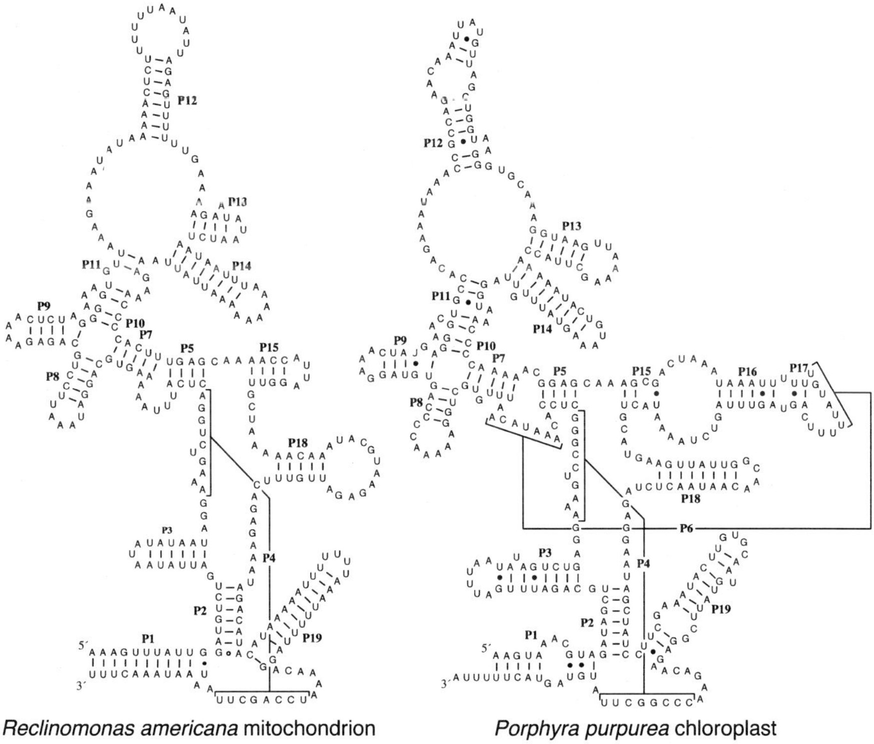

Fig. 2.5 Primitive organellar RNase P RNAs. The sequences and structures of RNase P RNAs encoded in the primitive mitochondrion of the jakobid *R. americana* and the primitive plastid of the rhodophyte *P. purpurea* reflect their bacterial ancestry. Helices are labeled P1-P17 according to Haas and Brown (1998). The structures RNase P RNAs in other, less primitive mitochondria and plastids are more divergent and less well characterized. RNase P in animal mitochondria and perhaps green plant chloroplasts lack an RNA component (see text). Structures are from the RNase P Database (Brown 1999)

in the yeast system, and what is clear in the human system is, not surprisingly, generally consistent with what is also found in yeast (at least in the case of the nuclear enzyme). Most other eukaryotic systems are known primarily from the sequences of the RNA and protein subunits identified by their similarity to those of yeast or humans. Here we will examine the nuclear and mitochondrial RNase P RNA holoenzymes and a closely related ribonucleoprotein complex, the RNase MRP holoenzyme.

2.4.1.1 Nuclear RNase P

The secondary structure of the nuclear RNase P in *S. cerevisiae* is consistent with much of the phylogenetically conserved core of the bacterial and archaeal enzymes, including P1, P2, and P3, P4, P7, P10/11, P12, P15, and conserved regions I-V

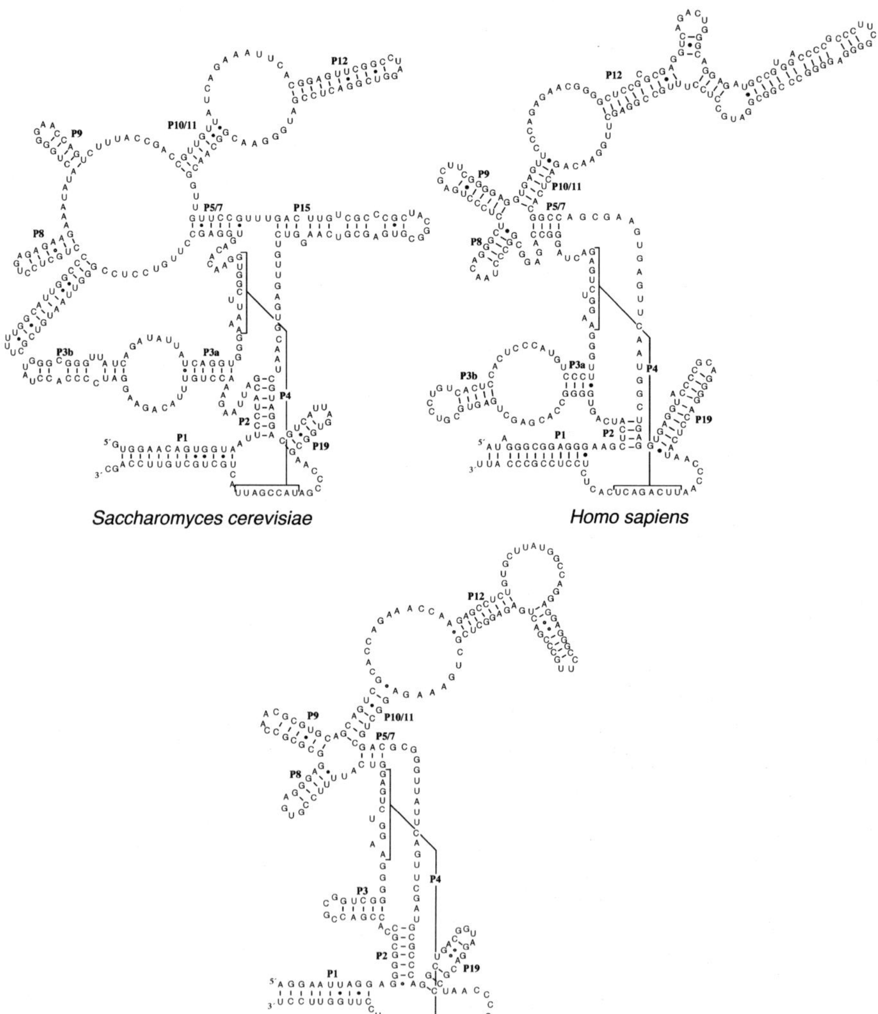

Fig. 2.6 Representative eukaryotic nuclear RNase P RNAs. Structural variation in nuclear RNase P RNAs is not yet sufficiently well studied to define "types" as in Bacteria and Archaea. Helices are labeled P1-P19 according to Haas and Brown (1998), but homology between some of these helices and those of their bacterial counterparts are questionable. For this reason, some helices are sometimes labeled "*e*P" (e.g., eP15), but the use of this nomenclature has not been consistent. Structures are from the RNase P Database (Brown 1999)

(Fig. 2.6) (Frank et al. 2000). Some of the RNase P RNA helical elements are found in the same location in *S. cerevisiae* as in bacteria and Archaea but do not otherwise share any obvious sequence or structural similarity. To denote this fundamental difference, they are sometimes described as "*e*ukaryal *p*air region" or "eP," although this has not been used consistently. These elements in the RNase P RNA of *S. cerevisiae* include eP8, eP9, eP15, and eP19. P3 in most Eukaryotes, but not in bacteria or Archaea, includes a large internal loop that is the binding site for the Pop6/7 protein heterodimer (Perederina et al. 2007). RNase MRP RNA shares both this loop and these proteins. The binding of Pop6/7 to P3 in the RNA may mediate the binding of other RNase P and RNase MRP protein subunits (Perederina et al. 2007). Nuclear RNase P RNAs notably lack L15, conserved in most bacterial and archaeal RNase P RNAs and the binding site for the 3′-NCCA tail of the pre-tRNA substrate. In addition, the P7/P8/P9/P10 cruciform found in most archaeal and bacterial RNase P RNAs is unusual in *S. cerevisiae* because of an extra RNA hairpin proximal of P8. However, other nuclear eukaryotic RNase P RNAs such as those found in *Homo sapiens*, *Schizosaccharomyces pombe*, and *Pichia strasburgensis* maintain the "standard" cruciform structure but lack the obvious sequence similarity to their homologs in Bacteria or Archaea (Frank et al. 2000).

The *S. cerevisiae* nuclear RNase P has no fewer than nine proteins associated with the holoenzyme complex (POP1, POP3, POP4, POP5, POP6, POP7, POP8, Rpp1, and Rpr2) (Fig. 2.7). Only Pop4, Pop1, and the heterodimer Pop6/7 appear to bind directly to the RNase P RNA (Houser-Scott et al. 2002; Srisawat et al. 2002; Perederina et al. 2007). Four of these proteins (POP4, Rpr2, Rpp1, and POP5) are homologous to the RNase P protein subunits found in Archaea (Ph1771p, Ph1601p, Ph1877p, Ph1481p in *P. horikoshii,* respectively). However, no similarity can be identified among any of the nuclear RNase P proteins subunits and the bacterial

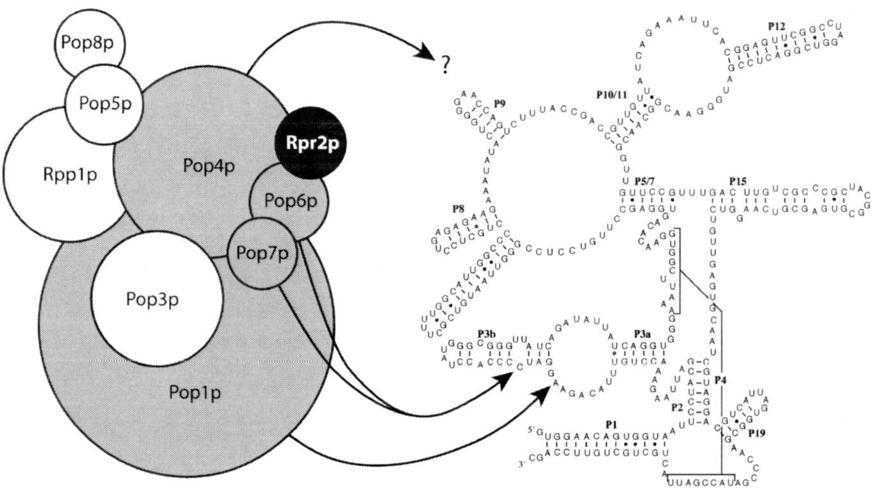

Fig. 2.7 *S. cerevisiae* nuclear RNase P protein:protein and protein:RNA interactions (Houser-Scott et al. 2002). Direct RNA-binding proteins are shaded grey. Pop1p binds to the P3 loop (between P3a and P3b). The Pop6p/Pop7b heterodimer also binds P3. The location of Pop4p binding in the RNA is not known. Prp2p, the single RNase P protein not also present in RNase MRP, is *shaded black*

rnpA protein. The specific roles of the nuclear proteins are not well-defined, but although these proteins are essential for catalysis, the catalytic center of the enzyme resides in the RNA rather than protein.

2.4.1.2 Mitochodrial RNase P

The RNase P of the *S. cerevisiae* mitochondrion is composed of an RNA subunit (Rpm1r) that is 490 nucleotides long and a single 105 kDa protein subunit (Rpm2p). The yeast mitochondrial RNase P RNA is so A + T-rich that the structure cannot be resolved by comparative analysis, but seems perhaps to contain only the core elements of the catalytic domain (Wise and Martin 1991). The Rpm2 protein is encoded in the nucleus and transported to the mitochondria and shares no obvious similarity to the RNase P protein subunit of bacteria or the nucleus (Morales et al. 1992). The protein is essential for catalytic activity (Morales et al. 1992). Surprisingly, the protein component (Rpm2p) is responsible for a host of functions in the nucleus and cytoplasm of the cell unrelated to the RNase P. For example, the C-terminal domain of Rpm2p is directly involved in the maturation of the RNase P RNA but even without this region of the protein the mitochondrial RNase P holoenzyme retains activity (Stribinskis et al. 2001b). Futhermore, Rpm2p is a transcriptional activator in the nucleus and interacts with the cytoplasmic processing bodies (P-bodies) along with Dcp2p (Stribinskis and Ramos 2007). Rpm2p activates transcription of many nuclear-encoded mitochondrial proteins, such as those involved in the import apparatus (TOM40, TOM6, TOM20, TOM22, and TOM37), and mitochondrial chaperons (HSP60 and HSP10) (Stribinskis et al. 2005). These multifunctional roles within the nucleus/cytoplasm and the lack of any conserved similarity with bacterial RNase P protein suggest that this was a pre-existing protein recruited to mitochondrial RNase P, perhaps to replace the original bacterial RNase P protein.

2.4.1.3 RNase MRP

RNase "*m*itochondrial *R*NA *p*rocessing" (MRP) is also an endoribonuclease, but is found only in eukaryotes. RNase MRP in *S. cerevisiae* has composed a single RNA molecule and at least 10 protein subunits (Pop1, Pop3, Pop4, Pop5, Pop6, Pop7, Pop8, Rmp1, Rpp1, and Snm1) (Fig. 2.8). Eight of these protein subunits are also subunits of RNase P. Although RNase MRP lacks the Rpr2p of RNase P, it contains two proteins unique to RNase MRP: Snm1p and Rmp1p. The RNA component of RNase MRP's catalytic domain (Domain 1) shares obvious similarity with the nuclear RNase P RNA in both sequence and secondary structure (Forster and Altman 1990; Li et al. 2002). These similarities in associated proteins and conserved RNA sequences/structures imply that the RNase MRP evolved from RNase P shortly after the three domains (Bacteria, Archaea, and Eukaryotes) diverged.

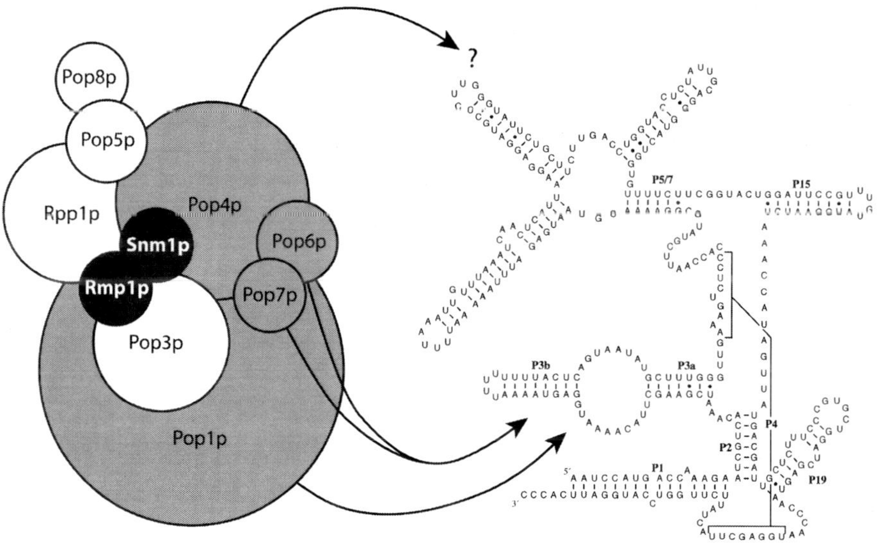

Fig. 2.8 *S. cerevisiae* RNase MRP secondary structure, protein:protein and protein:RNA interactions (Walker et al. 2005). Helices with clear homologs in RNase P RNA from the same organism are labeled P1-P19. Direct RNA-binding proteins are shaded *grey*. Pop1p binds to the P3 loop (between P3a and P3b). The Pop6p/Pop7b heterodimer also binds P3. The location of Pop4p binding in the RNA is not known. Snm1p and Rmp1p, the only RNase MRP proteins not shared with RNase P, are shaded black

RNase MRP was originally described as an endoribonuclease that cleaves primers for mitochondrial DNA replication (Chang and Clayton 1987). Later, RNase MRP was shown to be predominantly in the nucleus rather than the mitochondrion, and found to play a role in the processing of the 5.8 S ribosomal RNA precursor. It does this by cleaving the pre-rRNA at site A3 in the first internal transcribed spacer (ITS1). RNase MRP also cleaves the mRNA of CLB2 in its 5′-UTR (Gill et al. 2004). The cleavage of the 5′-UTR initiates the degradation of the transcript by Xrn1 nuclease. Cells with a defective RNase MRP suffer as a result from a late anaphase delay (Gill et al. 2004). Although both the RNA and protein subunits of RNase MRP are essential, neither the cleavage of pre-rRNA nor the cleavage of RNA primers associated with DNA replication in the mitochondria are essential for cell viability. Therefore, it seems likely that RNase MRP has other as yet unknown essential function(s).

2.5 Conclusion

Since the RNA component of RNase P is conserved among the three Domains, but the proteins are not, it is likely that the last common ancestor of these three Domains contained an RNA-only RNase P, and the associated proteins evolved later after the evolutionary divergence of bacteria on one hand and Archaea/eukaryotes on the other. The A type RNase P RNA common to most bacteria and Archaea gave rise to the other forms: type B and C in bacteria (and the organellar RNAs), type M and P in Archaea, and the various forms of the RNA (including that of RNase MRP), in the nucleus of eukaryotes. In the common ancestor of bacteria, the enzyme acquired the *rnpA* protein, which has remained essentially unchanged throughout this Domain. The common ancestor of Archaea and eukaryotes acquired a set of 4 RNase P proteins that remain (perhaps with some additions or loses) in Archaea. These four proteins along with the RNA serve as the core of the RNases P and MRP of the eukaryotic nucleus, to which many proteins have been added. Early in eukaryotic evolution (exactly where is unclear, but after the acquisition of most proteins), the RNase P RNA gene was duplicated, and one copy became specialized (including by some protein substitutions) for rRNA processing (at least) whereas the other retained the traditional RNase P function. In at least one archaeon, the need for RNase P seems to have been removed and the enzyme lost. In mitochondria and plastids generally, the RNA remains from their bacterial ancestry, but the *rnpA* protein has been replaced by other, nuclear-encoded proteins. In the animal mitochondrion, and perhaps green chloroplasts, RNase P has been replaced wholesale by an independent protein-only enzyme.

Acknowledgments This research was supported in part by the Intramural Research Program of the NIH, National Institute of Environmental Health Sciences.

References

Altman S, Wesolowski D et al (2005) RNase P cleaves transient structures in some riboswitches. Proc Natl Acad Sci U S A 102(32):11284–11289

Bai Y, Trang P et al (2008) Effective inhibition in animals of viral pathogenesis by a ribozyme derived from RNase P catalytic RNA. Proc Natl Acad Sci U S A 105(31):10919–10924

Ban N, Nissen P et al (2000) The complete atomic structure of the large ribosomal subunit at 2.4 A resolution. Science 289(5481):905–920

Brown JW (1999) The Ribonuclease P database. Nucleic Acids Res 27(1):314

Brown JW, Haas ES (1995) Ribonuclease P structure and function in Archaea. Mol Biol Rep 22(2–3):131–134

Brown JW, Haas ES et al (1991) Phylogenetic analysis and evolution of RNase P RNA in proteobacteria. J Bacteriol 173(12):3855–3863

Brown JW, Haas ES et al (1993) Characterization of ribonuclease P RNAs from thermophilic bacteria. Nucleic Acids Res 21(3):671–679

Brown JW, Nolan JM et al (1996) Comparative analysis of ribonuclease P RNA using gene sequences from natural microbial populations reveals tertiary structural elements. Proc Natl Acad Sci U S A 93(7):3001–3006

Buck AH, Dalby AB et al (2005) Protein activation of a ribozyme: the role of bacterial RNase P protein. EMBO J 24(19):3360–3368

Crary SM, Niranjanakunari S et al (1998) The protein component of Bacillus subtilis ribonuclease P increases catalytic efficiency by enhancing interactions with the 5′ leader sequence of pre-tRNAAsp. Biochemistry 37(26):9409–9416

Chang DD, Clayton DA (1987) A novel endoribonuclease cleaves at a priming site of mouse mitochondrial DNA replication. EMBO J 6(2):409–417
Chen JL, Pace NR (1997) Identification of a universally conserved core of ribonuclase P RNA. RNA 3:561–576
Christian EL, Harris ME (1999) The track of the pre-tRNA 5′ leader in the ribonuclease P ribozyme–substrate complex. Biochemistry 38(39):12629–12638
Coughlin DJ, Pleiss JA et al (2008) Genome-wide search for yeast RNase P substrates reveals role in maturation of intron-encoded box C/D small nucleolar RNAs. Proc Natl Acad Sci U S A 105(34):12218–12223
Di Giulio M (2006) *Nanoarchaeum equitans* is a living fossil. J Theor Biol 242:257–260
Feltens R, Gossringer M et al (2003) An unusual mechanism of bacterial gene expression revealed for the RNase P protein of Thermus strains. Proc Natl Acad Sci U S A 100(10):5724–5729
Forster AC, Altman S (1990) Similar cage-shaped structures for the RNA components of all ribonuclease P and ribonuclease MRP enzymes. Cell 62(3):407–409
Frank DN, Adamidi C et al (2000) Phylogenetic-comparative analysis of the eukaryal ribonuclease P RNA. RNA 6(12):1895–1904
Fukuhara H, Kifusa M et al (2006) A fifth protein subunit Ph1496p elevates the optimum temperature for the ribonuclease P activity from *Pyrococcus horikoshii* OT3. Biochem Biophys Res Commun 343(3):956–964
Guerrier-Takada C, Gardiner K et al (1983) The RNA moiety of ribonuclease P is the catalytic subunit of the enzyme. Cell 35(3):849–857
Gill T, Cai T et al (2004) RNase MRP cleaves the CLB2 mRNA to promote cell cycle progression: novel method of mRNA degradation. Mol Cell Biol 24(3):945–953
Gimple O, Schon A (2001) In vitro and in vivo processing of cyanelle tmRNA by RNase P. Biol Chem 382(10):1421–1429
Gösringer M, Hartmann R (2007) Function of heterologous and truncated RNase P proteins in *Bacillus subtilis*. Mol Microbiol 66:801–813
Haas ES, Brown JW (1998) Evolutionary variation in bacterial RNase P RNAs. Nucleic Acids Res 26(18):4093–4099
Haas ES, Banta AB et al (1996) Structure and evolution of ribonuclease P RNA in Gram-positive bacteria. Nucleic Acids Res 24(23):4775–4782
Hall TA, Brown JW (2002) Archaeal RNase P has multiple protein subunits homologous to eukaryotic nuclear RNase P proteins. RNA 8(3):296–306
Hall TA, Brown JW (2004) Interactions between RNase P protein subunits in archaea. Archaea 1(4):247–254
Hansen FG, Hansen EB et al (1982) The nucleotide sequence of the dnaA gene promoter and of the adjacent rpmH gene, coding for the ribosomal protein L34, of *Escherichia coli*. EMBO J 1(9):1043–1048
Hansen FG, Hansen EB et al (1985) Physical mapping and nucleotide sequence of the rnpA gene that encodes the protein component of ribonuclease P in *Escherichia coli*. Gene 38(1–3):85–93
Harris JK, Haas ES et al (2001) New insight into RNase P RNA structure from comparative analysis of the archaeal RNA. RNA 7(2):220–232
Hartmann E, Hartmann RK (2003) The enigma of ribonuclease P evolution. Trends Genet 19(10):561–569
Holzmann J, Frank P et al (2008) RNase P without RNA: identification and functional reconstitution of the human mitochondrial tRNA processing enzyme. Cell 135(3):462–474
Honda T, Kakuta Y et al (2008) Structure of an archaeal homolog of the human protein complex Rpp 21-Rpp29 that is a key core component for the assembly of active ribonuclease P. J Mol Biol 384(3):652–662
Hori Y, Tanaka T et al (2000) The catalytic RNA of RNase P from *Escherichia coli* cleaves Drosophila 2 S ribosomal RNA in vitro: a new type of naturally occurring substrate for the ribozyme. FEBS Lett 472(2–3):187–190
Houser-Scott F, Xiao S et al (2002) Interactions among the protein and RNA subunits of *Saccharomyces cerevisiae* nuclear RNase P. Proc Natl Acad Sci U S A 99(5):2684–2689

James BD, Olsen GJ et al (1988) The secondary structure of ribonuclease P RNA, the catalytic element of a ribonucleoprotein enzyme. Cell 52(1):19–26

Kakuta Y, Ishimatsu I et al (2005) Crystal structure of a ribonuclease P protein Ph1601p from *Pyrococcus horikoshii* OT3: an archaeal homologof human nuclear ribonuclease P protein Rpp 21. Biochemistry 44(36):12086–12093

Kawano S, Nakashima T et al (2006) Crystal structure of protein Ph1481p in complex with protein Ph1877p of archaeal RNase P from *Pyrococcus horikoshii* OT3: implication of dimer formation of the holoenzyme. J Mol Biol 357(2):583–591

Kazakov S, Altman S (1991) Site-specific cleavage by metal ion cofactors and inhibitors of M1 RNA, the catalytic subunit of RNase P from *Escherichia coli*. Proc Natl Acad Sci U S A 88(20):9193–9197

Kazantsev AV, Krivenko AA et al (2003) High-resolution structure of RNase P protein from *Thermotoga maritima*. Proc Natl Acad Sci U S A 100(13):7497–7502

Kazantsev AV, Krivenko AA et al (2005) Crystal structure of a bacterial ribonuclease P RNA. Proc Natl Acad Sci U S A 102(38):13392–13397

Kifusa M, Fukuhara H et al (2005) Protein–protein interactions in the subunits of ribonuclease P in the hyperthermophilic archaeon *Pyrococcus horikoshii* OT3. Biosci Biotechnol Biochem 69(6):1209–1212

Kirsebom LA, Svärd SG (1994) Base pairing between *Escherichia coli* RNase P RNA and its substrate. EMBO J. 13:4870–4876

Kouzuma Y, Mizoguchi M et al (2003) Reconstitution of archaeal ribonuclease P from RNA and four protein components. Biochem Biophys Res Commun 306(3):666–673

Kuhn JF, Tran EJ et al (2002) Archaeal ribosomal protein L7 is a functional homolog of the eukaryotic 15.5 kD/Snu13p snoRNP core protein. Nucleic Acids Res 30(4):931–941

Li Y, Altman S (2002) Partial reconstitution of human RNase P in HeLa cells between its RNA subunit with an affinity tag and the intact protein components. Nucleic Acids Res 30(17):3706–3711

Li Y, Altman S (2004) In search of RNase P RNA from microbial genomes. RNA 10(10): 1533–1540

Li X, Frank DN et al (2002) Phylogenetic analysis of the structure of RNase MRP RNA in yeasts. RNA 8(6):740–751

Lombo TB, Kaberdin VR (2008) RNA processing in *Aquifex aeolicus* involves RNase E/G and an RNase P-like activity. Biochem Biophys Res Commun 366(2):457–463

Loria A, Pan T (1996) Domain structure of the ribozyme from eubacterial ribonuclease P. RNA 2(6):551–563

Lowe T, Lai L, Chan P, Cozen A, Bernick D, Brown JW, and Gopalan V. Discovery of the Elusive Pyrobaculum RNase P; An Unexpected Form of an Ancient RNA. Nature (unpublished data)

Marszalkowski M, Willkomm DK et al (2008) 5′-end maturation of tRNA in *Aquifex aeolicus*. Biol Chem 389(4):395–403

Massire C, Jaeger L, Westhof E (1998) Derivation of the three-dimensional architecture of bacterial ribonuclease P RNAs from comparative sequence analysis. J Mol Biol 279:773–793

Morales MJ, Dang YL et al (1992) A 105-kDa protein is required for yeast mitochondrial RNase P activity. Proc Natl Acad Sci U S A 89(20):9875–9879

Nolan JM, Burke DH, Pace NR (1993) Circularly permuted tRNAs as specific photoaffinity probes of ribonuclease P RNA structure. Science 261:762–765

Niranjanakumari N, Stams T, et al (1998) Protein component of ribonuclease P alters substrate recognition by directly contacting precursor tRNA. Proc Natl Acad Sci U S A 95(26): 15212–15217

Numata T, Ishimatsu I et al (2004) Crystal structure of archaeal ribonuclease P protein Ph1771p from *Pyrococcus horikoshii* OT3: an archaeal homolog of eukaryotic ribonuclease P protein Rpp 29. RNA 10(9):1423–1432

Ogasawara N, Moriya S et al (1985) Conservation of genes and their organization in the chromosomal replication origin region of *Bacillus subtilis* and *Escherichia coli*. EMBO J 4(12): 3345–3350

Panagiotidis CA, Drainas D et al (1992) Modulation of ribonuclease P expression in *Escherichia coli* by polyamines. Int J Biochem 24(10):1625–1631

Pannucci JA, Haas ES et al (1999) RNase P RNAs from some Archaea are catalytically active. Proc Natl Acad Sci U S A 96(14):7803–7808

Perederina A, Esakova O et al (2007) Specific binding of a Pop6/Pop7 heterodimer to the P3 stem of the yeast RNase MRP and RNase P RNAs. RNA 13(10):1648–1655

Randau L, Munch R et al (2005b) *Nanoarchaeum equitans* creates functional tRNAs from separate genes for their 5′ and 3′ halves. Nature 433(7025):537–541

Randau L, Schroder I et al (2008) Life without RNase P. Nature 453(7191):120–123

Reyes-Darias JA, Sanchez-Luque FJ et al (2008) Inhibition of HIV-1 replication by RNA-based strategies. Curr HIV Res 6(6):500–514

Rozhdestvensky TS, Tang TH et al (2003) Binding of L7Ae protein to the K-turn of archaeal snoRNAs: a shared RNA binding motif for C/D and H/ACA box snoRNAs in Archaea. Nucleic Acids Res 31(3):869–877

Salazar L, Fsihi H et al (1996) Organization of the origins of replication of the chromosomes of *Mycobacterium smegmatis*, *Mycobacterium leprae* and *Mycobacterium tuberculosis* and isolation of a functional origin from *M. smegmatis*. Mol Microbiol 20(2):283–293

Seif E, Altman S (2008) RNase P cleaves the adenine riboswitch and stabilizes pbuE mRNA in *Bacillus subtilis*. RNA 14(6):1237–1243

Siegel RW, Banta AB, Haas ES, Brown JW, Pace NR (1996) *Mycoplasma fermentans* simplifies our view of the catalytic core of ribonuclease P RNA. RNA 2:452–462

Spitzfaden C, Nicholson N et al (2000) The structure of ribonuclease P protein from *Staphylococcus aureus* reveals a unique binding site for single-stranded RNA. J Mol Biol 295(1):105–115

Srisawat C, Houser-Scott F et al (2002) An active precursor in assembly of yeast nuclear ribonuclease P. RNA 8(10):1348–1360

Stams T, Niranjanakumari S et al (1998) Ribonuclease P protein structure: evolutionary origins in the translational apparatus. Science 280(5364):752–755

Stribinskis V, Ramos KS (2007) Rpm2p, a protein subunit of mitochondrial RNase P, physically and genetically interacts with cytoplasmic processing bodies. Nucleic Acids Res 35(4):1301–1311

Stribinskis V, Gao GJ et al (2001b) Rpm2p: separate domains promote tRNA and Rpm1r maturation in *Saccharomyces cerevisiae* mitochondria. Nucleic Acids Res 29(17):3631–3637

Stribinskis V, Heyman HC et al (2005) Rpm2p, a component of yeast mitochondrial RNase P, acts as a transcriptional activator in the nucleus. Mol Cell Biol 25(15):6546–6558

Svard SG, Kirsebom LA (1992) Several regions of a tRNA precursor determine the *Escherichia coli* RNase P cleavage site. J Mol Biol 227(4):1019–1031

Svard SG, Kagardt U et al (1996) Phylogenetic comparative mutational analysis of the base-pairing between RNase P RNA and its substrate. RNA 2(5):463–472

Takagi H, Watanabe M et al (2004) Crystal structure of the ribonuclease P protein Ph1877p from hyperthermophilic archaeon *Pyrococcus horikoshii* OT3. Biochem Biophys Res Commun 319(3):787–794

Torres-Larios A, Swinger KK et al (2005) Crystal structure of the RNA component of bacterial ribonuclease P. Nature 437(7058):584–587

Tous C, Vega-Palas MA et al (2001) Conditional expression of RNase P in the cyanobacterium *Synechocystis* sp. PCC6803 allows detection of precursor RNAs. Insight in the in vivo maturation pathway of transfer and other stable RNAs. J Biol Chem 276(31):29059–29066

Walker SC, Engelke DR (2008) A protein-only RNase P in human mitochondria. Cell 135(3):412–414

Walker SC, Aspinall TV, Gordon JM, Avis JM (2005) Probing the structure of *Saccharomyces cerevisiae* RNase MRP. Biochem Soc Trans 33:479–481

Waters E, Hohn MJ et al (2003) The genome of *Nanoarchaeum equitans*: insights into early archaeal evolution and derived parasitism. Proc Natl Acad Sci U S A 100(22):12984–12988

Waugh DS, Pace NR (1990) Complementation of an RNase P RNA (rnpB) gene deletion in *Escherichia coli* by homologous genes from distantly related eubacteria. J Bacteriol 172(11): 6316–6322

Waugh DS, Green CJ, Pace NR (1989) The design and catalytic properties of a simplified ribonuclease P RNA. Science 244:1569–1571

Wegmann U et al (2007) Complete genome sequence of the prototype lactic acid bacterium *Lactococcus lactis* subsp. *cremoris* MG1363. J Bacteriol 189:3256–3270

Westhof E, Wesolowski D et al (1996) Mapping in three dimensions of regions in a catalytic RNA protected from attack by an Fe(II)-EDTA reagent. J Mol Biol 258(4):600–613

Wise CA, Martin NC (1991) Dramatic size variation of yeast mictochondrial RNAs suggests that RNase P RNAs can be quite small. J Biol Chem 266:19154–19157

Chapter 3
Over a Decade of Bacterial Ribonuclease P Modeling

Benoît Masquida, Fabrice Jossinet, and Eric Westhof

3.1 Introduction

The prokaryotic ribonuclease P maturates precursor tRNAs (ptRNAs) by catalytically removing the 5′ leader sequence. The RNA subunit from this RNP carries out the catalytic activity. Hence, the RNase P is a ribozyme. The RNP builds up from a ~300-nt RNA and a ~100-aa protein. Two different scaffolds of the bacterial RNase P can be found. They consist of a common catalytic core surrounded by peripheral domains variable in size and structure. The A type (A stems for ancestor) is characteristic of *Escherichia coli* and cyanobacteria. The B type (B stems for *Bacillus*) is characteristic of *Bacillus subtilis*. tRNAs are substrates to many different biological processes including maturation, chemical modification, aminoacylation, protein translation and more that can be specific of a family of organisms (See (Giege 2008) for review), e.g., tRNA-dependent cell wall synthesis in firmicutes (Matsuhashi et al. 1965). This implies that selection pressures specific to all the pathways in which tRNAs intervene are simultaneously applied on the same molecular scaffold in a way characteristic to each organism. Conversely, it means that molecules acting on (p)tRNAs indirectly undergo selection pressures connected to all the pathways involving tRNAs. It is therefore not too surprising that the set of RNase P natural substrates differs slightly among bacteria (Altman et al. 1993; Hartmann et al. 1995; Peck-Miller and Altman 1991) and that RNase P RNAs have different secondary and tertiary structures.

From the molecular modeling point of view, one should thus distinguish the architectural features of the RNase P holoenzyme and substrates that are directly linked to the catalytic reaction from those linked to the other tRNA-processing pathways. Among the tRNA pathways, some are specific to a given tRNA or to a tRNA family (aminoacylation), and some use the whole pool of different tRNAs (tRNA processing, translation). For example, the anticodon loop is not recognized

B. Masquida (✉), F. Jossinet, and E. Westhof
Architecture et Réactivité de l'ARN, Université de Strasbourg, IBMC, CNRS, 15 rue René Descartes, Strasbourg 67084, France
e-mail: b.masquida@ibmc.u-strasbg.fr

F. Liu and S. Altman (eds.), *Ribonuclease P*, Protein Reviews 10,
DOI 10.1007/978-1-4419-1142-1_3, © Springer Science+Business Media, LLC 2010

by RNase P, making it likely that a substrate carrying a group I intron at this place (Rangan et al. 2004) is processed. Substrate recognition by RNase P relies on the acceptor stem and the T-arm (See (Kirsebom 2007) for review). It is known that 5′ and 3′ maturation events do not occur along a well-defined pathway, but under a process in which nucleases process the ptRNAs stochastically (Li and Deutscher 1996), which implies that RNase P can recognize pre-tRNAs bearing 3′ ends with different length and sequence (Gaur et al. 1996). The determinants for the recognition by both RNase P and EF-Tu during translation do not overlap, although they belong to the acceptor end and to the T-arm. Moreover, EF-Tu recognition implies that the tRNA has been fully maturated and amino-acylated. In each case, distinct subsets of nucleotides from these domains are used for proper recognition. In the case of RNase P, recognition mainly relies on the +1/72 base pair as well as on the discriminator base 73 (Brännvall et al. 2004) and C74 and 75 (See (Kirsebom 2007) for review). In the case of EF-Tu, the amino acid is the main binding element, and recognition of the T stem only affects the binding efficiency (Roy et al. 2007). Yet, contacts are observed between EF-Tu and the backbone of residues from the acceptor end of the amino-acylated tRNA that are also used by RNase P to recognize the ptRNA. This comparative observation means that while the same domains of tRNAs are used for recognition by both enzymes, the selection pressures occur on distinct residues in order to allow these two pathways to co-exist efficiently. The RNase P and EF-Tu, as well as other tRNA-using enzymes are thus evolutionarily linked through their substrates. This conclusion is also valid regarding the T-loop of the pre-tRNAs that participates in the recognition by RNase P through interactions with a specific region in a large loop (L11/12) from the specificity domain (See Chap. 8). Experimental evidence is in favor of discrete conformational changes and shows that RNase P and the substrate adapt to each other (or undergo induced-fit) in order to destabilize the scissile bond (Brännvall et al. 2007). This means that tRNA recognition by RNase P involves the dynamic processing of molecular information which implies the active detection of the leader sequence at the 5′ end of the substrate. Along the same lines, an active role for the tRNA has also been proposed during the elongation of protein translation by the ribosome (See (Daviter et al. 2006) for review). Finally, beyond the T-loop binding site (TBS) and the acceptor end binding site, the enzyme should be able to detect the presence of the 5′ leader sequence accurately.

RNase P is the only ribozyme acting as a turn-over enzyme on a *trans* substrate. This specific property means that its catalytic site cannot be easily circonscribed by detecting the nucleotide at which the cleavage occurs as is usually the case in self-cleaving ribozymes, such as group I and II introns and small ribozymes (hammerhead, hairpin, HDV, etc.). Studies dedicated to characterizing the catalytic site show the importance of the P4 pseudoknot apparently for binding metal ions required for the catalytic process (Christian et al. 2000). The cleavage products of the reaction are a 5′ leader stretch bearing a 3′ hydroxyl end and a tRNA bearing a 5′ phosphate. To perform this hydrolysis reaction, the ribozyme should share the catalytic ions with the pre-tRNA and prevent the 2′ OH group of the −1 residue of the pre-tRNA to participate in the reaction (by not allowing an in-line orientation between the phosphorus and O2′ atoms necessary for an SN2 mechanism). Indeed, such a reaction would yield a 5′ OH tRNA and a 2′,3′ cyclic phosphodiester. This may be a reason why the 2′ OH group of the −1 residue is not essential, although it strongly increases the speed of the reaction (Forster and Altman 1990), partly through the structuration of the catalytic pocket. Experimental evidence shows how the cleaved tRNAs lose affinity for the holoenzyme and also how the RNase P protein helps in promoting this stringency. The bacterial RNase P can carry out its reaction in the absence of the protein cofactor (Guerrier-Takada et al. 1983). However, without the protein, the required salt concentration is high, which allows the detection of aberrant cleavages at the −1 position (Brännvall and Kirsebom 1999). The protein cofactor binds the 5′ leader sequence and thus may help the RNA subunit identifying the scissile bond (Crary et al. 1998; Kurz et al. 1998; Loria et al. 1998). Interestingly, it has been difficult to demonstrate the catalytic properties of the eukaryotic or archaeal RNase P RNA but this has finally been done. In these cases, either a correct cleavage (Kikovska et al. 2007) or miscleavage was observed associated with a different chemistry (Stathopoulos et al. 2001). These observations show the critical role of the protein moiety in accelerating the reaction and accurately detecting the 5′ leader sequence. Moreover, an all-protein RNase P has been recently reconstructed from overexpressed individual human mitochondrial subunits and characterized (Holzmann et al. 2008), showing that the frontier of the RNA world is more blurry than previously thought (see for discussion (Cech 2009)).

3.1.1 Different Models

Several models have been built in more than a decade of research on RNase P. For each of them specific biochemical data have been used. However, since data have accumulated over time, the most recent models have shown a far better predictive power than the first ones. Consequently, models should be compared with caution after describing the type and extent of data on which they are specifically based. The best way of comparing models is to cluster them according to the publication date. Models were mainly proposed by the Pace and the Westhof laboratories. A model of a functionally important internal loop, the J15/16 loop, was proposed by the Harvey laboratory (Easterwood and Harvey 1997). The initial models of each team were published in 1994 (Harris et al. 1994; Westhof and Altman 1994). These models rationalized distinct sets of data of different nature (chemical probing vs. crosslink) and showed significantly different architectures. A revised version of the Westhof model essentially similar but incorporating new probing data, was published in 1996 (Westhof et al. 1996). As these models did not incorporate sufficient amount of biochemical data, the probability that they could reflect comparable solutions for the structure of the RNase P RNA was very low. Consequently, they will not be extensively discussed. With a growing corpus of data, a second set of models was published in 1997 by the Pace Laboratory (Harris et al. 1997), and by both laboratories in 1998 (Chen et al. 1998; Massire et al. 1998). In spite of different modeling strategies used by the two laboratories, convergence between the models was observed except in critical junctions around the central pseudoknot. The structural similarity resulted from the incorporation of biochemical data published by all the groups working on RNase P (they are difficult to acknowledge explicitly here). This situation remained unchanged until 2003 when a model tackling the interaction between the RNase P protein and RNA components was proposed (Tsai et al. 2003) based on a method in which the chemical probe tethered to the protein allowed for the exploring of its direct vicinity on the RNA (Heilek et al. 1995). This model was the last one that could not benefit from crystal structures of fragments or of the entire RNA from the RNase P. Between 2003 and 2005, two crystal structures of the S-domain (Krasilnikov et al. 2003, 2004;) and two crystal structures of the whole RNA were published (Kazantsev et al. 2005; Torres-Larios et al. 2005). This structural information provided a framework for the second generation of models from the Pace laboratory that aimed at proposing models for the prokaryotic ternary complex (the holoenzyme bound to the substrate) (Buck et al. 2005) and for the eukaryotic RNase P RNA (Marquez et al. 2006). The crystal structures of the RNA also provide a framework for the retrospective analysis of the models that will be the outline of this review.

3.1.2 Different Approaches

Ab initio modeling can be envisaged by applying and/or developing different methods. The method develops by the Westhof laboratory consists of transforming the structural information obtained by comparative sequence analysis into three-dimensional (3D) constraints derived from nucleic acid structural building blocks and stereochemistry. Secondary structure elements such as helices, single-stranded junctions and loops, are assembled interactively to build an initial model. Biochemical data resulting from enzymatic and chemical probing experiments are then incorporated step by step to lead to a model providing a correct stereochemical framework for data rationalization (Masquida and Westhof 2005). The model is then refined in order to avoid data outliers. This process is repeated until a best fit model is reached. This strategy is implemented in the MANIP package (Massire and Westhof 1998), which has been in use to build the models that will be discussed in this review. Hence, this method overlays two sets of independent constraints, the convergence of which is a good criterion to validate the model. The modeling methods in use in the Pace Laboratory are of two kinds. The first one aims at integrating data translated to distance contraints (for example crosslinking data) to derive a fullfilling model using molecular dynamics calculations implemented in the YAMMP-RNA program (Malhotra and Harvey 1994). This forcefield simulates pseudoatoms representing groups of nucleotides or helices and leads to a model made of a bundle of cylinders. A whole atom model can be further built on the obtained scaffold. This method has been used to generate the model published in (Harris et al. 1994) and in (Chen et al. 1998). Instead, the intermediate model from the Pace Laboratory (Harris et al. 1997) was built using a now obsolete version of the automated RNA modeling program MC-SYM (Major et al. 1991). This program is based on a constraintsatisfaction algorithm using constraints derived from the base pairs from the secondary structure and tertiary contacts. The set of solutions can be further analyzed by clustering.

3.1.3 Models as Answers to Biological Questions

In more than a decade of RNase P modeling, the aims have evolved from identifying the architecture of the RNA and its interactions with the ptRNA substrate and have finally focused on the structure of the ternary complex formed between the holoenzyme (both the protein and RNA subunits) bound to the substrate. Soon after the appearance of the initial models from each laboratory, important work showed that the RNase P RNA was actually built from two independently folding units, the specifity (S) and the catalytic (C) domains (Loria and Pan 1996; Pan 1995) that further interact to make up the RNA subunit (Fig. 3.1). Each domain accommodates a given region of the ptRNA substrate. The T-loop region and the acceptor end of the pre-tRNA are recognized by the S domain and the C domain, respectively. This constituted critical information because it showed that discrete subsets of secondary structure elements (helices, loops and single-stranded junctions) were interacting together before forming tertiary inter-domain contacts to sew the final RNA architecture. This observation reduced the complexity of the modeling problem considerably by reducing it to the modeling of two individual particles of medium size. In addition, the assembled RNA domains provide a scaffold for detecting the presence of the 5′ leader sequence in a process made accurate by the help of a tiny protein cofactor.

3.2 Models in a Historical Perspective

3.2.1 The Initial Models

Initial models have primarily relied either on chemical probing data (Westhof and Altman 1994) or on crosslinking data (Harris et al. 1994). The DMS and CMCT probes used in the initial Westhof–Altman model specifically target the Watson-Crick (WC) positions of the nucleotides (Brunel and Romby 2000). The use of such data relies on an accurate secondary structure and supports phylogenetic analysis in corroborating tertiary contacts, such as pseudoknots through the observation of protections in regions that are not base-paired. Perhaps because the model was not incorporating enough data, such as crosslinks, the model failed in detecting long-range interactions other than the ones that were already known from comparative sequence analysis (Darr et al. 1992; Haas et al. 1991). Yet, some local features such as the coupling of helices by stacking interactions were predicted successfully.

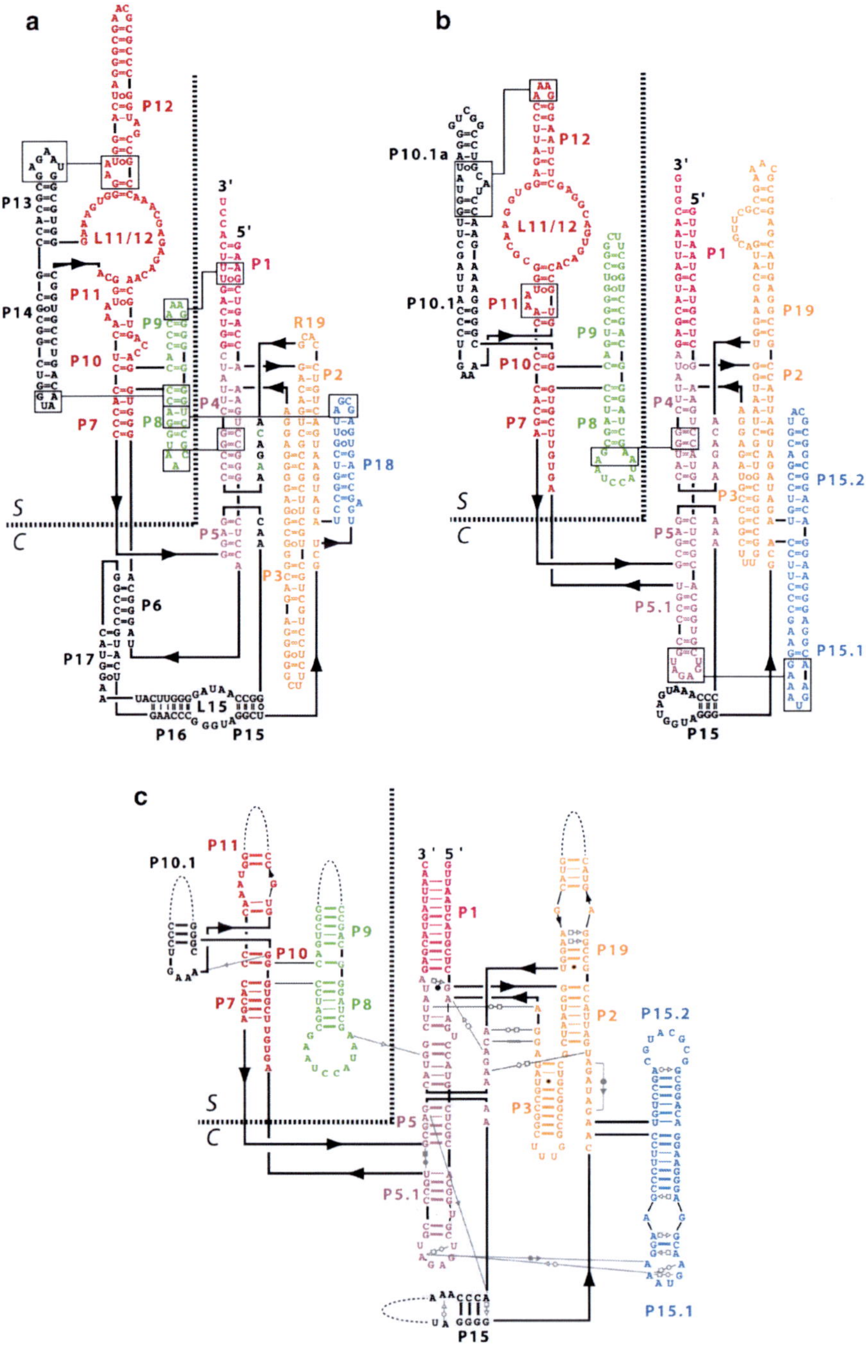

Fig. 3.1 Secondary structures of (**a**) the A (Ancestral) *E.coli* and (**b**) B (*Bacillus*) *Bacillus stearothermophilus* (*Bs*) RNase P RNA subtypes (Haas et al. 1994). The organization of the secondary structure elements (numbered P1–P19, P stems for Paired segment) is derived from the 3D model

This model was refined using iron–EDTA chemical probing (Westhof et al. 1996), but failed again at identifying the overall shape of the RNA particle because of the lack of tertiary structure information.

The initial model of the Pace Laboratory (Harris et al. 1994) was based on crosslinking data obtained using circularly permuted RNAse P RNAs incorporating the photoactivable probe APA (azidophenacyl) at the 5′ end or modified tRNAs. By evenly incorporating this probe at various locations within the RNA, it was possible to identify the tertiary structure organization of the different RNA domains and the relative position of the ptRNA. However, the precise structural features of the individual domains could not be tackled. The consequences were (1) a model barely showing the correct organization of the catalytic site and (2) its interaction with the ptRNA substrate. Nevertheless, constraints were still missing to accurately describe the architecture of the P11–P13 region from the S domain. The initial models from each laboratory had in common some structural features, such as the stacking between the P1, P2 and P3, and the interaction between the CCA 3′ end of the ptRNA substrate and the L15/16 loop of the RNase P RNA.

3.2.2 The Best Models in the Absence of RNase P Crystal Structures

The best models published by both laboratories in the absence of any crystal structure of the RNA of the RNase P were published in 1998 (Chen et al. 1998; Massire et al. 1998). The Massire model of both type A and B RNase P RNAs was based on a very careful comparative analysis of sequences that had been accumulating. This analysis led to the proposal of several tertiary interactions that were instrumental in building a correct 3D architecture. The final type A and B models consist of the assembly of four main helical stacks that pack together in order to form a highly compact structure (Fig. 3.1). In *E. coli*, these stacks are P1/P4/P5, P2/P3, P6/P7/P17/P10/P11/P12, and P13/P14. P15 and P16 are also stacked and pack along the P4/P5 elements. P18 packs along P4 on the side opposite to the ptRNA recognition site. Some differences arise in the *B. subtilis* model because of the different peripheral

Fig. 3.1 (continued) from Massire et al. (1998) and refined using the crystal structures from the A and B S-domains (Krasilnikov et al. 2004; Krasilnikov et al. 2003) and from the *Bs* whole RNA (Kazantsev et al. 2005) (**c**). These schemes give a better overview of the topology of the RNAs and are more helpful in comparing them when compared with the original secondary structures proposed earlier (Haas et al. 1994; Westhof and Altman 1994). Nucleotides involved in tertiary interactions are squared and linked by *thin lines*. The geometry of the base pairs is indicated using the Leontis–Westhof nomenclature, when possible (Leontis et al. 2002), on the secondary structure of the Bs. For each diagram, the *thick dotted line* between P5 and P7 indicates the separation between the specificity (S) and catalytic (C) domains (Loria and Pan 1996). Stems are color-coded according to the helical stacks originally proposed by Massire (1998).

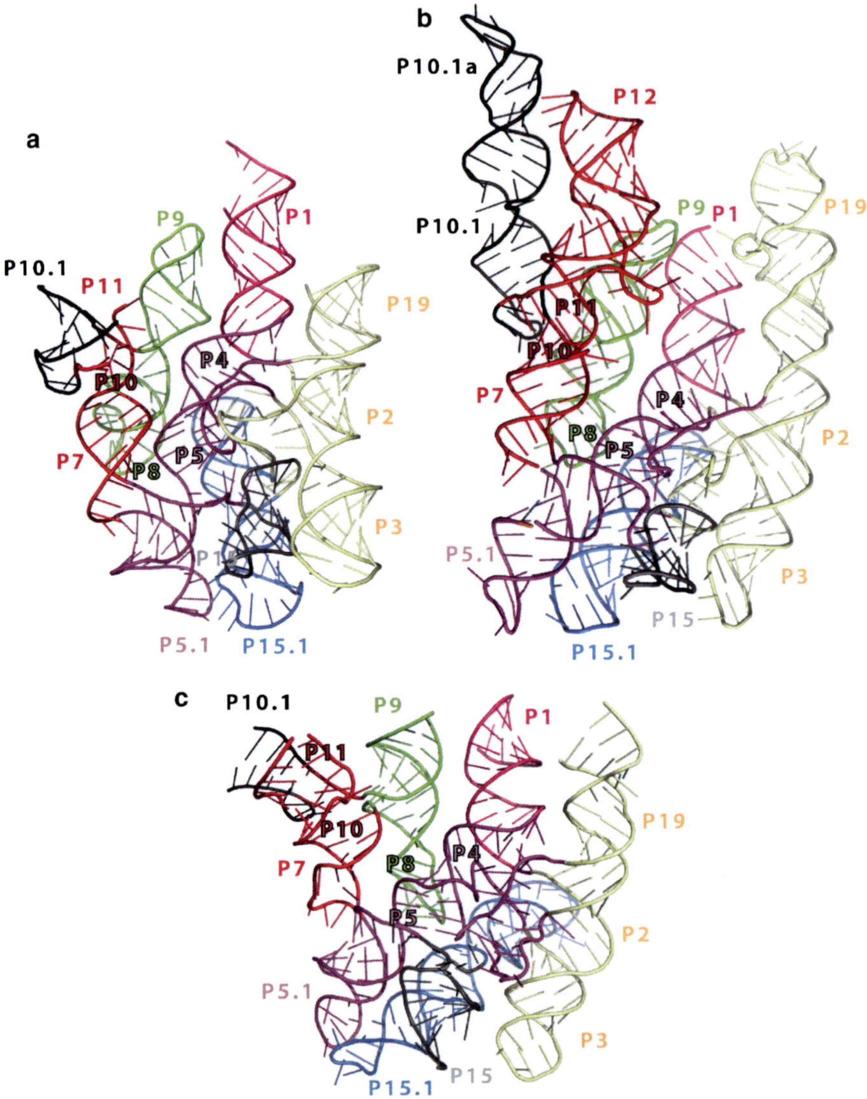

Fig. 3.2 Representation of the best 3D models of the B-type RNase P RNAs from the Pace (**a**) and the Westhof (**b**) laboratories. The color code is the same as in Fig. 3.1. The similar orientation of the two models underlines their high degree of similarity despite the different approaches undertaken to build them. Differences appear in the way P1 is stacked and in the orientation of several helical stacks. However, the elements on each face of the RNA are evenly distributed. The crystal structure of the whole *Bs* RNase P (Kazantsev et al. 2005) is represented in (**c**) as a standard for comparison of the two models

domains (Fig. 3.2). P2 is stacked on the P19 element that extends the corresponding R19 loop in *E. coli*. The P15.1/P15.2 stack occupies a location slightly different from the corresponding P18, albeit still lying on the face opposite to the ptRNA binding site. The P10.1 extension is roughly placed as the P13/P14 stack, although making different contacts with the P12 stem. As it is known that the RNA is composed of two self-folding domains (see above), the S and C domains were first built separately and then assembled together by taking advantage of tertiary interactions evidenced by comparative sequence analysis (Brown et al. 1996; Massire et al. 1997). The model is slightly curved with a concave internal side that binds the tRNA. The crystal structure of the *Bacillus stearothermophilus* RNase P RNA (Kazantsev et al. 2005) is rather flat, but has been solved without the ptRNA substrate. Residues supposed to be responsible for the binding of the protein subunit are clustered on the opposite convex side of the model.

The key tertiary interactions were contacts L14/P8, L18/P8 (Brown et al. 1996) and L9/P1 (Massire et al. 1997). The modeling led to the proposal of further tertiary interactions that were corroborated by comparative sequence analysis. These are L13/P12 (in *E. coli* or P10.1a/P12 in *B. subtilis*) interactions, the L5.1/L15.1 in the type B and the L8/P4. The L13/P12 interaction in type A RNase P was identified because it was recognized as occupying the same location as the evidenced interaction between P10.1a and P12 in type B which is similar to the tetraloop–tetraloop receptor interaction taking place in group I and II self-splicing introns (Cate et al. 1996; Costa and Michel 1997). The type-B-specific L5.1/L15.1 interaction was identified by the strong conservation of specific consensus sequences and by the fact that this loop-loop interaction motif was also swapping with a classical tetraloop in L5.1 that interacts in the shallow minor groove of an extended P15.1 stem. This interaction mimics the P6 pseudoknot observed in type A RNA. The L8/P4 interaction is a consequence of the previously identified L9/P1 interaction. As P8 and P9 are stacked, L8 is constrained by the geometry of the L9/P1 contacts to reach a location in the shallow minor groove of P4. Two strongly conserved A residues are present in good situation to interact with a tandem of conserved G=C pairs. In the type A ribozyme, P8 is thus the recipient of three tertiary interactions that lock the S domain onto the C domain. Notably, the crystal structure of the *Bs* RNase P RNA (Kazantsev et al. 2005) has confirmed this L8/P4 interaction and allowed for the observation of the other described tertiary contacts.

Extensive use of all published biochemical data allowed for the final validation of the model. Special attention was given to the crosslinking studies (Burgin and Pace 1990; Harris et al. 1994; Nolan et al. 1993). A majority of these crosslinks were compatible with the proposed model. Outliers were located in the L11/12 and P15 regions, which are poorly constrained and which are indeed disordered in the crystal structure (Kazantsev et al. 2005). Since P15 interacts both in vitro and in vivo with the ptRNA, the absence of the latter in the *Bs* crystal structure may be part of the reason for the apparent disorder of the former.

The architecture of the all-atom Westhof model is very similar to the architecture of the phosphate-only Pace model based on the crosslinking approach used to integrate APA in a circularly permuted form of the RNA (Harris et al. 1997). However, the sampling grid is finer than in the initial model (Harris et al. 1994). These models only present the core of the type A RNA. Comparative all-atom models of the two bacterial A and B types were proposed in order to document the arrangement of the different peripheral elements as supported by additional APA crosslinks (Chen et al. 1998). Residues encompassing elements P12, P13 and P14 in *E. coli* (P10.1 and P12 in *B. subtilis*) were not included, because the detected crosslinks were only local, leading to a lack of distance constraints between these elements and the core of the RNA subunit. This led to the conclusion that the P12–P14 region was protruding from the structure which is actually the case in the Westhof model (Massire et al. 1998). This view is supported by the *Bs* crystal structure in which the domain had to be modeled because of its absence from the electron density maps (Kazantsev et al. 2005). Interestingly, since the aim of the revised model (Chen et al. 1998) was mainly to integrate some tertiary contacts identified by comparative sequence analysis (Brown et al. 1996; Massire et al. 1997), a manual molecular modeling technique was used (Major et al. 1993) to build the final architecture of the 1997 Pace model. The Westhof and Pace models show a lot of similarity in the way the four helical domains are organized. The four-way junction P7-P9 is rather flat, with P7 and P10 coaxially stacked, as well as P7 and P8, as shown in the *Bs* crystal structure. The observed similarities also contribute to propose common interaction modes for the ptRNA substrate. The differences reside in the core, and specifically around P4 around which the topology of the joining region is different.

The *Bs* crystal structure showed that the topology of the Westhof model (Massire et al. 1998) was correct around the P4 region, whereas the Pace model (Chen et al. 1998) was not (Fig. 3.3). However, this observation is not of critical importance with

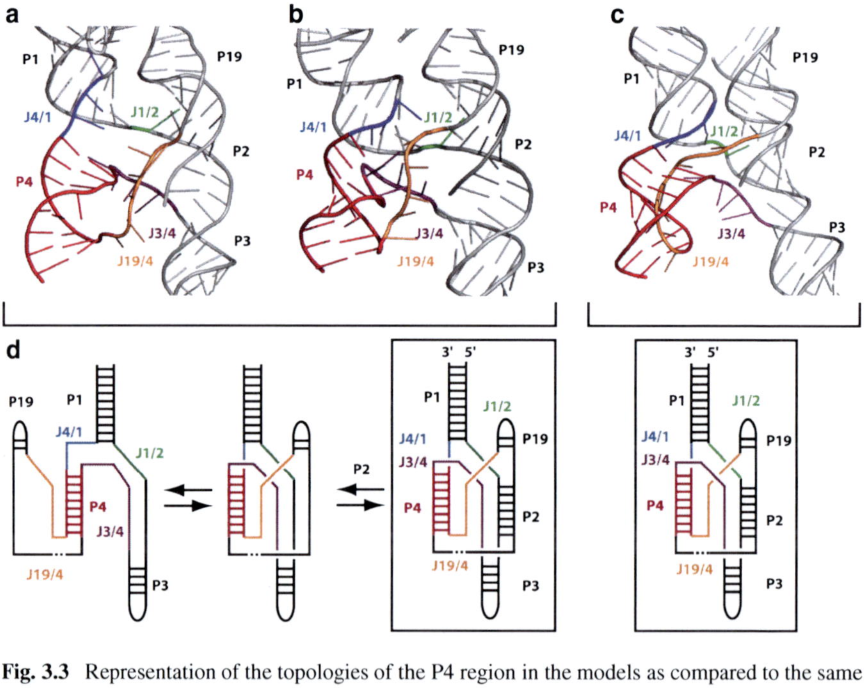

Fig. 3.3 Representation of the topologies of the P4 region in the models as compared to the same region in the *Bs* crystal structure (**b**) (Kazantsev et al. 2005). The path of the J19/4 junction is different in the two models leading to different topologies of the region containing the central pseudoknot P2/P4 (**d**). The *Bs* crystal structure reveals that the topology proposed by Westhof and colleagues (Massire et al. 1998) is correct (**a**) and not the topology of Pace and colleagues (Chen et al. 1998) (**b**). The consequence of this topology is to place this key junction on the outside of the enzyme that is responsible for the ptRNA recognition (See text)

regard to the correctness of the overall architecture; it is important regarding catalysis and folding pathways. The junction considered by the topological difference between the Westhof and Pace model is J19/4. It contains conserved residues that are important for catalysis; however, it is difficult to infer whether it interacts with the ptRNA substrate, the protein or both (Chen and Pace 1997). We have learnt from the *Bs* crystal structure (Kazantsev et al. 2005) that J19/4 actually weaves lots of contacts with the neighboring junctions J15.2/2 (corresponding to the *E. coli* J18/2) and J3/4, and that it is located on ptRNA substrate side. These junctions also contain conserved residues involved in catalysis and in ptRNA recognition. The identification of the correct topology was derived from a folding model of the RNase P RNA. This model is based on folding considerations of the junctions linking the central P2/P4 pseudoknot to the other regions of the RNA. Experimental evidence shows that P2 is a slowly forming element (Zarrinkar et al. 1996). Furthermore, P2 is shorter than P4, which means that P4 formation is energy-driven and precedes P2 formation (P2 is sometimes shorter or even absent in some RNase P). However, P4 is always present with, at most, one base pair less. Hence, P4 seems to fold before P2, although P2 is synthesized first during transcription. Therefore, the formation of P4 seems to promote the correct placement of the second strand of P2 in order to direct the recognition of the first strand. The application of this folding model to the Westhof 3D model leads to the positions of the P2-forming strands J1/2 and J2/4 against J19/4 and J18/2 so that the two sets of junctions do not cross each other. In other words, J1/2 and J2/4 on one hand, and J19/4 and J18/2 on the other hand form two sheets of junctions that pile against each other. Thus, the formation of P2 can be seen as a docking event. With such a topology, the early folding of P2 would prevent the folding of P4 by forming a real knot between the strands leading to P4. On the contrary, in the Pace model, the two sets of junctions are intricated resulting in the P3-forming strands entering the loop formed by the strands connecting the second strand of P2 to P4. With this topology, the folding of P2 cannot intervene prior to the folding of P4 in order to avoid the formation of a real knot in the RNA.

3.2.3 Models for the Holoenzyme

The validity of the RNase P RNA models complexed with the ptRNA substrate (Chen et al. 1998; Massire et al. 1998) was inferred from their degree of convergence. It is indeed clear that if the use of different experimental approaches coupled with different modeling techniques brings almost similar 3D models, it means that the amount of experimental and theoretical constraints of different origins is sufficient to describe the system adequately. Yet in 2002, because the protein subunit is dispensable in vitro, the complex formed between the protein and the RNA subunits had been somehow overlooked. The question then was determining the data on which a 3D model of the holoenzyme rely. The knowledge related to the holoenzyme or to the individual protein subunit can be summarized as follows. The structure of the RNase P protein from *B. subtilis* and *Staphylococcus aureus* were solved by crystallography (Stams et al. 1998) and NMR (Spitzfaden et al. 2000), respectively. Crosslinking studies had shown that the 5′ leader of the ptRNA was in contact with the central cleft of the protein subunit (Niranjanakumari et al. 1998). Iron–EDTA footprinting had also been carried out underlining potential regions of the RNase P RNA interacting with the protein (Loria et al. 1998; Rox et al. 2002; Talbot and Altman 1994; Vioque et al. 1988). These experiments showed three distinct footprint sites of the protein located in both the S and C domains in spite of the small size of the protein. This led to the conclusion that part of the footprints was probably due to conformational changes and/or to direct binding (Loria et al. 1998). This study also showed the lesser importance of the ptRNA T stem-loop for substrate recognition by the holoenzyme as compared with the naked RNA (Loria and Pan 1997), due to the compensation given by the stronger recognition of the RNA substrate proximal to the cleavage site. Moreover, mutational studies and phylogenetic studies had shown that the two C residues from the CCA ptRNA 3′ end were base-pairing to the well-conserved G292 and a 293 located in the internal loop separating stems P15 and P16 from the *E. coli* M1 RNA and from the reconstituted holoenzyme (Kirsebom and Svard 1994; Oh and Pace 1994; Svärd et al. 1996).

A model for the holoenzyme solely relying on the data described above could not be proposed with confidence. Neither the RNA domain nor the domain side interacting with the protein could be identified using such data. First, discriminating between direct and indirect footprints of the tiny 12-kDa P protein (*E. coli* C5) was critical in order to define a reasonable binding site between the two RNase P subunits. To achieve this goal, a probing method aiming at exploring the direct vicinity of specific C5 protein residues was implemented by tethering an Fe–EDTA probe to individual cystein residues engineered by mutagenesis in order to transform them into chemical nucleases. This technique has been successfully used to explore the direct environment of several ribosomal proteins (Heilek and Noller 1996a; Heilek and Noller 1996b; Heilek et al. 1995). Seven single Cys-substituted C5 protein mutants were derivatized with Fe–EDTA using EDTA-2-aminoethyl–pyridyl disulfide (EPD–Fe). These span over three different regions from the N-terminal, the centre and finally, to the C-terminal end of the protein. The seven cleavage patterns (Biswas et al. 2000; Tsai et al. 2003) when merged showed that all nucleotides in the direct vicinity of the EPD–Fe residues belonged to the C-domain. Strikingly, the footprints were located in the P3 and P4 stems, J18/2 and J19/4, reinforcing the topology of the junctions around the central pseudoknot presented in the Westhof model from (Massire et al. 1998). Thus, this model was used as a framework to map the protein-binding site (Fig. 3.4a). The crystal structure of the *B. subtilis* P protein was chosen to mimic the C5 protein from *Ec*. Positions of the EPD-Fe residues on C5 were deduced on the *B. subtilis* P protein from their sequence alignment. The Westhof model of the naked RNA (Massire et al. 1998) was adjusted in order to bring all the residues hit by the hydroxyl radical cleavage on the ptRNA substrate-binding face of the C domain. The data also allowed for defining the orientation of the protein with the help of the crosslinks between the protein and the 5′ end of the ptRNA (Niranjanakumari et al. 1998). The modeling was carried out on both the *E. coli* and the *B. subtilis* RNase P RNAs in order to understand the basis of the observation that RNase P holoenzyme formed from cross-species components are

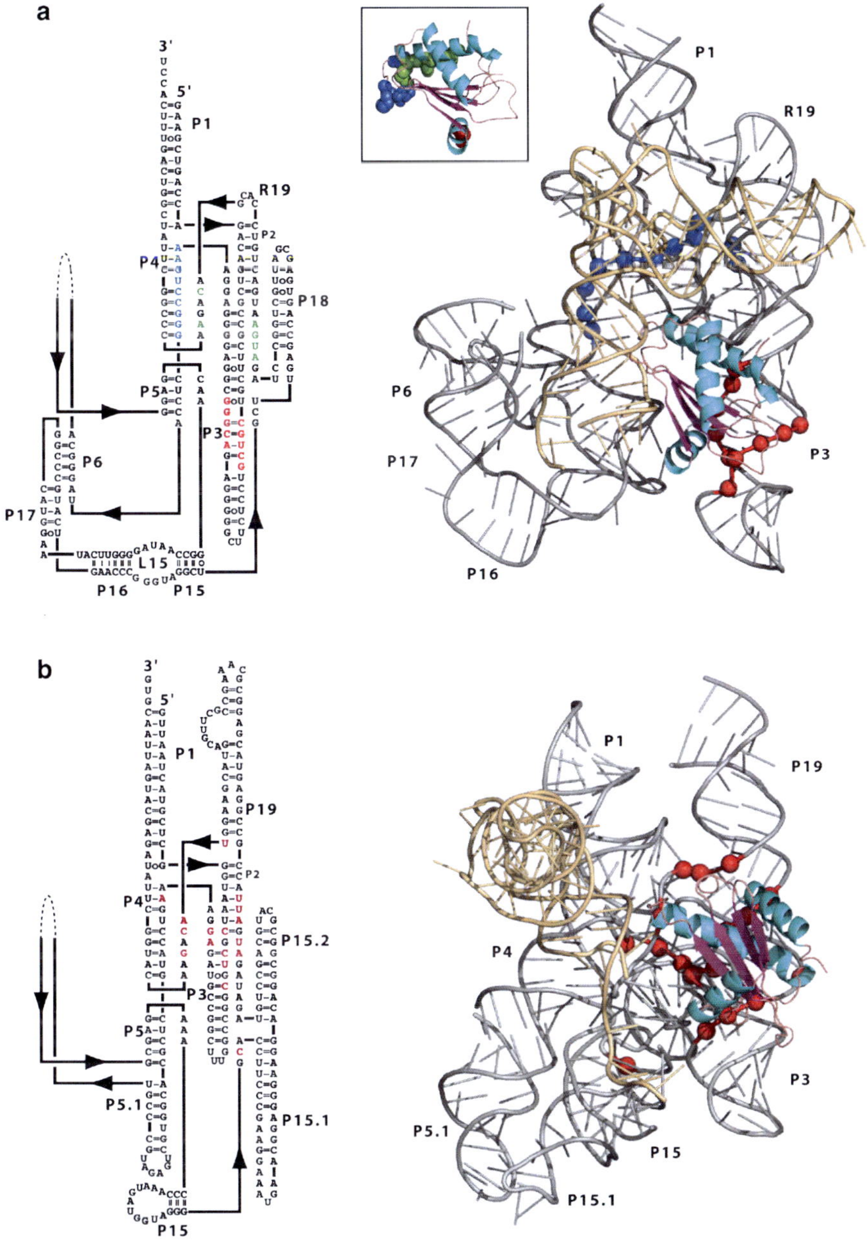

Fig. 3.4 Comparison between the holoenzyme models. (**a**) Secondary structure diagram of the Ec M1 RNA showing the cleavage pattern specifically due to the various EPD-Fe-substituted proteins (residues sensitive to Cys16, red; residues sensitive to Cys52, 54 or 59, blue; residues sensitive to Cys62, 66 or 70, green) (Left panel). Modified positions on the proteins are highlighted on the crystal structure of the *B.subtilis* RNase P protein (Stams et al. 1998) in the inset under the same orientation as in the model of the holoenzyme. (Right panel) 3D resulting model of the holoenzyme showing the position of the protein on the C-domain, as well as the position of the ptRNA. (**b**) Footprint of the RNase P protein mapped on the secondary structure of the *Bs* RNase P RNA (Left panel). (Right panel) 3D model of the holoenzyme showing the *B. subtilis* protein crystal structure (Stams et al. 1998) and the ptRNA substrate docked on the *Bs* RNase P RNA crystal structure (Kazantsev et al. 2005)

functional (Guerrier-Takada et al. 1983). Interestingly, the various architectural adjustments made to the RNA to fulfil the data have revealed a likely A-minor interaction (Doherty et al. 2001; Nissen et al. 2001) between the −1 and −2 residues from the 5′ leader of the ptRNA with the two consecutive A residues located in J5/15 (*Ec* A248 and 249). This interaction is proposed to constitute a sensor for the presence of the 5′ leader in a way helped by the protein. This interaction provides a rationale for the experimentally observed importance of J5/15 (Christian and Harris 1999; Christian et al. 1998; Siew et al. 1999). The final A and B type models (Tsai et al. 2003) render a picture of the holoenzymes with both the protein subunit and ptRNA lying on the same face of the C domain where they interact with a large part of the strongly conserved residues. Moreover, the protein is in close contact with the 5′ leader of the ptRNA. The arginine-rich consensus RNR motif (Stams et al. 1998) faces the RNA subunit, and more specifically, the residues hit by the EPD–Fe modified mutants. Furthermore, the models suggest that the RNR motif interacts with the RNase P RNA, whereas residues located in the cleft bind the 5′ leader of the ptRNA substrate. In our molecular model of the holoenzyme, residues A248 (*E. coli* numbering) and A249 from the J5/15 junction interact in the shallow groove of the −2/−1 step of the pre-tRNA (Tsai et al. 2003). These interactions intervene simultaneously with others in the course of the docking of the pre-tRNA. On the proximal side, its CCA 3′ end interacts with J16/15 and on the distal side, the D/T loop domain interacts with the specificity domain, and more precisely, in the L11/12 region. This AA–J5/15 interaction is based on A-minor motif (Doherty et al. 2001; Nissen et al. 2001) recognition of the shallow groove. Although, it was modeled slightly differently, it resembles the structure of the decoding site of the 16 S ribosomal subunit in contact with the codon/anticodon stem (Lescoute and Westhof 2006; Ogle et al. 2001; Vicens and Westhof 2001). Hence, such an interaction could occur and control the docking of the −2/−1 step of the pre-tRNA by the two A residues from the J5/15 junction in order to promote fast catalysis. However, the −2/−1 nucleotides are not known to form Watson–Crick base pairs and, thus, such a contact would be a novel type. Recently, it has been suggested that induced fit of the M1RNA indeed occurs upon substrate binding (Brännvall et al. 2007) reinforcing the analogy between the M1 RNA and the decoding site of the ribosome.

3.2.3.1 ptRNA Recognition and Crystal Packing

The crystal structure of the *B. subtilis* S domain (Krasilnikov et al. 2003) contains two S domain molecules in the asymmetric unit. The first one is less well ordered than the second one. Interestingly, the largest apparently unstructured part of the structure of the S domain is located in the J11/12 and J12/11 junctions where they form two T loop motifs (Krasilnikov and Mondragon 2003) which are typical of the tRNA structure. The J12/11 contains a G residue (G220) known to bind directly the T loop of the ptRNA substrate, as well as bulging As from P9 and P11 contacting the T stem (LaGrandeur et al. 1994; Odell et al. 1998). In the crystal structure, these residues in the second molecule clamp the highly non-canonical L8 loop of the first molecule.

Superimposition between the S domain of the holoenzyme model (Tsai et al. 2003) with the second molecule of the S domain shows that the S domain of the holoenzyme model is in very good agreement with the S domain crystal structure (the value of the calculated normalized rmsd is only 4.8 Å (Kleywegt and Jones 1994)). Furthermore, the first S domain molecule falls in place with the ptRNA substrate (Fig. 3.5). The T loop from the ptRNA docked in the model of the holoenzyme is located at the very same place where the L8 loop in the first molecule of the S domain crystal structure is located. Thus the S domain dimer in the asymmetric unit of the crystal structure may somehow validate the ptRNA recognition scheme by the holoenzyme in our model. This idea is indeed supported by the observation that the interactions leading to the formation of the S domain dimer do not correspond to the potential *B. subtilis* RNase P dimer that is observed in solution (Barrera et al. 2002).

Following the publication of the crystal structure of the complete *Bs* RNase P (Kazantsev et al. 2005), Pace and coworkers proposed another model of the holoenzyme complexed with the ptRNA substrate (Buck et al. 2005). This model was based on a comparative footprinting approach based on the isolation of folding intermediates.

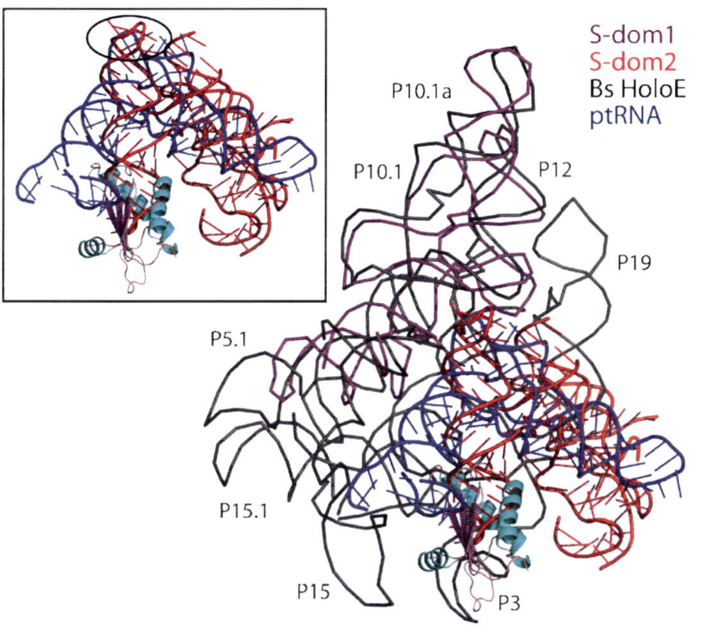

Fig. 3.5 Superimposition of the holoenzyme model on the dimer present in the symmetric unit of the *B. subtilis* S-domain crystal structure (Krasilnikov et al. 2003). The ribbon representation of the RNase P holoenzyme model is in *dark gray*. The associated pre-tRNA substrate is in blue and the protein moiety bound to the C-domain is represented according to the secondary structure to help in orienting the complex. The most complete S-domain monomer of the asymmetric unit (purple) is superimposed on the S-domain of the holoenzyme model (normalized rmsd 4.8 Å). The second S-domain monomer (red) is superimposed to the pre-tRNA. The inset shows the resulting superimposition for the pre-tRNA and the second S-domain monomer. The T loops are circled

An in-gel phosphorothioate–iodine footprint assay was develops to test the interactions between two RNase P RNAs (*E. coli* and *B. stearothermophilus*) and three RNase P proteins (*E. coli*, *B. subtilis* and *B. stearothermophilus*). This technique warrants the homogeneity of the probed complexes and allows specific probing of different conformational states as far as they can be isolated efficiently. In this way, the naked RNA corresponding to the native (N) and intermediate (I) folded forms were also probed. By substracting the probing patterns between the various forms of the RNA, it was possible to discriminate between protections because of the contacts between the protein and the RNA and the protections due to RNA structural rearrangements (Fig. 3.4b). In agreement with the Westhof holoenzyme model, all the protections due to direct contacts with the protein map to the C domain where they target the same nucleotide regions (P2, P3, P4, J19/4, J15.2/2 and J19/4). Subtle differences in the probing patterns are observed according to the RNA and protein used to form the complexes, but the overall patterns are very much similar. This eventually reflects differences in the interaction networks between the RNase P components. Notably, the A residues in J5/15, supposedly involved in the recognition of the −1 and −2 residues of the ptRNA substrate 5′ leader, are variably affected by the proteins. This holoenzyme model also suggests that the protein subunit binds the C domain and co-localizes on the same side as the ptRNA substrate. In both holoenzyme models, the protein is roughly oriented in the same way with respect to the RNR motif, facing the conserved junctions from the RNA core (J19/4 and J15.2/2) and P4, and the 5′ leader of the ptRNA substrate (which faces the hydrophobic cleft of the protein). The protein subunit is positions slightly more towards P4 in the Pace model than in the Westhof model. The structural differences between the model of the RNA subunit and the crystal structure of the *Bs* RNase P used in each study account for this consequence. Moreover, mild differences in the protection patterns of each footprinting study explain that different distance constraints have been used in each case. For example, a set of protections was located slightly closer to the bottom of P3 in the experiments carried out by Pace and coworkers resulting in the placement of the protein subunit closer to P15 rather than to P4 in the Westhof model.

3.3 Conclusions and Perspectives

As pointed out in the introduction, in order to play a predictive role, a model should answer defined questions and raise new ones. The RNase P models have perfectly fulfilled this role by rationalizing several points, such as the architecture and the topology of the RNase P RNA, the way the ptRNA substrate is recognized, and finally the way the protein docks onto the C domain and wraps the ptRNA 5′ leader. Importantly, the models have shown how a 3D architecture could be built from the individual S and C domains. A consequence of that was the identification of new complex tertiary interactions involving stem P8, showing that the core of the ribozyme was actually formed at the interface of the two domains. The models also rationalized the experimental observation that two regions of the ptRNA were specifically recognized, each by one domain, whereas the presence of 5′ leader was detected by a distinct junction. Furthermore, the models show that the situation is comparable to the group I intron case where a structurally very well-conserved catalytic core is surrounded by peripheral elements that vary a lot in sequence and shape, but still have a main role in stabilizing the structure of the core and the catalytic intermediates (Baird et al. 2005). Thus, the bacterial RNase P ribozymes apparently constitute an example of adaptive selection in which an ancestral RNase P ribozyme (type A) evolved to a type B typical of the Bacillus subdivision. Such an evolutionary mechanism can be explained by a neutral sequence drift based on the relative equilibrium between discrete folding intermediates of the RNA (Fig. 3.6). The ancestral (type A) RNase

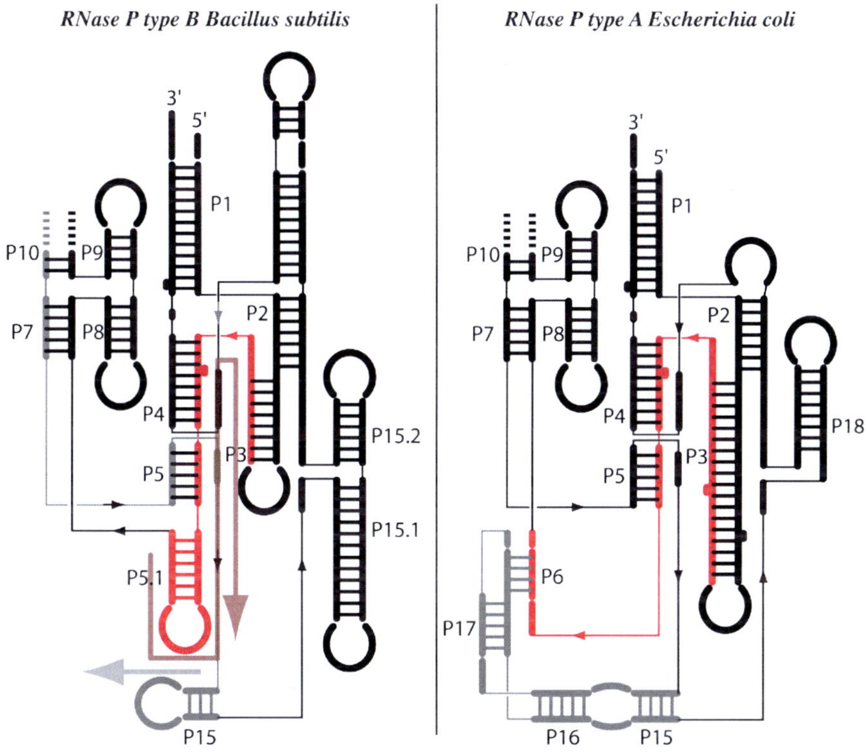

Fig. 3.6 Neutral sequence drift suggests a model to explain the evolutionary link between the A and B bacterial RNase P subtypes. Arbitrarily starting from the B type, enlengthen the red strand of P3 results in shortening of P5.1. Concomitantly, enlengthen P15 results in the shortening of P19. The loop of the newly formed P17 can interact with the remaining portion of P5.1 to form P6. Along the same line of idea, enlengthen P15 results in the appearance of the P16 and P17 elements allowing the formation of P6. These events should result in structural modification of the P7/P10 region, which is indeed observed. The directionality of this process is not known, although the B type derives from the A type. According to this scheme, it is likely that sequence drift favored misfolding of some key elements mentioned above, resulting in the loss of the P6 pseudo-knot and the stretching of the P19 hairpin. The new scaffold was selected because of its capacity to carry out the cleavage of the 5′ leader sequence of the ptRNA substrate

P comprises the P2/P4 and the P6/P17 pseudoknots. It is likely that a folding intermediate of a type A RNase P that undertook a critical amount of mutation may have been locked in a conformation showing alterations of the secondary structure (Beckert et al. 2008). Extending P19 by using residues building up the P16 and P17 elements would result in the shortening of the P15 peripheral domain. Similarly, if residues embedded in P6 form a hairpin taking advantage of base complementarity in downstream residues, P3 would be shortened. These concerted events may have led to the appearance of the two types of ribozymes.

In spite of all the progress made by crystallographic methods, there is still no crystal structure of the holoenzyme in the presence of the substrate. The current models thus constitute the most accomplished structural studies of the RNase P holoenzyme. In cases where crystals are difficult to obtain, models can be built on a critically shorter time scale and thus quickly provide efficient rationalization frameworks to help researchers interpret and design the experiments most adequate to target the important questions.

As a final note, one should consider the limitations of molecular modeling of RNA molecules. Clearly, for highly structured RNAs, such as the RNaseP or group I introns, rather well-defined architectures can be deduced with some confidence, especially if some key experimental constraints are available. Such architectures, although built from atomic fragments, cannot be considered at atomic resolution and they do contain numerous errors. Despite these limitations, such architectures, when understood properly, can be extremely valuable to the progress of the field. However, the recent crystal structures of fully active ribozymes, such as the hammerhead, hairpin or group II ribozymes, show pronounced deformations and propagations of induced-fits around the active sites (Michel et al. 2009). The prediction and modeling of such distorted states are totally out of reach of present computer modeling techniques.

References

Altman S, Kirsebom L, Talbot S (1993) Recent studies of ribonuclease P. FASEB J 7:7–14

Baird NJ, Westhof E, Qin H, Pan T, Sosnick TR (2005) Structure of a folding intermediate reveals the interplay between core and peripheral elements in RNA folding. J Mol Biol 352:712–722

Barrera A, Fang X, Jacob J, Casey E, Thiyagarajan P, Pan T (2002) Dimeric and monomeric *Bacillus subtilis* RNase P holoenzyme in the absence and presence of pre-tRNA substrates. Biochemistry 41:12986–12994

Beckert B, Nielsen H, Einvik C, Johansen SD, Westhof E, Masquida B (2008) Molecular modelling of the GIR1 branching ribozyme gives new insight into evolution of structurally related ribozymes. EMBO J 27:667–678

Biswas R, Ledman DW, Fox RO, Altman S, Gopalan V (2000) Mapping RNA–protein interactions in ribonuclease P from *Escherichia coli* using disulfide-linked EDTA–Fe. J Mol Biol 296:19–31

Brännvall M, Kirsebom LA (1999) Manganese ions induce miscleavage in the *Escherichia coli* RNase P RNA-catalyzed reaction. J Mol Biol 292:53–63

Brännvall M, Kikovska E, Kirsebom LA (2004) Cross talk between the +73/294 interaction and the cleavage site in RNase P RNA mediated cleavage. Nucleic Acids Res 32:5418–5429

Brännvall M, Kikovska E, Wu S, Kirsebom LA (2007) Evidence for induced fit in bacterial RNase P RNA-mediated cleavage. J Mol Biol 372:1149–1164
Brown JW, Nolan JM, Haas ES, Rubio MAT, Major F, Pace NR (1996) Comparative analysis of ribonuclease P RNA using gene sequences from natural microbial populations reveals tertiary structural elements. Proc Natl Acad Sci U S A 93:3001–3006
Brunel C, Romby P (2000) Probing RNA structure and RNA–ligand complexes with chemical probes. Methods Enzymol 318:3–21
Buck AH, Kazantsev AV, Dalby AB, Pace NR (2005) Structural perspective on the activation of RNase P RNA by protein. Nat Struct Mol Biol 12:958–964
Burgin AB, Pace NR (1990) Mapping the active site of ribonuclease P RNA using a substrate containing a photoaffinity agent. EMBO J 9:4111–4118
Cate JH, Gooding AR, Podell E, Zhou K, Golden BL, Kundrot CE, Cech TR, Doudna JA (1996) Crystal structure of a group I ribozyme domain: principles of RNA packing. Science 273:1678–1684
Cech TR (2009) Crawling out of the RNA world. Cell 136:599–602
Chen JL, Pace NR (1997) Identification of the universally conserved core of ribonuclease P RNA [letter]. RNA 3:557–560
Chen JL, Nolan JM, Harris ME, Pace NR (1998) Comparative photocross-linking analysis of the tertiary structures of *Escherichia coli* and *Bacillus subtilis* RNase P RNAs. EMBO J 17: 1515–1525
Christian EL, Harris ME (1999) The track of the pre-tRNA 5′ leader in the ribonuclease P ribozyme–substrate complex. Biochemistry 38:12629–12638
Christian EL, McPheeters DS, Harris ME (1998) Identification of individual nucleotides in the bacterial ribonuclease P ribozyme adjacent to the pre-tRNA cleavage site by short-range photo-cross-linking. Biochemistry 37:17618–17628
Christian EL, Kaye NM, Harris ME (2000) Helix P4 is a divalent metal ion binding site in the conserved core of the ribonuclease P ribozyme. RNA 6:511–519
Costa M, Michel F (1997) Rules for RNA recognition of GNRA tetraloops deduced by in vitro selection: comparison with in vivo evolution. EMBO J 16:3289–3302
Crary SM, Niranjanakumari S, Fierke CA (1998) The protein component of *Bacillus subtilis* ribonuclease P increases catalytic efficiency by enhancing interactions with the 5′ leader sequence of pre-tRNAAsp. Biochemistry 37:9409–9416
Darr SC, Brown JW, Pace NR (1992) The varieties of ribonuclease P. Trends Biochem Sci 17: 178–182
Daviter T, Gromadski KB, Rodnina MV (2006) The ribosome's response to codon–anticodon mismatches. Biochimie 88:1001–1011
Doherty EA, Batey RT, Masquida B, Doudna JA (2001) A universal mode of helix packing in RNA. Nat Struct Biol 8:339–343
Easterwood TR, Harvey SC (1997) Ribonuclease P RNA: models of the 15/16 bulge from *Escherichia coli* and the P15 stem loop of *Bacillus subtilis*. RNA 3:577–585
Forster AC and Altman S (1990) External guide sequences for an RNA enzyme. Science, 249:783–786
Gaur RK, Hanne A, Conrad F, Kahle D, Krupp G (1996) Differences in the interaction of *Escherichia coli* RNase P RNA with tRNAs containing a short or a long extra arm. RNA 2:674–681
Giege R (2008) Toward a more complete view of tRNA biology. Nat Struct Mol Biol 15: 1007–1014
Guerrier-Takada C, Gardiner K, Marsh T, Pace N, Altman S (1983) The RNA moiety of ribonuclease P is the catalytic subunit of the enzyme. Cell 35:849–857
Haas ES, Morse DP, Brown JW, Schmidt FJ, Pace NR (1991) Long-range structure in ribonuclease P RNA. Science 254:853–856
Haas ES, Brown JW, Pitulle C, Pace NR (1994) Further perspective on the catalytic core and secondary structure of ribonuclease P RNA. Proc Natl Acad Sci U S A 91:2527–2531

Harris ME, Nolan JM, Malhotra A, Brown JW, Harvey SC, Pace NR (1994) Use of photoaffinity crosslinking and molecular modeling to analyze the global architecture of ribonuclease P RNA. EMBO J 13:3953–3963

Harris ME, Kazantsev AV, Chen JL, Pace NR (1997) Analysis of the tertiary structure of the ribonuclease P ribozyme–substrate complex by site-specific photoaffinity crosslinking. RNA 3:561–576

Hartmann RK, Heinrich J, Schlegl J, Schuster H (1995) Precursor of C4 antisense RNA of bacteriophages P1 and P7 is a substrate for RNase P of *Escherichia coli*. Proc Natl Acad Sci U S A 92:5822–5826

Heilek G, Noller H (1996a) Directed hydroxyl radical probing of the rRNA neighborhood of ribosomal protein S13 using tethered Fe(II). RNA 2:597–602

Heilek G, Noller H (1996b) Site-directed hydroxyl radical probing of the rRNA neighborhood of ribosomal protein S5. Science 272:1659–1662

Heilek GM, Marusak R, Meares CF, Noller HF (1995) Directed hydroxyl radical probing of 16 S rRNA using Fe(II) tethered to ribosomal protein S4. Proc Natl Acad Sci U S A 92:1113–1116

Holzmann J, Frank P, Loffler E, Bennett KL, Gerner C, Rossmanith W (2008) RNase P without RNA: identification and functional reconstitution of the human mitochondrial tRNA processing enzyme. Cell 135:462–474

Kazantsev AV, Krivenko AA, Harrington DJ, Holbrook SR, Adams PD, Pace NR (2005) Crystal structure of a bacterial ribonuclease P RNA. Proc Natl Acad Sci U S A 102:13392–13397

Kikovska E, Svard SG, Kirsebom LA (2007) Eukaryotic RNase P RNA mediates cleavage in the absence of protein. Proc Natl Acad Sci U S A 104:2062–2067

Kirsebom LA (2007) RNase P RNA mediated cleavage: substrate recognition and catalysis. Biochimie 89:1183–1194

Kirsebom LA and Svärd SG (1994) Base pairing between Escherichia coli RNase P RNA and its substrate. Embo J, 13:4870–4876

Kleywegt GJ, Jones TA (1994) A super position. CCP4/ESF-EACBM Newsletter Protein Crystallogr 31:9–14

Krasilnikov AS, Mondragon A (2003) On the occurrence of the T-loop RNA folding motif in large RNA molecules. RNA 9:640–643

Krasilnikov AS, Yang X, Pan T, Mondragon A (2003) Crystal structure of the specificity domain of ribonuclease P. Nature 421:760–764

Krasilnikov AS, Xiao Y, Pan T, Mondragon A (2004) Basis for structural diversity in homologous RNAs. Science 306:104–107

Kurz JC, Niranjanakumari S, Fierke CA (1998) Protein component of *Bacillus subtilis* RNase P specifically enhances the affinity for precursor-tRNAAsp. Biochemistry 37:2393–2400

LaGrandeur TE, Huttenhofer A, Noller HF, Pace NR (1994) Phylogenetic comparative chemical footprint analysis of the interaction between ribonuclease P RNA and tRNA. EMBO J 13:3945–3952

Leontis NB, Stombaugh J, Westhof E (2002) The non-Watson–Crick base pairs and their associated isostericity matrices. Nucleic Acids Res 30:3497–3531

Lescoute A, Westhof E (2006) The A-minor motifs in the decoding recognition process. Biochimie 88:993–999

Li Z, Deutscher MP (1996) Maturation pathways for *E. coli* tRNA precursors: a random multienzyme process in vivo. Cell 86:503–512

Loria A, Pan T (1996) Domain structure of the ribozyme from eubacterial ribonuclease P. RNA 2:551–563

Loria A, Pan T (1997) Recognition of the T stem-loop of a pre-tRNA substrate by the ribozyme from *Bacillus subtilis* ribonuclease P. Biochemistry 36:6317–6325

Loria A, Niranjanakumari S, Fierke CA, Pan T (1998) Recognition of a pre-tRNA substrate by the *Bacillus subtilis* RNase P holoenzyme. Biochemistry 37:15466–15473

Major F, Turcotte M, Gautheret D, Lapalme G, Fillion E, Cedergren R (1991) The combination of symbolic and numerical computation for three-dimensional modeling of RNA. Science 253:1255–1260
Major F, Gautheret D, Cedergren R (1993) Reproducing the three-dimensional structure of a tRNA molecule from structural constraints. Proc Natl Acad Sci U S A 90:9408–9412
Malhotra A, Harvey SC (1994) A quantitative model of the *Escherichia coli* 16 S RNA in the 30 S ribosomal subunit. J Mol Biol 240:308–340
Marquez SM, Chen JL, Evans D, Pace NR (2006) Structure and function of eukaryotic Ribonuclease P RNA. Mol Cell 24:445–456
Masquida B, Westhof E (2005) Modeling the architecture of structured RNAs within a modular and hierarchical framework. In: Hartmann RK, Bindereif A, Schön A, Westhof E (eds) Handbook of RNA biochemistry. Wiley VCH Verlag Gmbh & Co, Weinheim, pp 536–545
Massire C, Westhof E (1998) MANIP: an interactive tool for modelling RNA. J Mol Graph Model 16:197–205, 255–257
Massire C, Jaeger L, Westhof E (1997) Phylogenetic evidence for a new tertiary interaction in bacterial RNase P RNA. RNA 3:553–556
Massire C, Jaeger L, Westhof E (1998) Derivation of the three-dimensional architecture of bacterial ribonuclease P RNAs from comparative sequence analysis. J Mol Biol 279:773–793
Massire C and Westhof E (1998) MANIP: an interactive tool for modelling RNA. J Mol Graph Model, 16:197–205, 255–197
Matsuhashi M, Dietrich CP, Strominger JL (1965) Incorporation of glycine into the cell wall glycopeptide in *Staphylococcus aureus*: role of sRNA and lipid intermediates. Proc Natl Acad Sci U S A 54:587–594
Michel F, Costa M, Westhof E (2009) The ribozyme core of group II introns: a structure in want of partners. Trends Biochem Sci 34:189–199
Niranjanakumari S, Stams T, Crary SM, Christianson DW, Fierke CA (1998) Protein component of the ribozyme ribonuclease P alters substrate recognition by directly contacting precursor tRNA. Proc Natl Acad Sci U S A 95:15212–15217
Nissen P, Ippolito JA, Ban N, Moore PB, Steitz TA (2001) RNA tertiary interactions in the large ribosomal subunit: the A-minor motif. Proc Natl Acad Sci U S A 98:4899–4903
Nolan JM, Burke DH, Pace NR (1993) Circularly permuted tRNAs as specific photoaffinity probes of ribonuclease P RNA structure. Science 261:762–765
Odell L, Huang V, Jakacka M, Pan T (1998) Interaction of structural modules in substrate binding by the ribozyme from *Bacillus subtilis* RNase P. Nucleic Acids Res 26:3717–3723
Ogle JM, Brodersen DE, Clemons WM Jr, Tarry MJ, Carter AP, Ramakrishnan V (2001) Recognition of cognate transfer RNA by the 30 S ribosomal subunit. Science 292:897–902
Oh BK and Pace NR (1994) Interaction of the 3'-end of tRNA with ribonuclease P RNA. Nucleic Acids Res, 22:4087–4094
Pan T (1995) Higher order folding and domain analysis of the ribozyme from *Bacillus subtilis* ribonuclease P. Biochemistry 34:902–909
Peck-Miller KA, Altman S (1991) Kinetics of the processing of the precursor to 4.5 S RNA, a naturally occurring substrate for RNase P from *Escherichia coli*. J Mol Biol 221:1–5
Rangan P, Masquida B, Westhof E, Woodson SA (2004) Architecture and folding mechanism of the Azoarcus Group I Pre-tRNA. J Mol Biol 339:41–51
Rox C, Feltens R, Pfeiffer T, Hartmann RK (2002) Potential contact sites between the protein and RNA subunit in the *Bacillus subtilis* RNase P holoenzyme. J Mol Biol 315:551–560
Roy H, Becker HD, Mazauric MH, Kern D (2007) Structural elements defining elongation factor Tu mediated suppression of codon ambiguity. Nucleic Acids Res 35:3420–3430
Siew D, Zahler NH, Cassano AG, Strobel SA, Harris ME (1999) Identification of adenosine functional groups involved in substrate binding by the ribonuclease P ribozyme. Biochemistry 38:1873–1883
Spitzfaden C, Nicholson N, Jones JJ, Guth S, Lehr R, Prescott CD, Hegg LA, Eggleston DS (2000) The structure of ribonuclease P protein from *Staphylococcus aureus* reveals a unique binding site for single-stranded RNA. J Mol Biol 295:105–115

Stams T, Niranjanakumari S, Fierke CA, Christianson DW (1998) Ribonuclease P protein structure: evolutionary origins in the translational apparatus. Science 280:752–755

Stathopoulos C, Tekos A, Zarkadis IK, Drainas D (2001) Extensive deproteinization of *Dictyostelium discoideum* RNase P reveals a new catalytic activity. Eur J Biochem 268:2134–2140

Svärd SG, Kagardt U and Kirsebom LA (1996) Phylogenetic comparative mutational analysis of the base-pairing between RNase P RNA and its substrate. Rna, 2:463–472

Talbot SJ, Altman S (1994) Gel retardation analysis of the interaction between C5 protein and M1 RNA in the formation of the ribonuclease P holoenzyme from *Escherichia coli*. Biochemistry 33:1399–1405

Torres-Larios A, Swinger KK, Krasilnikov AS, Pan T, Mondragon A (2005) Crystal structure of the RNA component of bacterial ribonuclease P. Nature 437:584–587

Tsai HY, Masquida B, Biswas R, Westhof E, Gopalan V (2003) Molecular modeling of the three-dimensional structure of the bacterial RNase P holoenzyme. J Mol Biol 325:661–675

Vicens Q, Westhof E (2001) Crystal structure of paromomycin docked into the eubacterial ribosomal decoding A site. Structure (Camb) 9:647–658

Vioque A, Arnez J, Altman S (1988) Protein–RNA interactions in the RNase P holoenzyme from *Escherichia coli*. J Mol Biol 202:835–848

Westhof E, Altman S (1994) Three-dimensional working model of M1 RNA, the catalytic RNA subunit of ribonuclease P from *Escherichia coli*. Proc Natl Acad Sci U S A 91:5133–5137

Westhof E, Wesolowski D, Altman S (1996) Mapping in three dimensions of regions in a catalytic RNA protected from attack by an Fe(II)–EDTA reagent. J Mol Biol 258:600–613

Zarrinkar PP, Wang J, Williamson JR (1996) Slow folding kinetics of RNase P RNA. RNA 2:564–573

Chapter 4
Structural Studies of Ribonuclease P

Alfonso Mondragón

4.1 Introduction

Ribonuclease P, or RNase P, was one of the first ribozymes identified (Guerrier-Takada et al. 1983) and has been found in all three domains of life. It is the only endonuclease responsible for processing the 5′ end of tRNA by cleaving a precursor and leading to tRNA maturation (for reviews see (Frank and Pace 1998; Altman and Kirsebom 1999; Hartmann and Hartmann 2003; Kazantsev and Pace 2006; Torres-Larios et al. 2006; Walker and Engelke 2006)). It also cleaves other substrates including some viral and phage RNAs (Mans et al. 1990; Hartmann et al. 1995), pre-4.5S RNA (Peck-Miller and Altman 1991), pre-tmRNA (Komine et al. 1994), a few mRNAs (Alifano et al. 1994; Li and Altman 2003), and riboswitches (Altman et al. 2005). RNase P contains one RNA component and at least one protein component. Unlike most other natural ribozymes, RNase P is a multiple turnover enzyme that recognizes and cleaves its substrate in *trans* through the recognition of secondary and tertiary structural elements in the substrate. There are clear similarities among the RNA component of all organisms (Siegel et al. 1996; Chen and Pace 1997), strongly suggesting evolutionary conservation and indicating that understanding the structure of RNase P of a particular organisms may help understand the architecture of RNase P of all organisms.

An atomic level understanding of RNase P function requires knowledge of the atomic structures of the RNA and protein components alone, of the holoenzyme, and of ternary complexes of the RNA, protein components, and substrate, product and intermediates. In recent years, significant progress in understanding the structure of RNase P has been made, primarily through crystallographic and NMR studies of the RNA and protein components of different bacteria and archaeons. The structures of the RNA component from two different bacteria provided information that help understanding the way the RNA folds into a complex three dimensional structure, the interactions responsible for stabilizing such a large molecule, and the location of

A. Mondragón (✉)
Department of Biochemistry, Molecular Biology, and Cell Biology,
Northwestern University, 2205 Tech Drive, Evanston, IL 60208, USA
e-mail: a-mondragon@northwestern.edu

F. Liu and S. Altman (eds.), *Ribonuclease P*, Protein Reviews 10,
DOI 10.1007/978-1-4419-1142-1_4, © Springer Science+Business Media, LLC 2010

many of the regions crucial for activity. Structural studies are also helping to further our understanding of the structure of the protein components. Overall, a much clearer picture of the structure of RNase P is emerging that starts to provide a solid foundation to understand its function. Nevertheless, the absence of high resolution structures of the RNA component and structural information on the mechanism of substrate recognition and processing are still major limitations that need to be overcome. Here we present a brief overview of the structures known to date of the RNA and protein components. It is hoped that in the future, additional structures will be elucidated to help continue bringing our understanding of RNase P to the atomic level.

4.2 Structural Studies of Bacterial RNase P

In bacteria, RNase P is composed of a large RNA subunit and a small protein. Based on sequence similarities, bacterial RNase P's can be divided into two major types: A and B, for Ancestral and Bacillus respectively. The RNase P RNA component (P RNA) consists of ~300–450 nucleotides and is formed by two independently folded domains (Pan 1995; Loria and Pan 1996) (delineated on the secondary structure diagrams in Fig. 4.1) whose discovery helped support the observation, first reported for the group I intron (Doudna and Cech 1995), that large RNA molecules

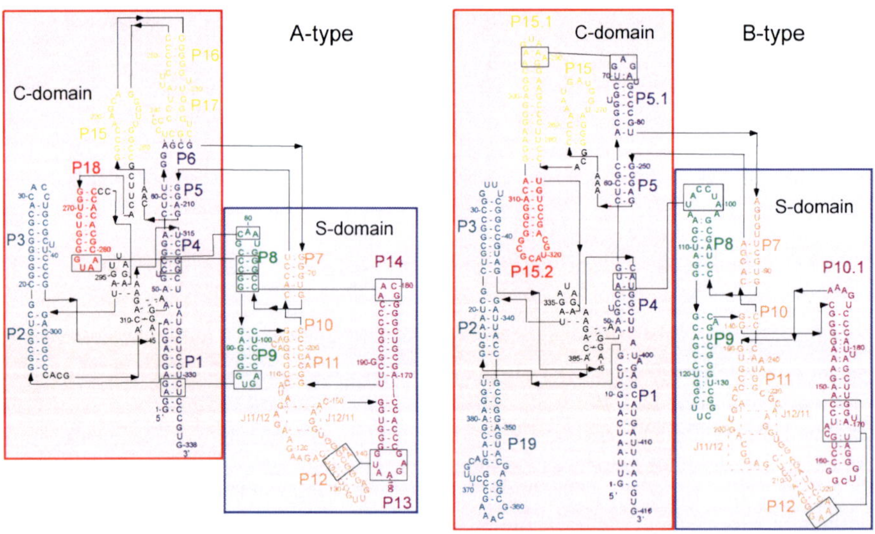

Fig. 4.1 Secondary structure diagrams of the RNA component of bacterial RNase P. The diagrams show the secondary structure for A- and B-type bacterial RNase P. The two structural domains are delimited by boxes, blue for the S-domain and red for the C-domain. Known tertiary interactions are indicated by connected boxes. Analogous secondary structural elements are colored identically in both diagrams. The diagrams were drawn based on the X-ray structures of the RNA component of *T. maritima* RNase P for the A-type (Torres-Larios et al. 2005) and *B. stearothermophilus* RNase P for the B-type (Kazantsev et al. 2005). In addition, the diagram of the S-domain of the B-type molecules was based on the structure of *B. subtilis* RNase P (Krasilnikov et al. 2003)

can form independently folded domains in a manner reminiscent of proteins. The two structural domains of bacterial P RNA are responsible for different activities: the specificity (S) domain recognizes the TΨC-loop of pre-tRNA while the catalytic (C) domain recognizes the acceptor stem and the 3′ CCA and catalyzes the hydrolysis of the 5′-leader. The protein component (P protein) is a small, basic protein of around 120 amino acids. Both RNA and protein components are needed for activity in vivo, but in vitro the P RNA can catalyze RNA cleavage in the absence of the protein, albeit with reduced efficiency.

In the absence of experimental structural information, models of the bacterial RNA component were proposed to account for the observed biochemical and phylogenetic data (Harris et al. 1994; Massire et al. 1998). A few years ago, structural studies of the different components of bacterial RNase P started to fill this void. Structures of the protein component from three different bacteria are now known (Stams et al. 1998; Spitzfaden et al. 2000; Kazantsev et al. 2003), but the structure of the protein without the RNA component provides limited information on the mechanism of RNase P. Until recently, the experimental information on the three dimensional structure of the RNA component was mostly confined to small regions of the molecule (for example (Schmitz and Tinoco 2000; Leeper et al. 2002; Getz et al. 2007)). This changed with the structural determination of one of the structural domains of the RNA component from two bacteria (Krasilnikov et al. 2003, 2004) and, later, with two structures at medium resolution of the entire RNA component (Kazantsev et al. 2005; Torres-Larios et al. 2005). These structures showed the architecture of large regions of the molecule and the way different structural elements come together and are stabilized. In addition, this collection of structures has provided important new information on the mechanism of RNase P catalysis and brings our understanding of this key ribozyme to the atomic level.

4.2.1 Structures of the Protein Component

Of the three structures of the protein component that have been solved, one is from the more common A-type (*Thermotoga maritima* (Kazantsev et al. 2003)), while the other two are from B-type RNase P (*Bacillus subtilis* (Stams et al. 1998) and *Staphylococcus aureus* (Spitzfaden et al. 2000)). Despite the primary sequence differences, when the three structures are compared they are extremely similar, which emphasizes the structural conservation amongst all bacterial RNase P regardless of type. The similarity is also consistent with the observed structural conservation of the RNA component (see below). In many ways, the structural similarity amongst bacterial RNase P proteins is not surprising, as it had been observed before that the protein component is interchangeable among different types of bacterial RNase P (Guerrier-Takada et al. 1983; Waugh and Pace 1990).

The structure of the protein component of *T. maritima* RNase P is shown in Fig. 4.2. This small protein folds into a compact domain formed by a central four-stranded β sheet with helices on both sides of the sheet. One of the helices contains a conserved sequence motif, termed the RNR motif, whose structure had been

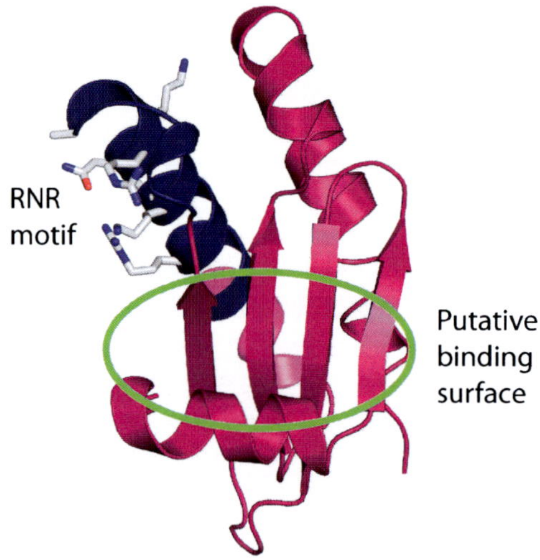

Fig. 4.2 Structure of the protein component of *T. maritima* RNase P. The ribbon diagram shows the high resolution structure of the protein component of an A-type bacterial RNase P (Kazantsev et al. 2003). The structures from two other bacteria are known and are similar (Stams et al. 1998; Spitzfaden et al. 2000). The diagram shows in blue the region spanning the RNR motif with some of the side chains that define it. A region that has been involved in interactions with RNA (Niranjanakumari et al. 1998; Spitzfaden et al. 2000) is *circled in green*

observed in other RNA binding proteins (Stams et al. 1998). In addition, one side of the β strand, which interacts with a short helix in the structure (Fig. 4.2), has been associated with RNA binding by a variety of approaches (Niranjanakumari et al. 1998; Spitzfaden et al. 2000). Although the structure of the protein in complex with RNA is not known, it is likely that the protein component will interact extensively with the RNA component and in a comparable fashion in all bacterial RNase Ps, regardless of type.

4.2.2 Structures of the S-Domain of Bacterial RNase P

The X-ray structures of the S-domain of *Thermus thermophilus* RNase P (Krasilnikov et al. 2004), an A-type RNase P, and *B. subtilis* RNase P (Krasilnikov et al. 2003), a B-type RNase P, represented the first structures of a large region of the RNA component of any RNase P elucidated (Fig. 4.3). The structures started to reveal the complexity and overall fold of the RNA component. The structures showed that the S-domains are stabilized by a variety of interactions which had been observed previously in other large RNA structures, such as tetraloop–tetraloop receptor interactions (Costa and Michel 1995) and A-minor motifs (Nissen et al. 2001). In addition, a very interesting finding was the first observation of the structure of a universally conserved region, in this case the region formed by universally

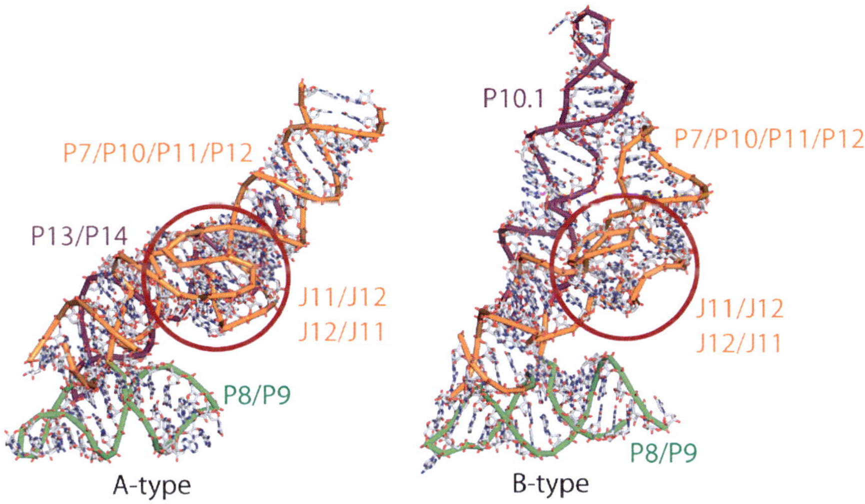

Fig. 4.3 Structure of the S-domain of bacterial RNase P. The diagram show the structure of the S-domain of A- and B-type bacterial RNase P (Krasilnikov et al. 2003, 2004). The S-domains of the two bacterial types of RNase P have different architectures and are stabilized differently, but the overall structure of the opening where the tRNA binding site is located is conserved. The opening is formed by non-helical region J11/12-J12/11 and stems P9, P10, and P11. Two interacting T-loops (*circled in red*) in J11/12-J12/11 are from universally conserved regions CR-II and CR-III and define a universally conserved module in the S-domain. The P7/P10/P11 region is shown in orange, the P8/P9 coaxial stem is shown in green, while the P13/P14 stem in the A-type or P10.1 stem in the B type are shown in purple

conserved regions (CR) II and III. These two universally conserved regions, which form part of the J11-12/J12-11 loops (see Fig. 4.1), fold in a complex fashion devoid of any canonical Watson–Crick base pairs. Instead, they form two T-loops motifs (Nagaswamy and Fox 2002; Krasilnikov and Mondragon 2003; Lee et al. 2003) that interlock with each other through a very extensive series of interactions. Furthermore, an analysis of the T-loop motifs revealed that the sequence conservation in these universally conserved regions probably arose due to the strong structural constraints imposed by the necessity of having two interacting T-loops (Krasilnikov and Mondragon 2003). Thus, these structures provided the first hint that structural conservation may be the driving force behind sequence conservation in the RNA component of RNase P. Comparison of the crystal structures of the S-domain from A- and B- type P RNAs (Krasilnikov et al. 2003, 2004) also revealed that while the secondary and tertiary structures are different, an opening in the S-domain lined by the nucleotides known to be involved in direct interactions with the tRNA TΨC-loop is structurally conserved. The structures thus showed for the first time how two different RNA molecules with the same function can form a similar core region by using different structural scaffolds, an observation that was corroborated later when comparing three different structures of group I introns (Adams et al. 2004; Guo et al. 2004; Golden et al. 2005; Vicens and Cech 2006; Woodson 2005).

4.2.3 Structures of the RNA Component of Bacterial RNase P

The crystal structures of the RNA component of an A-type and a B-type RNase P were solved to medium resolution. The structure of the RNA component of *T. maritima* RNase P was solved to 3.85 Å (Torres-Larios et al. 2005) while the RNA component of *Bacillus stearothermophilus* RNase P was solved to 3.3 Å (Kazantsev et al. 2005). To a large extent, the two structures are complementary as the *T. maritima* structure shows the two domains and the way they interact, albeit at low resolution, while the *B. stearothermophilus* structure shows greater detail, but only a small fraction of the S-domain is present in the structure.

The structures of the S-domains (Krasilnikov et al. 2003; Krasilnikov et al. 2004) had anticipated that, despite being classified as two different types, the RNA component from all bacteria would be structurally similar, particularly around the core functional regions. This was confirmed with the structures of the full length RNA components, which showed remarkable structural similarity for both bacterial types. Figure 4.4 shows a side by side comparison of structures of the type A and B

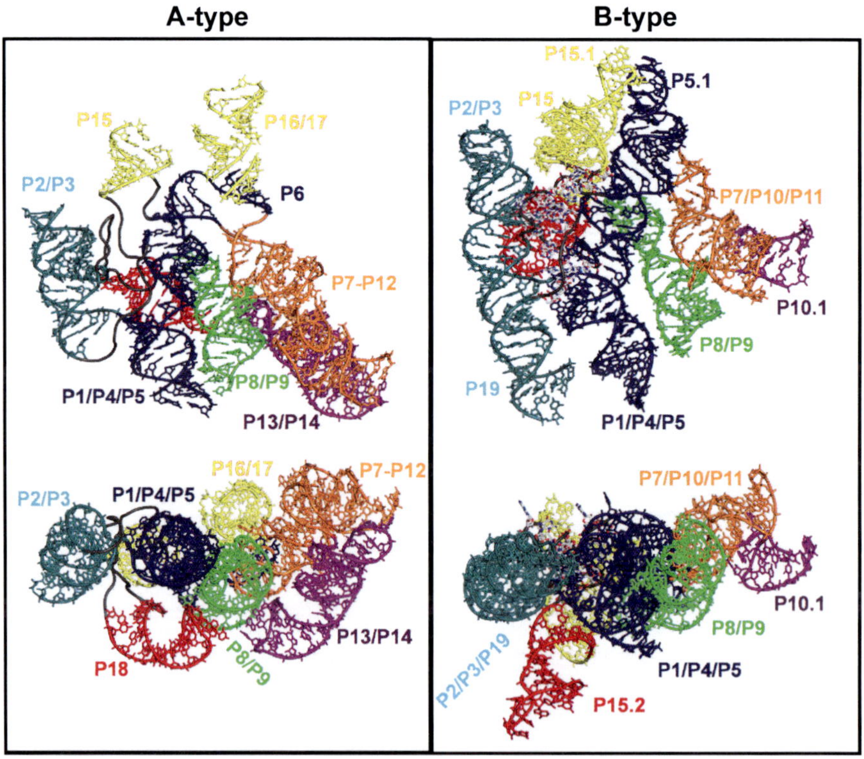

Fig. 4.4 Structures of the RNA component of bacterial RNase P. The diagrams show two orthogonal views of the structures of *T. maritima* RNase P (Torres-Larios et al. 2005), an A-type structure, and *B. stearothermophilus* RNase P (Kazantsev et al. 2005), a B-type structure. Analogous secondary structural elements are colored identically in both diagrams and correspond to the same color scheme used in Fig. 4.1

RNA molecules while Fig. 4.1 shows a schematic diagram of the secondary structure of both molecules that emphasizes the similarities in the secondary structure elements and also in the three dimensional arrangement of these elements. The overall secondary structure diagram is also in good agreement with previous predictions for the folding of the molecules (Massire et al. 1998). In both cases, the RNA molecule is composed of two one-helix thick layers with extensive interactions to stabilize the secondary structure elements.

The comparison of the two structures reveals an overall similar fold with a structurally conserved core (Fig. 4.5). The common structural core includes: (1) the coaxial stacks P1/P4/P5, P2/P3, P8/P9, and P15; (2) a non-canonical connector module located between stems P3 and P4; and (3) the J11/12–J12/11 module. In the structures, homologous features are structurally similar as are their relative orientations. For instance, a non-canonical connector region forms a distinctive structural module located in the same position relative to P1 and P4 (at a kink in the P1/P4 stack) in both structures. As in the case of the S-domain, the details of the way the core structures are stabilized are different in both cases. For example, in the A-type structure P18 is located behind and perpendicular to the P1/P4/P5 and P8/P9 stacks and acts as a buttress, in part by making a tetraloop–tetraloop receptor interaction between L18 and P8. B-type P RNA does not have P18, instead, the P15.2 stem is

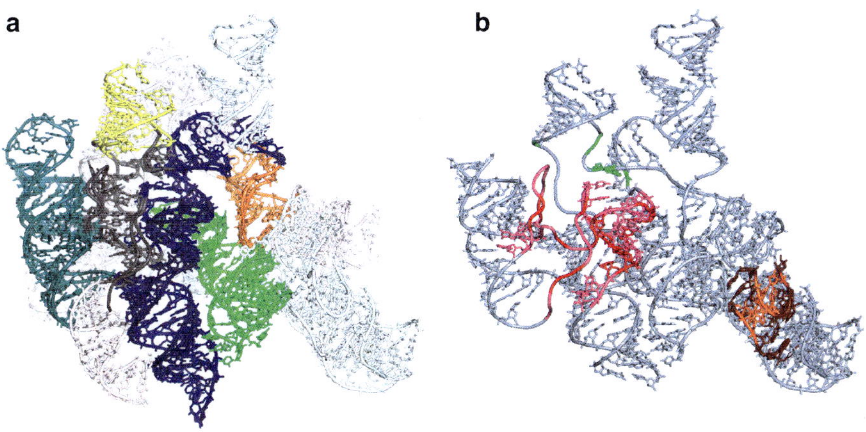

Fig. 4.5 Conserved regions in bacterial RNase P. (**a**) The diagram shows a superposition of the common structural core of bacterial RNase P. The common structural elements are colored following the same scheme used in Figs. 4.1 and 4.4. Non-common elements are colored light blue and light red for the A-and B-type molecules, respectively. The J11/12 and J12/11 module, which is part of this core, is not colored as it is disordered in the B-type structure. Note the marked structural similarity of the common core. This common core is supported by different peripheral secondary structure elements in both structures. (**b**) Location of the universally conserved regions in the structure of the A-type *T. maritima* RNase P. Regions CR-I, CR-IV and CR-V are shown in pink and red, with red corresponding to very highly conserved nucleotides. Regions CR-II and CR-III are shown in brown and orange, with orange corresponding to very highly conserved nucleotides. In addition, a region highly conserved in bacterial molecules is shown in green. In both diagrams, the orientation of the molecules is identical to the one in Fig. 4.4

found in a location equivalent to that of P18 in the A-type P RNA. In addition, the structures show that the S- and C-domains form independent domains that interact extensively with each other mainly through the P8 and P9 stems.

The structures also reveal the location of the universally conserved regions, shown in Fig. 4.5. The conserved regions cluster in two areas, which can be considered two universally conserved modules. Two of the universally conserved regions, CR-II and CR-III, form part of the S-domain while the other conserved regions mostly fold into loops joining secondary structure elements. CR-IV forms part of the linker joining stems P2 to P18, while CR-I and CR-V are located adjacent to the P4 stem. To a large extent, the conserved regions are formed by loops lacking an easily identifiable secondary or tertiary structure, which is highly indicative that sequence conservation is a reflection of strong structural constraints placed on these nucleotides. The fact that the universally conserved modules are present in RNase P from all domains of life suggests that the structures of the RNA component in all organisms are likely to have strong structural similarities in these regions.

Metals play an integral role in all RNA structures and in RNase P they play a crucial role in catalysis (Smith and Pace 1993; Beebe et al. 1996; Persson et al. 2003) and structure folding and substrate binding (Pan 1995; Fang et al. 1999; Kent et al. 2000). Knowledge of the location of the metals ions in the structure is thus important to understand its function. Structure determination of all the known structures involved incorporating heavy metals, such as osmium or barium, in the structure. The position of these heavier atoms may indicate where some magnesium ions are located. Due to the low resolution nature of the known structures, it is still impossible to identify all metal ions with confidence. However, a recent systematic study of metal binding to the C-domain of *B. stearothermophilus* RNase P helped identify some metal binding sites with high confidence (Kazantsev et al. 2009). Nevertheless, the number of metal binding sites identified is relatively small and it is not clear whether they play a role in catalysis and/or structure stabilization.

An important question that still remains unanswered relates to the recognition of the substrate by RNase P. Some of the regions and nucleotides that interact with the substrate have been mapped. Nucleotides in the opening formed by the S-domain interact with the TψC stem of tRNA (LaGrandeur et al. 1994; Loria and Pan 1997; Odell et al. 1998) while in the C-domain interactions have been mapped to regions around the P4 and P5 stems, the J5-15 linker, the J2-4 connector (reviewed in (Kurz and Fierke 2000; Christian et al. 2002)), and the L15 loop (Kirsebom and Svard 1994). When mapped onto the structures, these data indicate that there are two main regions of interaction between RNase P and tRNA: around the opening created by the P7–P11 cruciform and the J11-12/J12-11 conserved regions in the S-domain, and around the P4–P5 interface and L15 internal loop in the C-domain. Based on this information and on the structures, models of the RNase P/tRNA complex have been built (Buck et al. 2005; Torres-Larios et al. 2005). The level of detail in these models is quite limited due to the low resolution of the RNase P structures and also due to the relatively small number of constraints used to build the models. The remarkable thing about the models, which were built independently and using different approaches, is the extent of the overall agreement. Figure 4.6 shows both models

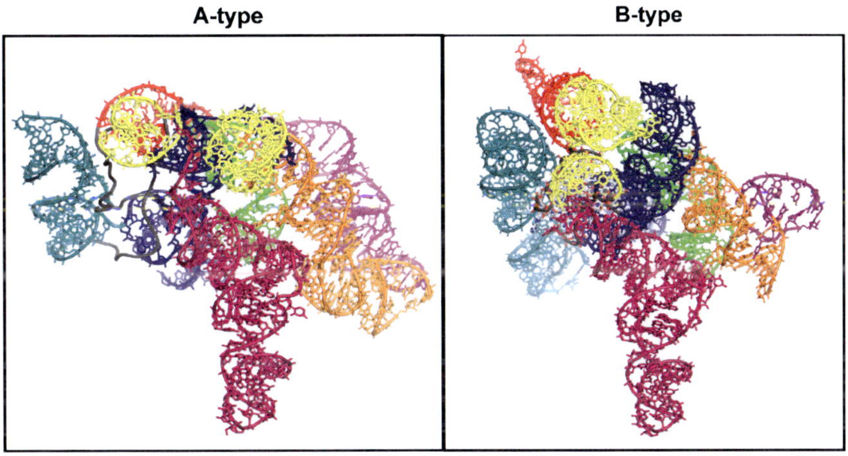

Fig. 4.6 Models of the tRNA–RNase P interaction. Diagrams of the models of complexes of tRNA with the A- and B-type RNA component of RNase P (Buck et al. 2005; Torres-Larios et al. 2005). Note the similarity in the manner in which the tRNA is placed in both models, which were built independently and using different constraints. In both models the tRNA is shown in pink and the P RNA is colored following the same scheme as in Fig. 4.1

side by side illustrating that the overall arrangement of the tRNA with respect to the RNA component is very similar in both cases and predicting similar interactions with the S- and C-domains. In the models, the TΨC loop of pre-tRNA interacts with nucleotides in the S-domain, the acceptor stem of tRNA is parallel to P4, and the cleavage site at the 5′ end of the tRNA is adjacent to a universally conserved module.

4.3 Structural Studies of Archaeal and Eukaryotic RNase P

Unlike bacterial RNase P, archaeal and eukaryotic RNase P contain several protein components. In both cases, there is a single RNA component with strong sequence resemblance to bacterial P RNA. As in the case of bacterial RNase P, the RNA component of archaea and eukaryotes is catalytic in the absence of proteins (Pannucci et al. 1999; Kikovska et al. 2007). A model for the RNA component of eukaryotic RNase P has been proposed based on the bacterial RNA structure and biochemical information (Marquez et al. 2006). Not surprisingly, the model suggests a molecule with a very similar core region to the bacterial ones. To date, no experimental structural information is yet available on the archaeal or eukaryotic RNA component that could help answer important questions regarding the similarity of the RNA component across all domains of life.

The number of protein components in archaeons and eukaryotes varies. Archaeal organisms have as few as four proteins, while yeast has at least nine

(Chamberlain et al. 1998) and humans have at least ten (Jarrous 2002) protein components. The function of the protein components is still not clear. Interestingly, none of the protein components from archaea or eukarya show any sequence similarity to the bacterial protein, posing an interesting question on the evolution of RNase P. In contrast, all four minimal proteins in archaea have an equivalent in eukarya (for reviews, see (Hartmann and Hartmann 2003; Evans et al. 2006)).

While there are no structures of eukaryotic RNase P proteins, there are now several structures of archaeal proteins, covering the full minimal set (Fig. 4.7). In addition, previous experiments had indicated that some of the proteins interact with each other (Jiang and Altman 2001; Houser-Scott et al. 2002; Hall and Brown 2004; Kifusa et al. 2005) while the reconstitution of archaeal RNase P shows that well-defined pairs of subunits are needed to recover activity (Tsai et al. 2003). Structures of complexes of pairs of proteins are now available (Kawano et al. 2006; Honda et al. 2008) that show the way some of the proteins interact with each other. A summary of the structures known to date is shown in Table 4.1.

From a structural perspective, the best characterized archaeal RNase P comes from a hyperthermophilic archaeon, *Pyrococcus horikoshii*. Five protein components have been identified, Ph1481p, Ph1877p, Ph1601p, Ph1771p, and Ph1496p. These proteins are also referred as *Pho*Pho5, *Pho*Rpp30, *Pho*Rpp21, *Pho*Rpp29, and *Pho*Rpp38 for their orthologs in humans: Pop5, Rpp30, Rpp21 Rpp29 and Rpp38. Of these proteins, Rpp21 and Rpp29 interact with each other to form a

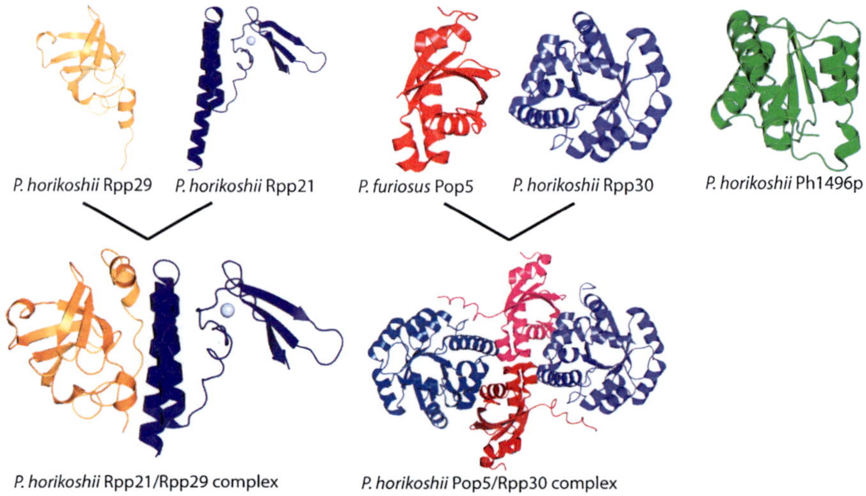

Fig. 4.7 Structure of the protein components of archaeal RNase P. The panel shows some of the structures of archaeal proteins known to date. The top row shows the structure of the individual proteins while the bottom row shows the structure of two protein heterodimers. In some cases, the structure of the same protein from different archaeal organisms is known, but only the structures from the *P. horikoshii* complexes have been elucidated. There is considerable structural conservation amongst the same protein from different archaeons, but none of these proteins show structural similarity to the bacterial protein. A compilation of all know archaeal structures is shown in Table 4.1

Table 4.1 Known structures of archaeal RNase P proteins

Structure	Organism	Human/yeast ortholog	References
Ph1496p	*P. hirokoshii*	None	Fukuhara et al. (2006)
*Pho*Rpp29 (Ph1771p)	*P. hirokoshii*	Rpp29/Pop4	Numata et al. (2004)
*mth*Rpp29 (Mth11)	*M. thermoautotrophicus*	Rpp29/Pop4	Boomershine et al. (2003)
*a*Rpp29	*A. fulgidus*	Rpp29/Pop4	Sidote and Hoffman (2003); Sidote et al. (2004)
Pfu Pop5	*P. furiosus*	Pop5/Pop5	Wilson et al. (2006)
*Pho*Rpp30 (Ph1877p)	*P. hirokoshii*	Rpp30/Rpp1	Takagi et al. (2004)
*Pho*Rpp21 (Ph1601p)	*P. hirokoshii*	Rpp21/Rpr2	Kakuta et al. (2005)
*pfu*Rpp21	*P. furiosus*	Rpp21/Rpr2	Amero et al. (2008)
*Pho*Rpp21–*Pho*Rpp29 complex (Ph1601p–Ph1877p complex)	*P. hirokoshii*	Rpp21-Rpp29/Rpr2-Pop4	Honda et al. (2008)
*Pho*Pop5–*Pho*Rpp30 complex (Ph1481p–Ph1877p complex)	*P. hirokoshii*	Pop5-Rpp30/Pop5-Rpp1	Kawano et al. (2006)

dimer, while Pop5 and Rpp30 interact to form another dimer. The five protein subunits and the RNA component can be reconstituted into an active RNase P (Kouzuma et al. 2003; Fukuhara et al. 2006). While no structure of the RNA component is available, the protein structures are providing a picture of the richness of the protein/protein interactions.

The structure of Pop5 is known from two organisms, *P. horikoshii* (Kawano et al. 2006) and *Pyrococcus furiosus* (Wilson et al. 2006). Pop5 is a small protein of around 120 amino acids. The two structures are basically identical, showing a central β strand with helices on both sides. The structure suggests that it belongs to the RRM (RNA Recognition Motif) domain of proteins (Kawano et al. 2006; Wilson et al. 2006), associated with RNA binding proteins. There is a weak structural resemblance to the bacterial RNase P protein, but the different topology and lack of sequence conservation makes it difficult to ascertain whether there is true structural similarity. The binding partner of Pop5 is Rpp30, whose structure is also known (Takagi et al. 2004). The *Pho*Rpp30 protein is much larger, around 210 amino acids, and folds into the familiar TIM barrel motif (Takagi et al. 2004). The structure of the complex of *Pho*Pop5 and *Pho*Rpp30 shows a heterotetramer formed by a dimer of *Pho*Pop5 and *Pho*Rpp30, but this tetramer may not represent a functional form.

The structures of the *Pho*Rpp21–*Pho*Rpp29 dimer (Honda et al. 2008) and of its individual components (Numata et al. 2004; Kakuta et al. 2005) are also known. In the case of Rpp29, structures from three archaeons, *P. horokoshii, Methanothermobacter thermoautotrophicus* and *Archaeoglobus fulgidus*, are known (Boomershine et al. 2003; Sidote and Hoffman 2003; Numata et al. 2004; Sidote et al. 2004; Kakuta et al. 2005). As expected, the structures from all three organisms are nearly identical in overall structure and topology, showing similarity to RNA binding proteins containing an Sm-like fold (Numata et al. 2004). The structures of Rpp21 from *P. horikoshii* (Kakuta et al. 2005) and *P. furiosus* (Amero et al. 2008) also show remarkable similarity. Rpp21 contains a zinc-ribbon binding domain linked to a long helical domain. The dimer of Rpp21 and Rpp29 shows that the two proteins associate through the long helical domain of Rpp21. The zinc-ribbon domain of Rpp21 and the Sm-like fold domain of Rpp29 are on opposite sides and one side of the dimer has a highly positively charged character, which suggests an RNA binding surface.

Overall, the knowledge of the structures of all these proteins helps us understand the overall structure of the protein components, the way the protein components interact with each other, and their evolutionary relationship to other proteins. Biochemical mapping of the protein/RNA interactions is also helping to paint a clearer picture of the position and possible role of the protein components. Nevertheless, the absence of a complex with the RNA component limits our structural understanding of archaeal and eukaryotic RNase P. Obtaining a complex with the RNA component remains a major future challenge.

4.4 Conclusions

The elucidation of the structures of various protein and RNA components from RNase P from different organisms helps to bring our understanding of the mechanism of action of this fundamental ribozyme to the atomic level. We now have a much more detailed view of the structure of the bacterial RNA component that provides an excellent picture of the overall architecture of the molecule and the way the different secondary structural elements and domains come together and are stabilized. Structures of the bacterial protein component show that this protein is identical in all bacterial RNase P types and that there are resemblances to other proteins known to interact with RNA. Finally, structures of several proteins from archaeal RNase P are now available and, together with biochemical data, are providing a clearer picture on how all these proteins interact with each other. Overall, the structural work on RNase P is changing our view of this important ribozyme and providing a solid framework to understand the mechanism by which RNase P recognizes and cleaves its substrates.

References

Adams PL, Stahley MR, Kosek AB, Wang J, Strobel SA (2004) Crystal structure of a self-splicing group I intron with both exons. Nature 430:45–50

Alifano P, Rivellini F, Piscitelli C, Arraiano CM, Bruni CB, Carlomagno MS (1994) Ribonuclease E provides substrates for ribonuclease P-dependent processing of a polycistronic mRNA. Genes Dev 8:3021–3231

Altman S, Kirsebom LA (1999) In: Gesteland RF, Cech TR, Atkins JF (eds) The RNA world. Cold Spring Harbor Laboratory Press, Cold Spring Harbor, NY, pp. 351–380

Altman S, Wesolowski D, Guerrier-Takada C, Li Y (2005) RNase P cleaves transient structures in some riboswitches. Proc Natl Acad Sci U S A 102:11284–11289
Amero CD, Boomershine WP, Xu Y, Foster M (2008) Solution structure of *Pyrococcus furiosus* RPP21, a component of the archaeal RNase P holoenzyme, and interactions with its RPP29 protein partner. Biochemistry 47:11704–11710
Beebe JA, Kurz JC, Fierke CA (1996) Magnesium ions are required by *Bacillus subtilis* ribonuclease P RNA for both binding and cleaving precursor tRNAAsp. Biochemistry 35:10493–10505
Boomershine WP, McElroy CA, Tsai HY, Wilson RC, Gopalan V, Foster MP (2003) Structure of Mth11/Mth Rpp 29, an essential protein subunit of archaeal and eukaryotic RNase P. Proc Natl Acad Sci U S A 100:15398–15403
Buck AH, Kazantsev AV, Dalby AB, Pace NR (2005) Structural perspective on the activation of RNase P RNA by protein. Nat Struct Mol Biol 12(11):958–964
Chamberlain JR, Lee Y, Lane WS, Engelke DR (1998) Purification and characterization of the nuclear RNase P holoenzyme complex reveals extensive subunit overlap with RNase MRP. Genes Dev 12:1678–1690
Chen J-L, Pace NR (1997) Identification of the universally conserved core of ribonuclease P RNA. RNA 3:557–560
Christian EL, Zahler NH, Kaye NM, Harris ME (2002) Analysis of substrate recognition by the ribonucleoprotein endonuclease RNase P. Methods 28:307–322
Costa M, Michel F (1995) Frequent use of the same tertiary motif by self-folding RNAs. EMBO J 14:1276–1285
Doudna JA, Cech TR (1995) Self-assembly of a group I intron active site from its component tertiary structural domains. RNA 1:36–45
Evans D, Marquez SM, Pace NR (2006) RNase P: interface of the RNA and protein worlds. Trends Biochem Sci 31:333–341
Fang XW, Pan T, Sosnick TR (1999) Mg^{2+}-dependent folding of a large ribozyme without kinetic traps. Nat Struct Biol 6:1091–1095
Frank DN, Pace NR (1998) Ribonuclease P: unity and diversity in a tRNA processing ribozyme. Annu Rev Biochem 67:153–180
Fukuhara H, Kifusa M, Watanabe M, Terada A, Honda T, Numata T, Kakuta Y, Kimura M (2006) A fifth protein subunit Ph1496p elevates the optimum temperature for the ribonuclease P activity from *Pyrococcus horikoshii* OT3. Biochem Biophys Res Commun 343:956–964
Getz MM, Andrews AJ, Fierke CA, Al-Hashimi HM (2007) Structural plasticity and Mg^{2+} binding properties of RNase P P4 from combined analysis of NMR residual dipolar couplings and motionally decoupled spin relaxation. RNA 13:251–266
Golden BL, Kim H, Chase E (2005) Crystal structure of a phage Twort group I ribozyme-product complex. Nat Struct Mol Biol 12:82–89
Guerrier-Takada C, Gardiner K, Marsh T, Pace N, Altman S (1983) The RNA moiety of ribonuclease P is the catalytic subunit of the enzyme. Cell 35:849–857
Guo F, Gooding AR, Cech TR (2004) Structure of the Tetrahymena ribozyme: base triple sandwich and metal ion at the active site. Mol Cell 16:351–362
Hall TA, Brown JW (2004) Interactions between RNase P protein subunits in archaea. Archaea 1:247–254
Harris ME, Nolan JM, Malhotra A, Brown JW, Harvey SC, Pace NR (1994) Use of photoaffinity crosslinking and molecular modeling to analyze the global architecture of ribonuclease P RNA. EMBO J 13:3953–3963
Hartmann E, Hartmann RK (2003) The enigma of ribonuclease P evolution. Trends Genet 19:561–569
Hartmann RK, Heinrich J, Schlegl J, Schuster H (1995) Precursor of C4 antisense RNA of bacteriophages P1 and P7 is a substrate for RNase P of *Escherichia coli*. Proc Natl Acad Sci U S A 92:5822–5826
Honda T, Kakuta Y, Kimura K, Saho J, Kimura M (2008) Structure of an archaeal homolog of the human protein complex Rpp 21-Rpp29 that is a key core component for the assembly of active ribonuclease P. J Mol Biol 384:652–662

Houser-Scott F, Xiao S, Millikin CE, Zengel JM, Lindahl L, Engelke DR (2002) Interactions among the protein and RNA subunits of *Saccharomyces cerevisiae* nuclear RNase P. Proc Natl Acad Sci U S A 99:2684–2689

Jarrous N (2002) Human ribonuclease P: subunits, function, and intranuclear localization. RNA 8:1–7

Jiang T, Altman S (2001) Protein–protein interactions with subunits of human nuclear RNase P. Proc Natl Acad Sci U S A 98:920–925

Kakuta Y, Ishimatsu I, Numata T, Kimura K, Yao M, Tanaka I, Kimura M (2005) Crystal structure of a ribonuclease P protein Ph1601p from *Pyrococcus horikoshii* OT3: an archaeal homologue of human nuclear ribonuclease P protein Rpp 21. Biochemistry 44:12086–12093

Kawano S, Nakashima T, Kakuta Y, Tanaka I, Kimura M (2006) Crystal structure of protein Ph1481p in complex with protein Ph1877p of archaeal RNase P from *Pyrococcus horikoshii* OT3: implication of dimer formation of the holoenzyme. J Mol Biol 357:583–591

Kazantsev AV, Pace NR (2006) Bacterial RNase P: a new view of an ancient enzyme. Nat Rev Microbiol 4:729–740

Kazantsev AV, Krivenko AA, Harrington DJ, Carter RJ, Holbrook SR, Adams PD, Pace NR (2003) High-resolution structure of RNase P protein from *Thermotoga maritima*. Proc Natl Acad Sci U S A 100:7497–7502

Kazantsev AV, Krivenko AA, Harrington DJ, Holbrook SR, Adams PD, Pace NR (2005) Crystal structure of a bacterial ribonuclease P RNA. Proc Natl Acad Sci U S A 102:13392–13397

Kazantsev AV, Krivenko AA, Pace NR (2009) Mapping metal-binding sites in the catalytic domain of bacterial RNase P RNA. RNA 15:266–276

Kent O, Chaulk SG, MacMillan AM (2000) Kinetic analysis of the M1 RNA folding pathway. J Mol Biol 304:699–705

Kifusa M, Fukuhara H, Hayashi T, Kimura M (2005) Protein–protein interactions in the subunits of ribonuclease P in the hyperthermophilic archaeon *Pyrococcus horikoshii* OT3. Biosci Biotechnol Biochem 69:1209–1212

Kikovska E, Svard SG, Kirsebom LA (2007) Eukaryotic RNase P RNA mediates cleavage in the absence of protein. Proc Natl Acad Sci U S A 104:2062–2067

Kirsebom LA, Svard SG (1994) Base pairing between *Escherichia coli* RNase P RNA and its substrate. EMBO J 13:4870–4876

Komine Y, Kitabatake M, Yokogawa T, Nishikawa K, Inokuchi H (1994) A tRNA-like structure is present in 10Sa RNA, a small stable RNA from *Escherichia coli*. Proc Natl Acad Sci U S A 91:9223–9227

Kouzuma Y, Mizoguchi M, Takagi H, Fukuhara H, Tsukamoto M, Numata T, Kimura M (2003) Reconstitution of archaeal ribonuclease P from RNA and four protein components. Biochem Biophys Res Commun 306:666–673

Krasilnikov AS, Mondragon A (2003) On the occurrence of the T-loop RNA folding motif in large RNA molecules. RNA 9:640–643

Krasilnikov AS, Yang X, Pan T, Mondragón A (2003) Crystal structure of the specificity domain of ribonuclease P. Nature 421:760–764

Krasilnikov AS, Xiao Y, Pan T, Mondragon A (2004) Basis for structural diversity in homologous RNAs. Science 306:104–107

Kurz JC, Fierke CA (2000) Ribonuclease P: a ribonucleoprotein enzyme. Curr Opin Chem Biol 4:553–558

LaGrandeur TE, Huttenhofer A, Noller HF, Pace NR (1994) Phylogenetic comparative chemical footprint analysis of the interaction between ribonuclease P RNA and tRNA. EMBO J 13:3945–3952

Lee JC, Cannone JJ, Gutell RR (2003) The lonepair triloop: a new motif in RNA structure. J Mol Biol 325:65–83

Leeper TC, Martin MB, Kim H, Cox S, Semenchenko V, Schmidt FJ, Van Doren SR (2002) Structure of the UGAGAU hexaloop that braces *Bacillus* RNase P for action. Nat Struct Biol 9:397–403

Li Y, Altman S (2003) A specific endoribonuclease, RNase P, affects gene expression of polycistronic operon mRNAs. Proc Natl Acad Sci U S A 100:13213–13218

Loria A, Pan T (1996) Domain structure of the ribozyme from eubacterial ribonuclease P. RNA 2:551–563

Loria A, Pan T (1997) Recognition of the T stem-loop of a pre-tRNA substrate by the ribozyme from *Bacillus subtilis* ribonuclease P. Biochemistry 36:6317–6325

Mans RM, Guerrier-Takada C, Altman S, Pleij CW (1990) Interaction of RNase P from *Escherichia coli* with pseudoknotted structures in viral RNAs. Nucleic Acids Res 18:3479–3487

Marquez SM, Chen JL, Evans D, Pace NR (2006) Structure and function of eukaryotic ribonuclease P RNA. Mol Cell 24:445–456

Massire C, Jaeger L, Westhof E (1998) Derivation of the three-dimensional architecture of bacterial ribonuclease P RNAs from comparative sequence analysis. J Mol Biol 279:773–793

Nagaswamy U, Fox GE (2002) Frequent occurrence of the T-loop RNA folding motif in ribosomal RNAs. RNA 8:1112–1119

Niranjanakumari S, Stams T, Crary SM, Christianson DW, Fierke CA (1998) Protein component of the ribozyme ribonuclease P alters substrate recognition by directly contacting precursor tRNA. Proc Natl Acad Sci U S A 95:15212–15217

Nissen P, Ippolito JA, Ban N, Moore PB, Steitz TA (2001) RNA tertiary interactions in the large ribosomal subunit: the A-minor motif. Proc Natl Acad Sci U S A 98:4899–4903

Numata T, Ishimatsu I, Kakuta Y, Tanaka I, Kimura M (2004) Crystal structure of archaeal ribonuclease P protein Ph1771p from *Pyrococcus horikoshii* OT3: an archaeal homolog of eukaryotic ribonuclease P protein Rpp 29. RNA 10:1423–1432

Odell L, Huang V, Jakacka M, Pan T (1998) Interaction of structural modules in substrate binding by the ribozyme from *Bacillus subtilis* RNase P. Nucleic Acids Res 26:3717–3723

Pan T (1995) Higher order folding and domain analysis of the ribozyme from *Bacillus subtilis* ribonuclease P. Biochemistry 34:902–909

Pannucci JA, Haas ES, Hall TA, Harris JK, Brown JW (1999) RNase P RNAs from some Archaea are catalytically active. Proc Natl Acad Sci USA 96:7803–7808

Peck-Miller KA, Altman S (1991) Kinetics of the processing of the precursor to 4.5 S RNA, a naturally occurring substrate for RNase P from *Escherichia coli*. J Mol Biol 221:1–5

Persson T, Cuzic S, Hartmann RK (2003) Catalysis by RNase P RNA: unique features and unprecedented active site plasticity. J Biol Chem 278:43394–43401

Schmitz M, Tinoco I Jr (2000) Solution structure and metal-ion binding of the P4 element from bacterial RNase P RNA. RNA 6:1212–1225

Sidote DJ, Hoffman DW (2003) NMR structure of an archaeal homologue of ribonuclease P protein Rpp 29. Biochemistry 42:13541–13550

Sidote DJ, Heideker J, Hoffman DW (2004) Crystal structure of archaeal ribonuclease P protein aRpp 29 from *Archaeoglobus fulgidus*. Biochemistry 43:14128–14138

Siegel RW, Banta AB, Haas ES, Brown JW, Pace NR (1996) *Mycoplasma fermentans* simplifies our view of the catalytic core of ribonuclease P RNA. RNA 2:452–462

Smith D, Pace NR (1993) Multiple magnesium ions in the ribonuclease P reaction mechanism. Biochemistry 32:5273–5281

Spitzfaden C, Nicholson N, Jones JJ, Guth S, Lehr R, Prescott CD, Hegg LA, Eggleston DS (2000) The structure of ribonuclease P protein from *Staphylococcus aureus* reveals a unique binding site for single-stranded RNA. J Mol Biol 295:105–115

Stams T, Niranjanakumari S, Fierke CA, Christianson DW (1998) Ribonuclease P protein structure: evolutionary origins in the translational apparatus. Science 280:752–755

Takagi H, Watanabe M, Kakuta Y, Kamachi R, Numata T, Tanaka I, Kimura M (2004) Crystal structure of the ribonuclease P protein Ph1877p from hyperthermophilic archaeon *Pyrococcus horikoshii* OT3. Biochem Biophys Res Commun 319:787–794

Torres-Larios A, Swinger KK, Krasilnikov AS, Pan T, Mondragon A (2005) Crystal structure of the RNA component of bacterial ribonuclease P. Nature 437:584–587

Torres-Larios A, Swinger KK, Pan T, Mondragon A (2006) Structure of ribonuclease P – a universal ribozyme. Curr Opin Struct Biol 16:327–335

Tsai HY, Masquida B, Biswas R, Westhof E, Gopalan V (2003) Molecular modeling of the three-dimensional structure of the bacterial RNase P holoenzyme. J Mol Biol 325:661–675

Vicens Q, Cech TR (2006) Atomic level architecture of group I introns revealed. Trends Biochem Sci 31(1):41–51

Walker SC, Engelke DR (2006) Ribonuclease P: the evolution of an ancient RNA enzyme. Crit Rev Biochem Mol Biol 41:77–102

Waugh DS, Pace NR (1990) Complementation of an RNase P RNA (rnpB) gene deletion in *Escherichia coli* by homologous genes from distantly related eubacteria. J Bacteriol 172:6316–6322

Wilson RC, Bohlen CJ, Foster MP, Bell CE (2006) Structure of Pfu Pop5, an archaeal RNase P protein. Proc Natl Acad Sci U S A 103:873–878

Woodson SA (2005) Structure and assembly of group I introns. Curr Opin Struct Biol 15: 324–330

Chapter 5
Folding of Bacterial RNase P RNA

Tao Pan

5.1 Introduction

Bacterial RNase P RNA (P RNA) is among the best studied in RNA folding. The attractiveness of P RNA for folding studies includes its large collection of natural sequences (http://www.mbio.ncsu.edu/RNaseP/home.html), two types of distinct tertiary structures and four available crystal structures (Krasilnikov et al. 2003; Krasilnikov et al. 2004; Torres-Larios et al. 2005; Kazantsev et al. 2005). This rich information in structural knowledge has facilitated biophysical studies of P RNA folding as well as the analysis of P RNA folding during transcription. The natural sequences are derived from bacteria that live at a wide range of environmental conditions including temperature, salinity and nutrients. This wide distribution allows a fundamental question in RNA folding to be addressed uniquely for RNase P RNA studies: how does tertiary RNA structure and folding co-evolve with environmental adaptation?

Bacterial P RNA is catalytically active in the absence of its protein subunit, therefore, it folds on its own. In contrast, eukaryotic P RNA is catalytically inactive without their protein subunits, and no evidence has emerged that it can fold into a well-defined tertiary structure by itself (Xiao et al. 2002). Therefore, folding studies of RNase P RNA has so far been confined to bacterial P RNAs. It is conceivable however, that the eukaryotic RNase P can be excellent study objects of assembly and folding of large ribonucleo-protein particles (RNP) in the future.

RNA folding studies can be conceptually divided into three types: biophysical, co-transcriptional, and cellular. The biophysical studies aim to understand how RNA folds as a polymer chain with a four letter code. In general, biophysical studies examine the pathway formation and energetic landscapes upon the variation of solvent conditions (e.g., [Mg^{2+}], [urea]) and temperature. Co-transcriptional studies aim to understand how RNA folds during transcription. Co-transcriptional studies also examine the folding properties of RNA as related to properties of the RNA polymerase.

T. Pan (✉)
Department of Biochemistry and Molecular Biology, University of Chicago,
929 East 57th street, Chicago, IL, 60637, USA
e-mail: taopan@uchicago.edu

F. Liu and S. Altman (eds.), *Ribonuclease P*, Protein Reviews 10,
DOI 10.1007/978-1-4419-1142-1_5,

Cellular studies aim to understand how RNA folding occurs in the cell in the presence of RNA chaperon proteins (Herschlag 1995), RNA helicases (Russell 2008), and RNA polymerase co-factors (Mooney et al. 1998; Pan et al. 1999a). This review focuses on the biophysical and co-transcriptional studies of P RNA.

5.2 Experimental Techniques and Data Analysis to Study P RNA Folding

Five basic techniques have been used to study the folding of the full-length P RNA.

1. *Catalytic activity*: The appearance of the catalytic activity for the pre-tRNA substrate is the most decisive method to demonstrate that the folding of this ribozyme has reached its fully native state. For folding thermodynamics, a general protocol involves folding of the P RNA alone at varying [Mg^{2+}]. The folding mixture is then combined with radio-labeled pre-tRNA substrate to determine the amount of P RNA folded. For folding kinetics, a general protocol involves initiating folding of the P RNA alone (e.g., upon the addition of Mg^{2+}). Aliquots from the folding mixture are taken at various times (t_1) and mixed with radio-labeled pre-tRNA substrate; new aliquots from catalysis reaction are taken at various times (t_2) and analyzed by gel electrophoresis. The effectiveness of this assay requires $t_2 \ll t_1$ so that additional folding during the catalytic assay is negligible. The time resolution of this technique is >15 s.
2. *Hydroxyl radical footprinting*: The appearance of protected residues generally correlates with the formation of the tertiary structure. This assay has a high, residue-specific resolution. For folding thermodynamics, a general protocol involves folding of the radio-labeled P RNA alone at varying [Mg^{2+}]. Hydroxyl radical generating reagents (Fe-EDTA or peroxonitrous acid) are added to determine the amount of P RNA folded. For folding kinetics, a general protocol involves initiating folding of the radio-labeled P RNA alone (e.g., upon the addition of Mg^{2+}). Aliquots from the folding mixture are taken at various times and mixed with peroxonitrous acid (which generates hydroxyl radical in a matter of seconds or less), and analyzed by gel electrophoresis. The time resolution of this technique is <5 s.

By using synchrotron radiation to general hydroxyl radicals as in the folding studies of group I intron ribozymes (Sclavi et al. 1998; Brenowitz et al. 2002), this technique can have a time resolution of milliseconds.

3. *Oligonucleotide hybridization*: the accessible regions during the folding reaction are rapidly hybridized with various, complementary oligonucleotides of ~10 residues. The hybridized oligonucleotides are detected by RNase H cleavage of the radio-labeled P RNA (Zarrinkar and Williamson 1994; Treiber and Williamson 2000). For folding thermodynamics, a general protocol involves folding of the

radio-labeled P RNA alone at varying [Mg^{2+}]. The folding mixture is then combined with a large excess of an oligonucleotide plus RNase H to determine the amount of P RNA folded. For folding kinetics, a general protocol involves folding of the radio-labeled P RNA alone (e.g., upon the addition of Mg^{2+}). Aliquots from the folding mixture are taken at various times (t_1) and mixed with a large excess of an oligonucleotide plus RNase H. The reaction proceeds for a fixed time (t_2) and analyzed by gel electrophoresis. Again, the effectiveness of this assay requires $t_2 \ll t_1$ so that additional folding during the oligo-hybridization/RNase H cleavage is negligible. The time resolution of this technique is >15 s.

4. *Optical spectroscopy*: folding is monitored by changes in hypochromicity (UV absorbance) or differences in the UV absorption of polarized light (circular dichroism = CD (Pan and Sosnick 1997; Sosnick et al. 2000)). These techniques have no site-specific resolution, but they are convenient to do and provide fast time resolution (<5 ms for UV) or continuous measurement of folding transitions in a semi-automated fashion (CD and UV). For folding thermodynamics, a general protocol involves monitoring the change of CD/UV signals at several wavelengths upon the addition of varying [Mg^{2+}] using a syringe-pump driven titrator (from the same sample). For folding kinetics, a general protocol involves injecting a high [Mg^{2+}] solution into a P RNA solution using the titrator, followed by monitoring the change of CD/UV signal at one fixed wavelength over time.
5. *Small-angle X-ray scattering (SAXS)*: SAXS provides information on the global size and shape of an RNA under any solution condition (Lipfert and Doniach 2007). Simulated SAXS data can be generated from any structural model; in turn, experimental SAXS data can be converted into global shapes (Petoukhov and Svergun 2007; Koch et al. 2003). Therefore, SAXS enables the comparison between structural models of folding states to experimental data. The only limitation of this method is that the experiments have to be performed at synchrotrons. For example, the SAXS data for P RNA studies were obtained at the Advanced Photon Source (APS) at the Argonne National Laboratory.

5.2.1 Data Analysis

For folding thermodynamics, a simple analysis uses the Hill-type equation:

$$\text{Fraction native } (f_{\mathrm{N}}) = \left[\mathrm{Mg}^{2+}\right]^{\mathrm{n}} / \left(K_{Mg}{}^{\mathrm{n}} + \left[\mathrm{Mg}^{2+}\right]^{\mathrm{n}}\right) \tag{1}$$

where $K_{\mathrm{Mg}} = [\mathrm{Mg}^{2+}]$ is the transition midpoint at which 50% of the RNA is folded and n is the Hill constant which describes the degree of folding cooperativity.

For folding kinetics, a simple analysis uses the single-exponential description:

$$\text{Fraction native } (f_{\mathrm{N}}) = 1 - \mathrm{e}^{-k_f t} \tag{2}$$

where k_{f} is the folding rate, and folding half-time is obtained by $t_{1/2} = \ln 2/k_{\mathrm{f}}$.

5.3 Tertiary Structure of P RNA

Bacterial P RNA consists of two types that have a common core, but different peripheral structures. The A and B types are present in distinct evolutionary branches: the A (ancestral) type is present in over 90% of all P RNAs, whereas the B (Bacillus) type is confined within branches of Gram-positive bacteria. At the structural level, each A or B type P RNA is made of two independently folding domains that have different functional roles: the catalytic domain (C-domain) contains the active site of this ribozyme, whereas the specificity domain (S-domain) binds the T stem-loop of pre-tRNA substrate to confer the substrate specificity (Loria and Pan 1996; Massire et al. 1998).

Four crystal structures have been solved for the bacterial P RNA (Torres-Larios et al. 2006; Kazantsev and Pace 2006): an A-type S-domain, a B-type S-domain, an A-type full-length, and a B-type C-domain + a fraction of S-domain. These structures show that both types of P RNA contain a common core, but different peripheral structures, as indicated by their phylogenic conservation (Fig. 5.1). For instance, the common core structure for the S-domain is made of a four-way junction plus a two-loop structure made of numerous non-canonical base pairs. In the B-type S-domain, this core is supported by one peripheral, tetraloop-receptor structure that allows the correct orientation between the structural elements in the core. In the A-type S-domain, this core is supported by two peripheral structures, a tetraloop-helix and a heptaloop-receptor to support the core. Having two peripheral structures in the A-type likely increases the modularity and evo-ability of this type of structure as suggested by the wide-spread occurrence of the A-type S-domain in the bacterial branches. Folding of various P RNA domains, B-type C-domain, A and B-type S-domains has been extensively and comprehensively reviewed in a 2007 article (Baird et al. 2007). This review focuses on folding studies of the full-length, bacterial RNase P RNA.

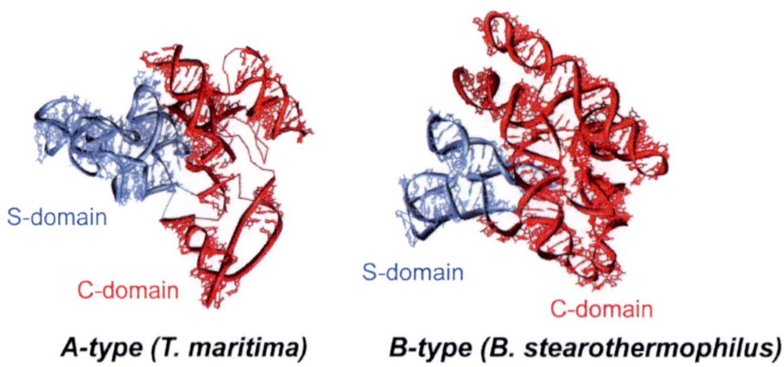

Fig. 5.1 Crystal structure of an A- and a B-type P RNA. The catalytic domain (C) is in *red* and the specificity domain (S) is in *blue*. A challenging feat for these molecules is to fold all secondary structures correctly. Misfold that significantly slows folding generally involves the formation of a short region of non-native secondary structure

The presence of two domains in P RNA elevates the importance of folding relationships among the domains and possible roles of domain-domain interactions in folding of the full-length P RNA. For a B-type P RNA (Table 5.1), the domains are able to fold to their native structures at significantly different levels of [Mg^{2+}] (as indicated by the K_{Mg} value = [Mg^{2+}] where 50% of the RNA is folded) and cooperativity (as indicated by the Hill constant n of the folding transition). This large disparity in stability allows several simplistic models to be proposed for the folding pathway of the full-length P RNA.

Model I: Both domains fold independently from each other. For the full-length P RNA folding, the thermodynamics should follow that of the C-domain, the kinetics should follow that of the S-domain.

Model II: Inter-domain, non-native structure forms during folding. Here, the thermodynamics and kinetics should follow neither of the domains, this non-native structure is absent in the folding of the domains.

Model III: Both domains fold independently followed by domain-domain interactions. If the domain-domain interactions are weaker (i.e., require higher [Mg^{2+}]) than domain folding themselves, then one additional folding step would be observed at $K_{Mg} > K_{Mg}$ (C-domain). If the domain-domain interactions are stronger than domain folding themselves, then the folding should have $K_{Mg} < K_{Mg}$ (C-domain). The folding kinetics should be either limited by S-domain folding, or the additional rearrangement needed for domain-domain interactions.

5.4 Folding of the B-type, *B. subtilis* P RNA

The B-type, *B. subtilis* P RNA has been the more extensively studied than the A-type, *E. coli* P RNA in folding, even though the *E. coli* P RNA has been the major subject for studies of catalysis. The reason for this is due to their differences in folding behavior: the *B. subtilis* P RNA is well-behaved in Mg^{2+}-dependent refolding as a vast majority (>90%) of this RNA folds along a single, well-defined pathway. In contrast, re-folding of *E. coli* P RNA often exhibits a multi-pathway behavior (unpublished results), so that the biophysical characterization has been more difficult.

Thermodynamics: The *B. subtilis* P RNA folds in a minimal three-state fashion, U-to-I_{eq}-to-N (U = unfolded, I_{eq} = equilibrium folding intermediate, N = native) at increasing Mg^{2+} concentration (Pan and Sosnick 1997). The U-to-I_{eq} transition occurs at lower [Mg^{2+}] and has an empirical Hill constant of one. This transition may contain multiple, overlapping folding transitions. The I_{eq}-to-N transition is of greater interest since it leads to the formation of the native structure. At 37°C, this transition has a Hill constant of ~4, and a K_{Mg} value between 0.8 and 3.2 mM in the literature. The variation of K_{Mg} value is largely due to the differences in monovalent ion (NaCl, KCl or NH_4Cl (Zarrinkar et al. 1996)) or buffer concentration (Tris–HCl (Pan et al. 1999b)). The low value (0.8 mM) is at no monovalent ion and low

Table 5.1 Folding parameters of the *B. subtilis* P RNA domains in 20 mM Tris–HCl, pH 8, 37°C

	$[Mg^{2+}]$ needed for 50% folding (K_{Mg}, mM)	Cooperativity of folding transition (Hill constant, *n*)	Approximate time needed for 50% folding ($t_{1/2}$, s)	Reference
S-domain	0.2	7	10	(Baird et al. 2005, 2007)
C-domain	1.2	3	0.1	(Fang et al. 1999a, b)
Full-length (S+C)	0.8	4	100	(Pan and Sosnick 1997; Zarrinkar et al. 1996)

(20 mM) buffer, whereas the high value (3.2 mM) is at high monovalent ion (up to 150 mM) and high (50 mM) buffer. In spite of the variation in K_{Mg} value, the Hill constant is unchanged, as is typical for folding of many tertiary RNAs.

Kinetics: Folding of the U-to-I_{eq} transition is very fast, occurring in less than 1 ms. Furthermore, the I_{eq} state contains ~5/6th of the amount of the native structure, as discussed below. The speed and amount of structure formation in this transition suggests that RNA folding can be inherently very fast. In contrast, the folding half-life of the I_{eq}-to-N transition is on the order of ~100 s at 37°C (Pan and Sosnick 1997). Therefore, the slow time scale of the I_{eq}-to-N transition may not be a necessary property of the large RNA folding.

Structural changes in folding: At the I_{eq} state, the vast majority of the secondary structure is already formed, because no hypochromicity change was observed between I_{eq} and N. A useful structural parameter in quantifying the amount of structural formation in a folding transition is the urea *m*-value (unit of kcal $mol^{-1} M^{-1}$ (Shelton et al. 1999)). The *m*-value is proportional to the amount of surface area buried in a folding transition. For *B. subtilis* P RNA, the calculated total *m*-value from completely unfolded (a polymer chain without any structure, this is experimentally un-attainable for the full-length P RNA, however) is ~15 kcal $mol^{-1} M^{-1}$. The I_{eq}-to-N transition has *m*-value of ~1.7 kcal $mol^{-1} M^{-1}$ (Pan and Sosnick 1997), indicating that 85–90% of the native structure (which includes secondary and tertiary structures) is already present in the I_{eq} state.

Additional kinetic intermediates are present on the pathway between I_{eq} and the N state. Oligonucleotide hybridization studies indicate that an I_k intermediate contains the structure of the P4 helix, but not the P7 and P2 helices (Zarrinkar et al. 1996). Furthermore, the 3′ portion of the P7 helix is inaccessible, while the 5′ portion is accessible to oligo-hybridization. This difference suggests that the 3′ portion of the P7 helix is involved in a non-native structure. This non-native structure has to be disrupted for the folding to proceed to the native state. This disruption likely accounts for the slow folding rate of this P RNA. Structural disruption in the folding transition state was validated by an increased exposure of surface area at this step (folding rate is accelerated upon addition of urea with $m = -0.3$ kcal $mol^{-1} M^{-1}$ (Pan and Sosnick 1997)). The involvement of the 3′ portion of P7 in this non-native structure is supported by the finding that disruption of the P5–P7 bond in a circularly permuted P RNA resulted in >10-fold faster folding (Pan et al. 1999b).

Only one inter-domain interaction is present in the B-type P RNA: the L8 loop interacts with the P4 helix (Kazantsev et al. 2005). This interaction is observed in the crystal structure of the *B. stearothermophilus* P RNA. However, it is likely to be very weak for the *B. subtilis* P RNA in solution as indicated by the extreme flexibility of the global shape of this P RNA as measured by SAXS (Fang et al. 2001). In the absence of monovalent ions, this interaction is most likely absent and not accounted for in many folding studies.

In summary, the *B. subtilis* P RNA folds through a minimal three-state pathway. Folding occurs at [Mg^{2+}] (<3.2 mM) that is significant lower than that required for its optimal catalytic activity (>50 mM). Folding kinetics is limited by the disruption of a non-native structure containing the 3′ portion of P7. Multiple lines of evidence indicate that this non-native structure contains interacting residues from both domains, consistent with the folding model II described above. A model of this non-native structure is shown in Fig. 5.2.

5.5 Folding of the A-type, *E. coli* P RNA

Thermodynamics and kinetics: Only the folding transition leading to the native state has been studied for *E. coli* P RNA (Loria and Pan 1996; Zarrinkar et al. 1996; Kent et al. 2000). At 37°C, this transition has a Hill constant of ~2.5, and a K_{Mg} value between 1.0 and 2.4 mM in the literature. Again, the variation of K_{Mg} value is likely due to the changes in the monovalent ions or buffer concentration. The folding half-life of this transition leading to N state is also on the order of ~100 s at 37°C (Zarrinkar et al. 1996; Kent et al. 2000).

Structural changes in folding: Oligonucleotide hybridization studies indicate that a major kinetic folding intermediate lacks the native-like P7 and P2 helices (Zarrinkar et al. 1996). The 5′ portion of P7 is accessible to oligo-hybridization, but the accessibility of the 3′ portion of the P7 helix is not tested. Assuming that the 3′ portion of P7 is also inaccessible in the kinetic intermediate, a non-native structure would have to be disrupted for the folding to proceed to the native state, consistent with the slow, observed folding rate.

A rapid hydroxyl radical protection assay using peroxonitrous acid provides a more detailed picture on the structural formation in folding kinetics (Kent et al. 2000). The formation of the P12–L13 interactions and P7–P11 four-way junctions in the S-domain, as well as the P4 and P5 regions in the C-domain occurs much faster than the rate-limiting step. The rate-limiting step involves the formation of the P8–L14 interactions in the S-domain and the P6 pseudoknot in the C-domain. Because hydroxyl radical protection primarily detects the formation of buried structures, it is unclear whether the rate-limiting step involves the disruption of a non-native structure.

Two inter-domain interactions are present in the A-type P RNA: the L8 loop-P4 helix and the L9 loop–P1 helix interactions (Massire et al. 1998). These interactions are thought to be stronger than the single interdomain interaction in a B-type P RNA.

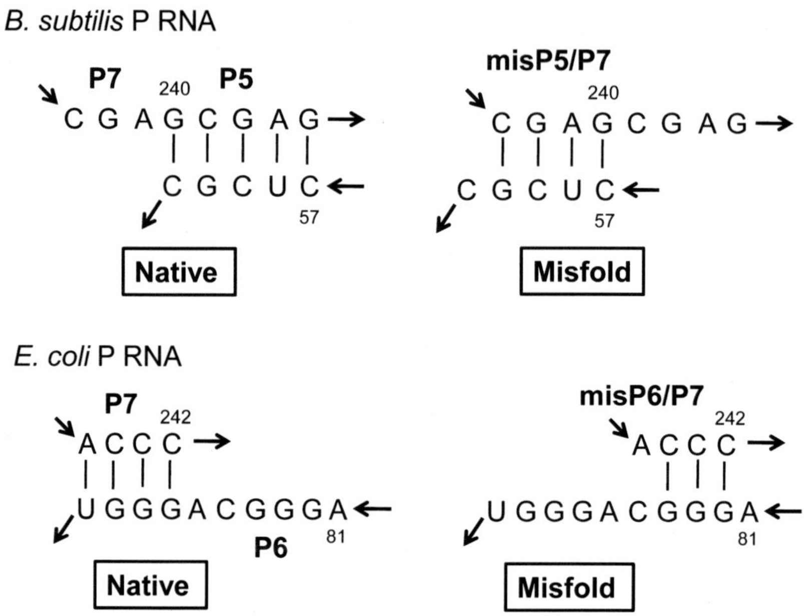

Fig. 5.2 Models on non-native structures in P RNA folding. *B. subtilis* P RNA has a CGAG repeat in the P5/P7 region that can base pair correctly (native) or incorrectly (misfold). Similarly, *E. coli* P RNA has an AGGG repeat in the P6/P7 region that can base pair correctly (native) or incorrectly (misfold)

Therefore, domain–domain interactions probably play a more significant role in the folding of an A-type P RNA.

In summary, the *E. coli* P RNA also folds at [Mg^{2+}] (<2.4 mM) that is significant lower than that required for its optimal catalytic activity (>25 mM). Folding kinetics is slow, and may also be limited by the disruption of a non-native structure. It is unclear whether this non-native structure contains residues from both domains, or is confined within a domain, as folding kinetics of either domain alone has not been reported. A model of this non-native structure is shown in Fig. 5.2.

5.6 P RNA Folding During Transcription

The process of co-transcriptional folding occurs with three fundamentally different components compared to Mg^{2+}-induced refolding:

1. Transcription has a 5–3′ polarity, so that the 5′ portion of the nascent RNA can fold into another structure before the 3′ portion is transcribed. This sequentiality is particularly important for the formation of long-range helices because their 5′ and 3′ portions are far apart in sequence; the 5′ half is present on the nascent RNA without its native partner for some time.

Therefore, the formation of native secondary structure is a very significant issue in co-transcriptional folding.

2. Bacterial or phage RNA polymerases have different elongation speeds, ranging from very fast (T7 RNA polymerase, high [NTP], hundreds of nucleotides/second) to intermediate (*E. coli* RNA polymerase, high [NTP], tens of nucleotides/second) to very slow (*E. coli* RNA polymerase, low [NTP], a few nucleotides/second). Therefore, varying time windows exist between the synthesis of an upstream portion (e.g., 5′ portion of a long-range helix) and the downstream interaction partner (3′ portion of this long-range helix).
3. Bacterial RNA polymerases does not elongate at a uniform speed. Rather, it transcribes some portions of the RNA very rapidly and temporarily pauses at distinct locations during elongation. Pausing is a distinct property for bacterial RNA polymerases and its location and duration depend on the interaction of the nascent RNA with the transcribing RNA polymerase (Mooney et al. 1998; Artsimovitch and Landick 2000). As discussed below, pausing plays a very significant role in the co-transcriptional folding of bacterial P RNA.

When compared with the biophysical studies of RNA folding, the available tools to study co-transcriptional folding are more limited. Because transcription and folding occurs continuously, it is difficult to monitor folding behavior spectroscopically. Co-transcriptional folding can be monitored by oligonucleotide hybridization, although this assay cannot readily distinguish the folding of a local region versus global folding to the native state. Nevertheless, oligonucleotide hybridization is useful in comparing folding under different conditions or for mutant RNAs (Wong et al. 2007). Even though single molecule tools have been applied extensively to study the transcription and even pausing (Davenport et al. 2000; Tolic-Norrelykke et al. 2004; Herbert et al. 2006), these studies focus on monitoring the RNA polymerase or the template DNA during transcription. To my knowledge, no single molecule study has been published that directly monitors the folding of the nascent RNA molecule during transcription.

The recovery of catalytic activity has been used to study P RNA folding during transcription. For both *B. subtilis* and *E. coli* P RNAs, the standard pre-tRNA substrate is cleaved only when both domains of the P RNA are folded. For *B. subtilis* P RNA, an in vitro selected substrate is also available, and its cleavage requires only the C-domain (Pan et al. 1999a). This difference permits the identification of a specific folding intermediate that contains the folded C-domain, but not the S-domain. Unfortunately, the in vitro selected substrate is not sufficiently active for *E. coli* P RNA for it to be useful in studying co-transcriptional folding of *E. coli* P RNA (unpublished results).

The polarity of *B. subtilis* P RNA can be changed to study the importance of the order of structural formation in co-transcriptional folding (Pan et al. 1999a; Wong et al. 2005). The wild-type P RNA starts with ~1/3 of the C-domain, followed by the S-domain and then the remaining ~2/3 of the C-domain. In this arrangement, folding during transcription by its cognate, *B. subtilis* RNA polymerase and non-cognate, *E. coli* RNA polymerase and even the phage T7 RNA polymerase occurs at a half time of ~100 s at 37°C. Folding is limited by the S-domain formation, because an intermediate containing folded C-domain can be detected using the in vitro selected substrate.

To study the importance of connectivity and folding, the polarity of P RNA is changed by making circularly permuted (CP) versions. Two CP versions are made, one starts with the entire C-domain followed by an S-domain (termed CP240 because the 5′ nucleotide is the 240th nucleotide in the wild-type P RNA), and the other starts with the entire S-domain followed by a C-domain. Both alterations in the order of structure formation, however, do not initially change the folding behavior of P RNA.

Unexpectedly, CP240 folding during transcription by *E. coli* RNA polymerase changes completely when the transcriptional elongation factor NusA was added (Pan et al. 1999a). NusA protein is known to change the pausing locations and durations of the *E. coli* RNA polymerase. For *B. subtilis* P RNA, NusA makes a particular pause site much stronger (at the 55th nucleotide of the wild-type P RNA). This longer duration in pausing accelerates folding of the S-domain during transcription by at least tenfold, while the folding rate of the C-domain remains unchanged. Since the entire C-domain is transcribed prior to the S-domain CP240, the pause site is near the end of the C-domain. It is amusing to see that the NusA enhanced pausing at this site only changes folding of the S-domain which has not yet been transcribed when pausing occurred! Subsequent study (Wong et al. 2005) rationalizes this result. Without the pause, the immediate 5′ region of the S-domain (5′P7) is entangled with a hairpin loop (L15) of the C-domain, and somehow, the NusA induced pause eliminates this misfold in the nascent RNA. It remains unclear how this works at the molecular level. One idea is that when pausing occurs, the NusA protein blocks the access of L15–5′P7 either through direct binding to L15 or making this interaction impossible topologically.

If pausing can significantly influence P RNA folding, shouldn't nature already use this to accelerate folding? Indeed, a very strong pause is present in the transcription of *E. coli* P RNA by its cognate, *E. coli* RNA polymerase and NusA is not needed here (Wong et al. 2007). Pausing at this site is much weaker during transcription by a non-cognate, *B. subtilis* RNA polymerase, most likely because these two polymerases are known to have very distinct pausing properties (Artsimovitch et al. 2000). This pause is very important for the folding of *E. coli* P RNA. Its reduction by any means (non-cognate polymerase, pausing-deficient *E. coli* RNA polymerase, P RNA mutants) all results in significantly slower folding. This pause site is located strategically just downstream of the 5′ portion of all six long-range helices in the *E. coli* P RNA. Structural mapping of the nascent RNA in this paused complex indicates that these 5′ portions fold into a defined, non-native structure, and the most likely function of this non-native structure is to sequester these segments in such a way as to facilitate folding once their 3′ base-paring partners are transcribed. Exactly how pausing leads to the formation of this beneficial non-native structure is unclear (the nascent RNA structure in the absence of the pause is unstable and hence cannot be mapped for comparison). One idea is that when pausing occurs, the RNA polymerase acts as a "folding chaperon" to help the formation of this structure. A model of how this non-native structure helps folding is shown in Fig. 5.3.

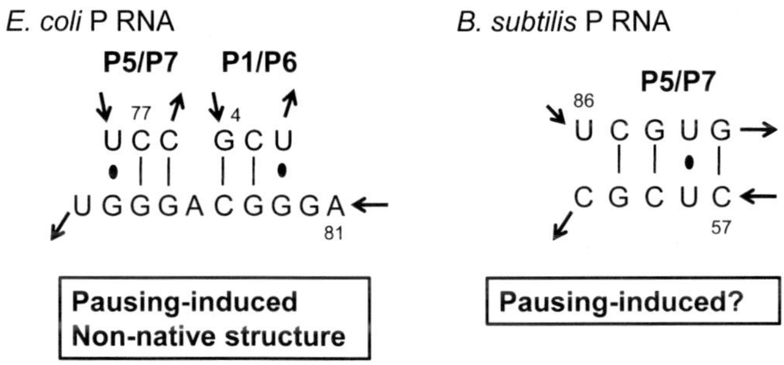

Fig. 5.3 Model on how *E. coli* P RNA escapes the P6/P7 misfold during transcription. Pausing at nucleotide 119 allows the formation of this non-native structure in the nascent RNA (all residues are derived from the 5′ portion of helices P1, P5, P6 and P7). This non-native structure is also set-up in such a way to facilitate the formation of native P7, P5 and P6 helices when their 3′ portions are transcribed in that order. A similar set-up for *B. subtilis* P RNA to escape the P5/P7 misfold during transcription may involve a non-native structure with just the 5′ portion of P5 and P7 helices. There is yet no experimental evidence that this structure of *B. subtilis* P RNA indeed forms during transcription

5.7 Future Directions and Questions for P RNA Folding

1. *Detailed structural understanding of folding pathways.* This route involves structural determination of key folding intermediates. Because these intermediates are meta-stable and P RNA is large, it will be difficult to apply standard high resolution techniques of crystallography and NMR. A recent advance shows that combining solution analysis and small-angle X-ray scattering can generate atomic structural models of a S-domain folding intermediate (Baird et al. 2005). Furthermore, we find that the molecular envelop of RNA folding intermediates can be determined by cryo-electron microscopy (cryo-EM). The cryo-EM image combined with all-atom simulations allow a more precise determination of the intermediate structure (Baird et al. submitted). These types of techniques can be readily applied to solve the structures of folding intermediates of the full-length P RNA.
2. *How is stability related to folding?* A-type P RNA is present in bacteria that live at a wide-range of environments and temperature. These P RNAs should all have similar tertiary structures but their stabilities are likely to match their ecological niche. Understanding how natural RNAs with the same native structure adjust their stability should reveal the stabilizing (or destabilizing) hotspots or may even enable rational design of RNA structures with appropriate stabilities.
3. *How does folding affect the assembly of bacterial RNase P holoenzyme?* The bacterial protein binds to the C-domain and significantly helps RNase P catalysis. The interaction of this protein with an intermediate form or the nascent RNA transcript could alter the downstream folding behavior. A possibility remains that the holoenzyme assembly may be faster or more specific (in the presence of other cellular RNAs) if the RNase P protein selectively binds to a non-native structure present in a folding intermediate.

4. *How does RNase P RNA folding influence cellular function?* We already know that P RNA folding during transcription by its cognate RNA polymerase is a very complex process involving specific non-native structures and biphasic folding kinetics. Does more than one structural form of RNase P RNA exist in the cell? Are there other cellular proteins that affect the P RNA folding? How do P RNA folding speed and partitioning into fast and slow folding fractions affect the cell fitness?

Acknowledgments The RNA folding work in my lab was supported by NIH (R01GM57880). I thank Tobin Sosnick for long-term collaborations, comments and discussions on this chapter.

References

Artsimovitch I, Landick R (2000) Pausing by bacterial RNA polymerase is mediated by mechanistically distinct classes of signals. Proc Natl Acad Sci USA 97(13):7090–7095

Artsimovitch I et al (2000) RNA polymerases from *Bacillus subtilis* and *Escherichia coli* differ in recognition of regulatory signals in vitro. J Bacteriol 182(21):6027–6035

Baird NJ et al (2005) Structure of a folding intermediate reveals the interplay between core and peripheral elements in RNA folding. J Mol Biol 352(3):712–722

Baird NJ et al (2007) Folding of a universal ribozyme: the ribonuclease P RNA. Q Rev Biophys 40(2):113–161

Brenowitz M et al (2002) Probing the structural dynamics of nucleic acids by quantitative time-resolved and equilibrium hydroxyl radical "footprinting". Curr Opin Struct Biol 2(5):648–653

Davenport RJ et al. (2000) Single-molecule study of transcriptional pausing and arrest by *E. coli* RNA polymerase. Science 287(5462):2497–2500.

Fang X, Pan T, Sosnick TR (1999a) A thermodynamic framework and cooperativity in the tertiary folding of a Mg2+-dependent ribozyme. Biochemistry 38(51):16840–16846

Fang X, Pan T, Sosnick TR (1999b) Mg2+-dependent folding of a large ribozyme without kinetic traps. Nat Struct Biol 6(12):1091–1095

Fang XW et al (2001) The Bacillus subtilis RNase P holoenzyme contains two RNase P RNA and two RNase P protein subunits. RNA 7(2):233–241

Herbert KM et al (2006) Sequence-resolved detection of pausing by single RNA polymerase molecules. Cell 125(6):1083–1094

Herschlag D (1995) RNA chaperones and the RNA folding problem. J Biol Chem 270(36):20871–20874

Kazantsev AV, Pace NR (2006) Bacterial RNase P: a new view of an ancient enzyme. Nat Rev Microbiol 4(10):729–740

Kazantsev AV et al (2005) Crystal structure of a bacterial ribonuclease P RNA. Proc Natl Acad Sci USA 102(38):13392–13397

Kent O, Chaulk SG, MacMillan AM (2000) Kinetic analysis of the M1 RNA folding pathway. J Mol Biol 304(5):699–705

Koch MH, Vachette P, Svergun DI (2003) Small-angle scattering: a view on the properties, structures and structural changes of biological macromolecules in solution. Q Rev Biophys 36(2):147–227

Krasilnikov AS, Yang X, Pan T, Mondragon A (2003) Crystal structure of the specificity domain of ribonuclease P. Nature 421(6924):760–764

Krasilnikov AS et al (2004) Basis for structural diversity in homologous RNAs. Science 306(5693):104–107

Lipfert J, Doniach S (2007) Small-angle X-ray scattering from RNA, proteins, and protein complexes. Annu Rev Biophys Biomol Struct 36:307–327

Loria A, Pan T (1996) Domain structure of the ribozyme from eubacterial ribonuclease P. RNA 2(6):551–563

Massire C, Jaeger L, Westhof E (1998) Derivation of the three-dimensional architecture of bacterial ribonuclease P RNAs from comparative sequence analysis. J Mol Biol 279(4):773–793

Mooney RA, Artsimovitch I, Landick R (1998) Information processing by RNA polymerase: recognition of regulatory signals during RNA chain elongation. J Bacteriol 180(13):3265–3275

Pan T, Sosnick TR (1997) Intermediates and kinetic traps in the folding of a large ribozyme revealed by circular dichroism and UV absorbance spectroscopies and catalytic activity. Nat Struct Biol 4(11):931–938

Pan T et al (1999a) Folding of a large ribozyme during transcription and the effect of the elongation factor NusA. Proc Natl Acad Sci USA 96(17):9545–9550

Pan T, Fang X, Sosnick TR (1999b) Pathway modulation circular permutation and rapid RNA folding under kinetic control. J Mol Biol 286(3):721–731

Petoukhov MV, Svergun DI (2007) Analysis of X-ray and neutron scattering from biomacromolecular solutions. Curr Opin Struct Biol 17(5):562–571

Russell R (2008) RNA misfolding and the action of chaperones. Front Biosci 13:1–20

Sclavi B et al (1998) RNA folding at millisecond intervals by synchrotron hydroxyl radical footprinting. Science 279(5358):1940–1943

Shelton VM, Sosnick TR, Pan T (1999) Applicability of urea in the thermodynamic analysis of secondary and tertiary RNA folding. Biochemistry 38(51):16831–16839

Sosnick TR, Fang X, Shelton VM (2000) Application of circular dichroism to study RNA folding transitions. Methods Enzymol 317:393–409

Tolic-Norrelykke SF et al (2004) Diversity in the rates of transcript elongation by single RNA polymerase molecules. J Biol Chem 279(5):3292–3299

Torres-Larios A et al (2005) Crystal structure of the RNA component of bacterial ribonuclease P. Nature 437(7058):584–587

Torres-Larios A et al (2006) Structure of ribonuclease P – a universal ribozyme. Curr Opin Struct Biol 16(3):327–335

Treiber DK, Williamson JR (2000) Kinetic oligonucleotide hybridization for monitoring kinetic folding of large RNAs. Methods Enzymol 317:330–353

Wong T, Sosnick TR, Pan T (2005) Mechanistic insights on the folding of a large ribozyme during transcription. Biochemistry 44(20):7535–7542

Wong TN, Sosnick TR, Pan T (2007) Folding of noncoding RNAs during transcription facilitated by pausing-induced nonnative structures. Proc Natl Acad Sci USA 104(46):17995–18000

Xiao S et al (2002) Eukaryotic ribonuclease P: a plurality of ribonucleoprotein enzymes. Annu Rev Biochem 71:165–189

Zarrinkar PP, Williamson JR (1994) Kinetic intermediates in RNA folding. Science 265(5174):918–924

Zarrinkar PP, Wang J, Williamson JR (1996) Slow folding kinetics of RNase P RNA. RNA 2(6):564–573

Chapter 6
Kinetic Mechanism of Bacterial RNase P

Kristin S. Koutmou, John Hsieh, and Carol A. Fierke

6.1 Introduction

Following the discovery of catalytically active RNAs, there has been great interest in elucidating the catalytic mechanism of the RNA component of ribonuclease P, P RNA, and comparing it with the mechanisms of proteins that perform similar reactions (Guerrier-Takada et al. 1983; Zaug et al. 1988; Marshall et al. 2008). In vivo both the RNA and protein components of ribonuclease P are required for function (Gossringer et al. 2006). However, in vitro P RNA alone is sufficient to catalyze the maturation of the 5′ end of precursor tRNAs (pre-tRNAs) under high salt and divalent metal ion concentrations, making RNase P a ribozyme (Guerrier-Takada et al. 1983; Gossringer et al. 2006). Unlike catalysis by small ribozymes that produce a 2′, 3′-cyclic phosphate and a 5′ hydroxyl group, RNase P catalyzes the cleavage of a specific phosphodiester bond to yield 5′ phosphate and 3′ hydroxyl end groups (Fig. 6.1) (Hanna and Doudna 2000; DeRose 2002; Doudna and Cech 2002; Harris and Christian 2003; Hsieh et al. 2004). Since RNase P was one of the first identified ribozymes, initial investigations into the mechanism of this enzyme focused almost exclusively on the steady-state kinetics of pre-tRNA cleavage catalyzed by bacterial P RNA, in particular M1 from *E. coli* and P RNA from *B. subtilis* (Guerrier-Takada et al. 1983). These investigations revealed that the cleavage of pre-tRNA catalyzed by RNase P RNA is pH independent, has a divalent metal ion dependence (Mg^{2+}, Mn^{2+} or Ca^{2+}), does not accumulate any covalent intermediates, and is increased by high monovalent salt concentrations (Altman 1989; Pace and Smith 1990; Smith et al. 1992; Warnecke et al. 1999). However, these steady-state studies have been of limited use in understanding the RNase P catalytic mechanism because the rate limiting step under steady-state turnover at saturating substrate concentrations is tRNA product release and not pre-tRNA cleavage (Reich et al. 1988; Tallsjo and Kirsebom 1993; Beebe and Fierke 1994; Beebe et al. 1996; Kurz and Fierke 2002).

K.S. Koutmou, J. Hsieh, and C.A. Fierke (✉)
Department of Chemistry, University of Michigan, 930 N. University Ave,
Ann Arbor, MI 48109, USA
e-mail: fierke@umich.edu

F. Liu and S. Altman (eds.), *Ribonuclease P*, Protein Reviews 10,
DOI 10.1007/978-1-4419-1142-1_6, © Springer Science+Business Media, LLC 2010

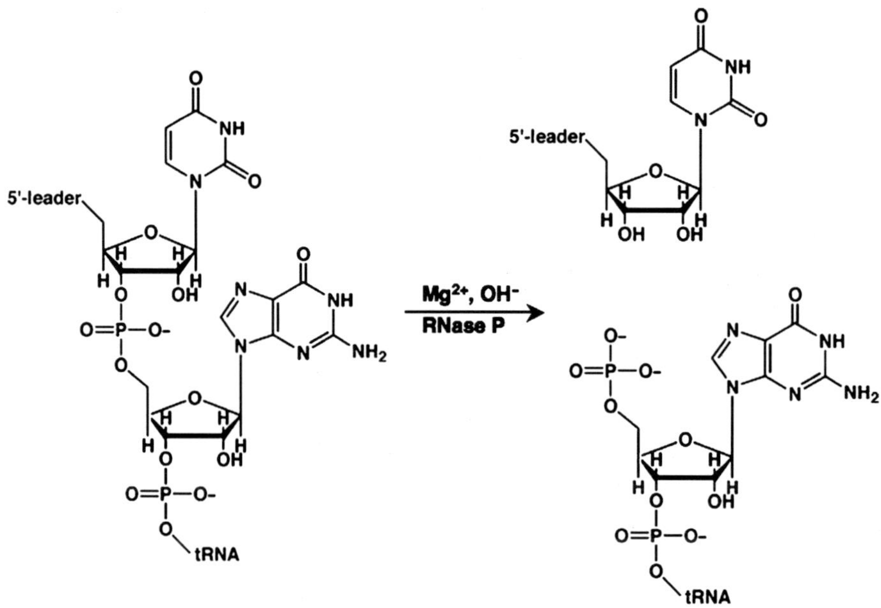

Fig. 6.1 Reaction catalyzed by RNase P. RNase P catalyzes cleavage of a phosphodiester bond in pre-tRNA between a guanosine (+1) and a uracil (−1) which are the nucleotides most frequently observed at these positions in bacterial pre-tRNAs. The nucleophile in this reaction is proposed to be a magnesium hydroxide coordinated to RNase P

As such, all the other reaction steps are masked under steady-state turnover, and thus presteady state kinetic investigations have been crucial to understanding both the kinetic and the catalytic mechanisms of RNase P. These kinetic and mutagenesis studies can now be understood in the context of the P RNA structure since the crystal structures of *T. maritima* and *B. stearothermophilus* have been solved.

Recently, the mechanism of the bacterial RNase P holoenzyme (P RNA•protein complex) has been extensively probed with a combination of steady-state and pre-steady state studies. These investigations reveal that the protein cofactor (C5 in *E. coli*, and P protein in *B. subtilis*) enhances RNase P catalysis, and suggest a number of distinct roles for the bacterial protein in the kinetic mechanism (Kurz et al. 1998; Niranjanakumari et al. 1998; Buck et al. 2005; Sun et al. 2006; Sun and Harris 2007). It is also important to note that there are several examples of the site-specific cleavage of non-tRNA substrates by RNase P (Bothwell et al. 1976; Peck-Miller and Altman 1991; Alifano et al. 1994; Komine et al. 1994; Hartmann et al. 1995; Gimple and Schon 2001; Altman et al. 2005; Seif and Altman 2008). Although the rate constants for RNase P-catalyzed cleavage for these substrates differ from those reported for pre-tRNA, it is unknown whether the rate-limiting step and the kinetic mechanism are the same as previously observed for pre-tRNA substrates. However, these investigations suggest that the role of the P protein in enhancing the activity of RNase P may be larger for cleavage of non-tRNA substrates when compared with the pre-tRNA (Peck-Miller and Altman 1991; Hartmann et al. 1995; Gimple and Schon 2001). In this chapter, we discuss the chemical and kinetic mechanism of bacterial RNase P, with special emphasis on highlighting the contributions of metal and protein cofactors to RNase P catalysis.

6.2 P RNA Coordinates a Metal-Hydroxyl That Functions as a Nucleophile to Catalyze Hydrolysis of the Scissile Phosphate

The hydrolysis of phosphodiester bonds is exceedingly slow even though it is thermodynamically favorable (Westheimer 1987). In an aqueous solution without a catalyst present, these cleavage reactions are estimated to take centuries to occur ($t_{1/2} > 100{,}000$ years) (Radzicka and Wolfenden 1995; Cassano et al. 2004b). However, RNA and protein enzymes make phosphodiester bond hydrolysis possible in biology by enhancing the reaction rate by 10^{12}–10^{14}-fold. Therefore, RNase P is required to catalyze the cleavage of the 5′ leader cleavage of pre-tRNA to form tRNA (Serpersu et al. 1986; Beebe and Fierke 1994; Cassano et al. 2004a).

Magnesium hydroxide has been proposed as a catalytic species required for the phosphodiester cleavage catalyzed by RNase P (Cassano et al. 2004a). In general, hydrolysis can be catalyzed by hydroxide ions acting either as a nucleophile or as a general base. The ability to distinguish between these two mechanisms is of interest because they provide two potential strategies for enzymes to activate a water nucleophile (Cassano et al. 2002). ^{18}O solvent isotope effect studies of the reaction catalyzed by *E. coli* RNase P are most consistent with bond cleavage having a concerted S_N2 mechanism, where the water nucleophile is activated by metal ion coordination (Cassano et al. 2004a). A concerted S_N2-like mechanism for phosphodiester cleavage has also been proposed for reactions catalyzed by the other metal-dependent ribozymes (Frank and Pace 1998; Kurz and Fierke 2000; Shan et al. 2001). In conjunction with the requirement for a divalent cation, such as magnesium, to activate catalysis by RNase P, it is likely that the metal-bound hydroxide functions as the nucleophile in the 5′ leader cleavage reaction.

6.2.1 P RNA Helix P4 is Important for the Binding and Positioning of Metals Required for Catalysis

Currently it is still unclear precisely which groups in RNase P coordinate the metal ion(s) involved in catalysis. Nonetheless, several individual nucleotides and regions of P RNA have been identified as making contributions to the RNase P ribozyme catalysis. RNase P RNA is comprised of two independently folding domains, the specificity (S-domain), and catalytic domain (C-domain) (Pan 1995; Loria and Pan 1996). The C-domain of P RNA (nucleotides 1–85, 240–409 in *B. subtilis*) is the region where nucleotides required for both P protein binding and pre-tRNA cleavage are located (Pan 1995; Loria and Pan 2001). Crystal structures of P RNAs from *T. maritima* and *B. stearothermophilus* reveal that the RNase P catalytic core is formed by coaxial stacked helices P1/P4/P5, P2/P3, and P8/P9, and this spatial arrangement is highly similar between A-type and B-type RNAs (Kazantsev et al. 2005; Torres-Larios et al. 2005). An exhaustive deletion analysis of *B. subtilis* P RNA demonstrated that most of the RNase P RNA sequence is important for high catalytic activity (Waugh and Pace 1993). However, deletion of nucleotides 225–270 are particularly deleterious to the catalytic activity of RNase P; in *B. subtilis* P RNA this region spans from J12/11, P11, P7, P5, J5/15 to P15 in the secondary structure (Fig. 6.2) (Waugh and Pace 1993; Kazantsev et al. 2005; Torres-Larios et al. 2005). The results from these deletion studies are unsurprising given that J5/15 and P15 loop regions of the ribozyme form Watson–Crick base pairs with pre-tRNA, including the nucleotide on the 5′ side of the cleavage site (N(−1)), and the CCA at the 3′ end of many pre-tRNAs (Kirsebom and Svard 1994; Svard et al. 1996; Brannvall et al. 1998, 2003; Zahler et al. 2005). Additionally, short range crosslinking experiments demonstrated that J5/15 and J18/2 of P RNA are in close proximity to the pre-tRNA cleavage site (Christian et al. 1998; Christian and Harris 1999). Furthermore, deletion of helix P4 from *B. subtilis* RNase P RNA (nucleotides G46-U55 in *B. subtilis*) results in an RNA sequence that is unable to fold into a native structure, bind substrate, or catalyze pre-tRNA 5′ cleavage demonstrating the importance of this helix (Crary et al. 2002). These studies highlight regions of the catalytic domain of P RNA that are important for the formation of the active site of this enzyme.

Several lines of evidence suggest that catalytically important metal ions bind to the P4 helix of P RNA (Frank and Pace 1998). Helix P4 possesses more than half of the 21 universally conserved nucleotides in RNase P RNA, and is thus the most highly conserved region of the enzyme (Pace and Brown 1995). Although there are substantial sequence and secondary structural differences between the bacterial P RNAs and their archael and eucaryal counterparts, P RNAs from all three domains of life possess P4 homologs (Forster and Altman 1990; Nieuwlandt et al. 1991; LaGrandeur et al. 1993; Tranguch and Engelke 1993). Additionally, single nucleotide point mutations in the P4 helix can reduce RNase P activity by over 100-fold in vitro, and inhibit cell growth in vivo under conditions of magnesium starvation (Waugh and Pace 1993; Frank and Pace 1997; Crary et al. 2002; Rasmussen and Nolan 2002).

Further evidence that important metal ions interact with helix P4 of RNase P comes from the phosphorothioate-rescue experiments; these data suggest that metal binding to helix P4 leads to specific, but possibly indirect, stabilization of the catalytic transition state (Christian et al. 2000, 2006). Sulfur substitution at the nonbridging oxygens of the phosphodiester bonds between nucleotides A49 and G50 in *B. subtilis* P4 helix (A67 and G68 in *E. coli*) decrease the cleavage rate constant by > 700-fold without affecting the affinity of pre-tRNA, and some of these reductions in activity can be rescued by manganese and/or cadmium ions (Hardt et al. 1995; Harris and Pace 1995; Christian et al. 2000, 2006; Crary et al. 2002). Additionally, a point mutation that moves the position of the conserved U-bulge in P4 enhances the catalytic activity of *E. coli* P RNA in the presence of calcium (Fig. 6.2) (Frank and Pace 1997). These observations illustrate the potentially profound effects of the coupling metal ion binding

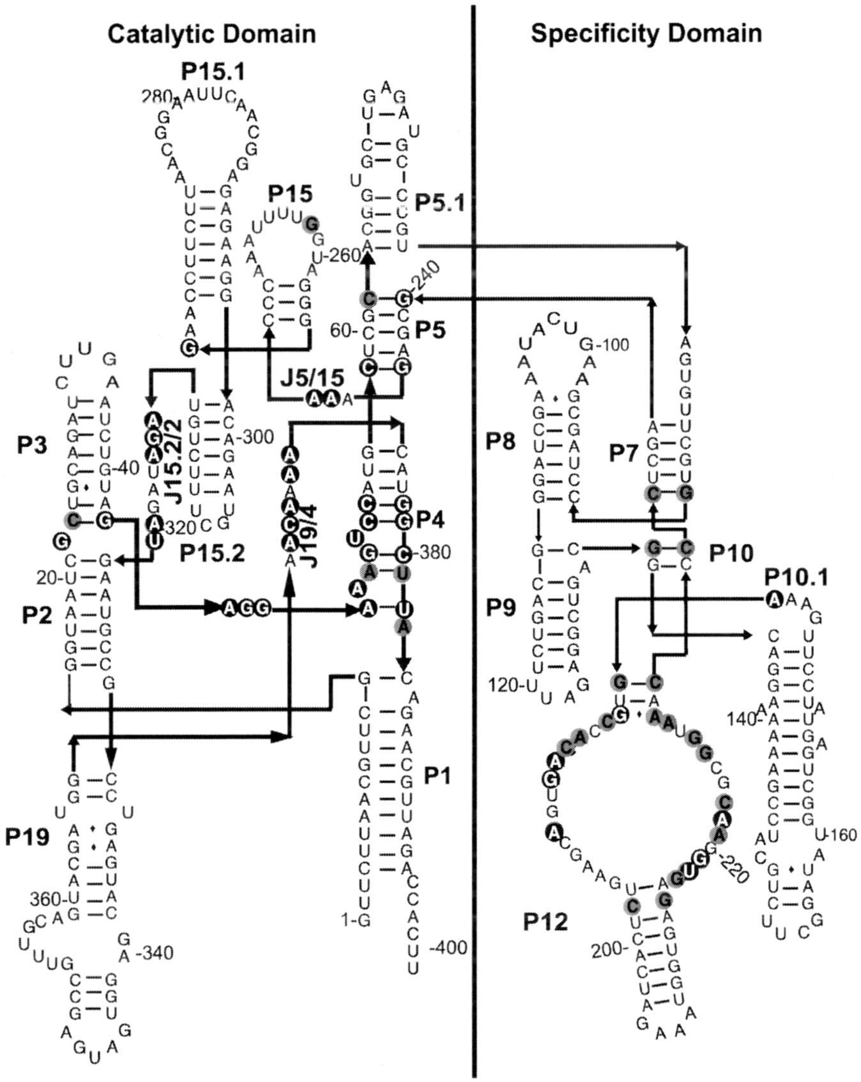

Fig. 6.2 Secondary structure of *B. subtilis* P RNA based on topology from the X-ray structure of the *B. stearothermophilus* P RNA (Kazantsev et al. 2005). Conserved nucleotides identified from phylogenetic analysis of bacterial P RNA (Pace and Brown 1995) are highlighted: universally conserved nucleotides are in white on filled circles (8-Ball format), while those are presented in at least 80% of sequences are in bold on *gray circles*

with P RNA structure and function (Christian et al. 2006). Corroboration for metal binding in P4 comes from NMR studies that demonstrate the specific magnesium binding sites in a stem-loop derived from the *B. subtilis* P4 helix sequence (Schmitz and Tinoco 2000; Getz et al. 2007). These results strongly suggest that the P4 helix binds and/or positions the metal ions that are essential for activity.

Several investigations have evaluated the nucleotide functional groups that are important for RNase P RNA catalytic activity. Nucleotide analog interference mapping (NAIM) and site-specific modification experiments have been used to probe the effects of modifications of nucleotides on substrate binding, metal affinity, and catalysis by the *E. coli* M1 ribozyme (Heide et al. 1999, 2001a, b; Kaye et al. 2002; Rox et al. 2002). Multiple nucleotide alterations decrease the activity of M1 RNA at low Mg(II) concentration (10 mM), although these effects are rescued by addition of 100 mM Mg(II) (Kaye et al. 2002). The majority of the nucleotide positions detected by NAIM are located within the P1–P4 multihelix junction, which agrees well with the suggestion that this region comprises the ribozyme active site. These results reveal that 2′ hydroxyl and purine N7 functional groups within the C-domain are particularly important for catalysis of pre-tRNA cleavage (Kaye et al. 2002). The metal sensitivity of the NAIM studies highlights the importance of the relationship between the RNase P RNA catalytic activity and the metal ion-dependent structure of the P1–P4 multihelix junction.

6.3 A Minimal Kinetic Scheme for RNase P

The first detailed kinetic scheme for bacterial P RNA was described by Fierke and coworkers for *B. subtilis* P RNA (Scheme 6.1) (Beebe and Fierke 1994). This mechanism includes four distinct events: (1) rapid and essentially irreversible binding of pre-tRNA to P RNA (k_1); (2) fast phosphodiester bond cleavage (k_2); (3) rapid dissociation of the 5′ leader RNA, and slow, rate-limiting dissociation of the mature tRNA (k_6); and (4) a slow conformational change in P RNA before binding a second substrate molecule (k_8) that is observed at high magnesium concentrations (Beebe and Fierke 1994). An alternative reaction scheme that only includes the initial binding, bond cleavage, and product release steps is sufficient to fit the most kinetic data (Scheme 6.2) (Beebe and Fierke 1994; Beebe et al. 1996). Scheme 6.2 can be used to compare the observed kinetic rate constants for *B. subtilis* RNase P RNA, *B. subtilis* holoenzyme (P RNA•P protein complex), and *S. cerevisiae* RNase P (Table 6.1). The differences between the kinetic mechanisms of the bacterial P RNA, holoenzyme, and yeast RNase P will be discussed in a later section.

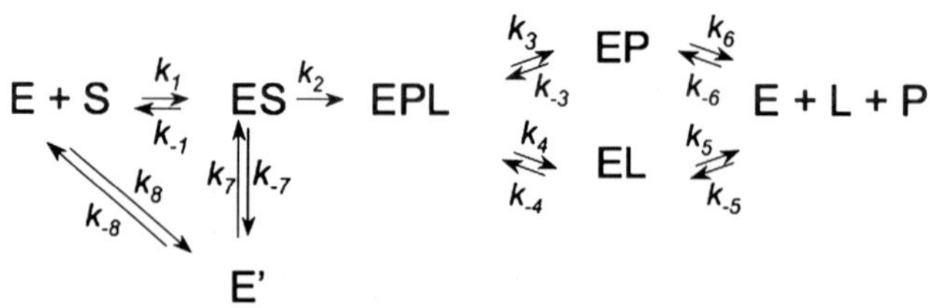

Scheme 6.1 Kinetic mechanism of *B. subtilis* P RNA at 800 mM NH_4Cl, 100 mM $MgCl_2$, 50 mM Tris (pH 8), 0.05% nonidet 40 (NP40), 0.1% SDS at 37°C (Beebe and Fierke 1994). *E* is P RNA; *S* is pre-$tRNA^{Asp}$; *ES* is P RNA•pre-RNA; *P* is mature tRNA; *L* is 5′ leader; *EP* is P RNA•mature tRNA; *EL* is P RNA•5′ leader

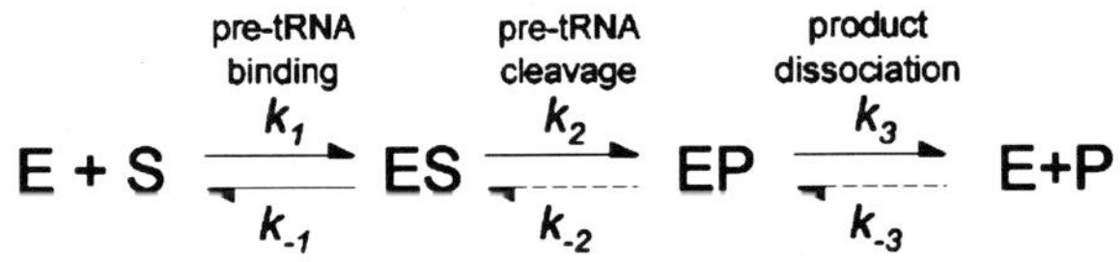

Scheme 6.2 Minimal kinetic mechanism for cleavage of pre-tRNA catalyzed by RNase P. *E* (RNase P); *S* (pre-tRNA); *ES* (RNase P•pre-tRNA complex); *EP* (RNase P•5′ leader•mature tRNA complex); *P* (5′ leader and mature tRNA)

Table 6.1 Kinetic parameters for pre-tRNA processing catalyzed by RNase P

	B. subtilis P RNA	*B. subtilis* holoenzyme	Yeast holoenzyme
k_{cat} low pH	$0.03 \pm 0.01\ s^{-1}$ [a]	$0.27 \pm 0.02\ s^{-1}$ [a]	–
k_{cat} high pH	$0.5 \pm 0.06\ s^{-1}$ [a]	$0.83 \pm 0.05\ s^{-1}$ [a]	$0.080 \pm 0.005\ s^{-1}$ [b]
K_M low pH	21 ± 9 mM [a]	0.08 ± 0.02 mM [a]	–
K_M high pH	27 ± 8 mM [a]	0.23 ± 0.06 mM [a]	$0.10 \pm 0.02\ \mu M$ [b]
k_{cat}/K_M low pH	$0.0014 \pm 0.0001\ \mu M^{-1} s^{-1}$ [a]	$3.4 \pm 0.6\ \mu M^{-1} s^{-1}$ [a]	–
k_{cat}/K_M high pH	$0.019 \pm 0.004\ \mu M^{-1} s^{-1}$ [a]	$3.5 \pm 0.7\ \mu M^{-1} s^{-1}$ [a]	$0.8 \pm 0.1\ \mu M^{-1} s^{-1}$ [b]
k_1	$6.3 \pm 0.2\ \mu M^{-1} s^{-1}$ [c]	$6.7 \pm 0.2\ \mu M^{-1} s^{-1}$ [a]	$1.1 \pm 0.2\ \mu M^{-1} s^{-1}$ [b]
k_{-1}	$0.6 \pm 0.2\ s^{-1}$ [c]	$0.04 + 0.03\ s^{-1}$ [a]	$0.05 \pm 0.02\ s^{-1}$ [b]
k_2	$0.09 \pm 0.04\ s^{-1}$ [a]	$0.33 \pm 0.01\ s^{-1}$ [a]	$0.37 \pm 0.03\ s^{-1}$ [b]
K_D tRNA, low pH (Mg^{2+})	$12 \pm 3\ \mu M$ [d]	$1.2 \pm 0.1\ \mu M$ [d]	–
K_D tRNA, high pH (Mg^{2+})	$3.2 \pm 0.5\ \mu M$ [d]	$0.47 \pm 0.08\ \mu M$ [d]	–
K_D tRNA, low pH (Ca^{2+})	0.3 ± 0.1 mM [d]	0.2 ± 0.06 mM [d]	$29.4 \pm 0.8\ \mu M$ [e]
K_D tRNA, high pH (Ca^{2+})	0.11 ± 0.02 mM [d]	0.06 ± 0.02 mM [d]	–
K_D pre-tRNA (Mg^{2+})	$60 \pm 20\ \mu M$ [d]	$0.006 \pm 0.005\ \mu M$ [d]	–
K_D pre-tRNA (Ca^{2+})	$4 \pm 1\ \mu M$ [d]	$0.0004 \pm 0.0002\ \mu M$ [d]	$0.05 \pm 0.02\ \mu M$ [f]

[a] Kurz et al. 1998. Measured at 100 mM NH_4Cl, 10 mM $MgCl_2$, 50 mM Tris, 50 mM MES, pH 6.1 (low pH) or 7.8 (high pH), at 37°C, and pre-$tRNA^{Asp}$ from *B. subtilis*. Steady-state parameters for the P RNA catalyzed reaction were extrapolated from kinetic measurements of up to 10 µM pre-$tRNA^{Asp}$

[b] Yeast nuclear RNase P holoenzyme, from Hsieh et al. 2009. Measured at 100 mM KCl, 10 mM $MgCl_2$, 10 mM HEPES, pH 8.0 at 20°C, and yeast pre-$tRNA^{Tyr}$

[c] Beebe and Fierke 1994. Measurements at 800 mM NH_4Cl, 100 mM $MgCl_2$, 50 mM Tris (pH 8), 0.05% nonidet P40, 0.1% SDS, at 37°C, and pre-$tRNA^{Asp}$ from *B. subtilis*

[d] Kurz et al. 1998. Affinity measurements at 100 mM NH_4Cl, 10 mM $CaCl_2$ or 10 mM $MgCl_2$, 50 mM Tris, 50 mM MES, pH 6.1 (low pH) or 7.8 (high pH), at 37°C, and pre-$tRNA^{Asp}$ from *B. subtilis*

[e] Ziehler et al. 2000

[f] Calculated from kinetic data in Hsieh et al. (2009)

Information about the kinetic and catalytic mechanism of RNase P has been derived from a combination of the transient and steady-state kinetic experiments (for examples, see Beebe and Fierke 1994; Beebe et al. 1996; Crary et al. 1998; Kurz et al. 1998; Loria et al. 1998; Ziehler et al. 2000, 2003; Kurz and Fierke 2002; Zahler et al. 2005, 2010 (Unpublished data); Sun et al. 2006; Sun and Harris 2007). In comparison to steady-state kinetics, transient kinetic experiments analyze either the cleavage of less than one molecule of substrate per enzyme active site using excess enzyme (single-turnover), or the formation of the enzyme-bound intermediates that accumulate under steady-state turnover (transient kinetics). The use of transient kinetic techniques has facilitated the distinction between the macroscopic properties described by steady-state parameters (k_{cat}, K_M, and k_{cat}/K_M), and microscopic rate constants for discrete steps in the kinetic mechanism presented in Schemes 6.1 and 6.2 (see below; Beebe and Fierke 1994; Fierke and Hammes 1995). A brief discussion of how to assess the microscopic rate constants in Scheme 6.2 through the use of both single turnover and transient kinetic techniques is provided below (for examples, see Beebe and Fierke 1994; Beebe et al. 1996; Crary et al. 1998; Kurz et al. 1998; Loria et al. 1998; Ziehler et al. 2000; Kurz and Fierke 2002; Zahler et al. 2003, 2005, 2010 (Unpublished data); Sun et al. 2006; Sun and Harris 2007).

For kinetic investigations of RNase P, 5′-^{32}P labeled pre-tRNA is generally incubated with enzyme and the time-dependent production of the 5′ leader product is monitored. Cleavage reactions are typically initiated by mixing the RNase P and substrate, and terminated by the addition of EDTA (at least two-fold molar excess over the divalent metal ion concentration) since metal chelation is fast relative to the RNase P catalyzed cleavage. Under pseudo-first order and single-turnover conditions ($[E] >> [S]$), an observed rate constant, k_{obs}, for product formation (the sum of $[EP]$ and $[P]$) can be calculated from a single exponential fit to the data. The observed cleavage rate is limited by: (a) at low $[E]$ by the ES complex formation ($[E][S]k_1 = [E]k_1'$), and (b) at high $[E]$, the cleavage rate constant, k_2, when the binding step is no longer rate-limiting ($[E]k_1 > 10x\ k_2$). The time course for the appearance of product formation can be fit to a mechanism of either a single first-order reaction (at high and low $[E]$) or two consecutive first-order reactions (at intermediate concentrations of E), consistent with $k_2 > k_{-1}$ under many solution conditions (Beebe and Fierke 1994; Beebe et al. 1996; Crary et al. 1998; Kurz et al. 1998). The bi-molecular binding rate constant, k_1, and the cleavage rate constant, k_2, can thus be measured by performing single turnover experiments over a range of the enzyme concentrations.

Transient kinetic experiments can also provide information about the rate constant for product release (k_3 in Scheme 6.2). In these experiments, a rapid formation of products (the burst phase) is followed by a slower linear phase, and the linear region extrapolates to a nonzero *y*-intercept (for example, see (Beebe and Fierke 1994) and reference therein). Burst kinetics indicate that the RNase P•product complex is formed rapidly followed by a slow step, presumed to be the product release (k_3 in Scheme 6.2). Therefore, unlike under single turnover conditions, the observed rate constant (k_{obs}) for the burst phase includes the binding and cleavage of pre-tRNA to form steady concentrations of the ES and EP complexes while subsequent turnovers are limited by the rate constant for formation of free enzyme. Furthermore, the rate constant for dissociation of product from RNase P can be measured by rapidly separating the bound and unbound product using a gel filtration centrifuge column. Together, these nonsteady state approaches allow a better understanding of the kinetic mechanism of both the RNase P ribozyme and holoenzyme (Beebe and Fierke 1994; Beebe et al. 1996; Crary et al. 1998; Kurz et al. 1998), as well as providing a clearer picture of the molecular recognition of pre-tRNA substrates (Loria et al. 1998; Loria and Pan 1998; Zahler et al. 2003, 2005; Sun et al. 2006).

Table 6.1 summarizes the rate constants obtained from the studies of *B. subtilis* ribozyme and holoenzyme, as well as yeast nuclear RNase P under similar reaction conditions (Kurz et al. 1998; Hsieh et al. 2009). The bulk of the microscopic rate constants for pre-tRNA cleavage catalyze by RNase P reported in the literature can not be directly compared because the studies are performed under different solution conditions and with different substrates. No standard conditions for measurement of enzyme activity have been determined, making it difficult to compare the rate constants for this enzyme since the activity depends on the concentration of mono- and divalent cations and pH. The values reported in Table 6.1 include both steady-state and pre-steady state kinetic parameters. Equations (6.1)–(6.3) describe the observed steady-state kinetic parameters derived from Scheme 6.2 assuming that cleavage is essentially irreversible ($k_{-2}<k_2$, k_3). Consistent with multiple kinetic steps prior to the rate-limiting step in catalytic mechanism, the value of k_{cat}/K_M represents the *lower-limit* for the bimolecular association rate constant (Fersht 1985) and equals this rate constant only when k_{-1} is much less than k_2 (6.1). Under most conditions K_M (6.3) is *not* equal to the substrate dissociation constant ($K_{D\,(pre\text{-}tRNA)}=k_{-1}/k_1$). Only under solution conditions in which $k_{-1}>k_2$, such as low salt, low pH and/or substitution of calcium for magnesium, does the value of K_M approximate the value of K_D (Smith et al. 1992; Beebe and Fierke 1994; Beebe et al. 1996). Under in vivo conditions RNase P turnover might be limited by binding of substrate (k_{cat}/K_M conditions) instead of the product release (k_{cat}) due to the low concentrations of individual pre-tRNAs (Beebe and Fierke 1994; Ziehler et al. 2000).

$$\frac{k_{cat}}{K_M}=\frac{k_1k_2k_3}{k_2k_3+k_{-1}(k_{-2}+k_3)}\sim k_1\left(\frac{k_2}{k_{-1}+k_2}\right), \tag{6.1}$$

$$k_{cat}=\frac{k_2k_3}{k_2+k_{-2}+k_3}\sim\frac{k_2k_3}{k_2+k_3}, \tag{6.2}$$

$$K_M=\frac{k_2k_3+k_{-1}(k_{-2}+k_3)}{k_1(k_2+k_{-2}+k_3)}\sim\frac{(k_2+k_{-1})k_3}{k_1(k_2+k_3)}. \tag{6.3}$$

Although RNase P RNA alone is catalytically active, the protein cofactor enhances enzyme catalysis. At moderate salt, addition of the *B. subtilis* protein to form the holoenzyme decreases the value of K_M by a factor of 120, and the k_{cat}/K_M value by 180-fold for pre-tRNA^{Asp} (Table 6.1) with modest effects on the value of $k_{cat.}$. However, under different conditions (low pH) the enhancement of the value of k_{cat}/K_M is as large as 2,500-fold (Kurz et al. 1998). Under these conditions, the rate-limiting steps for P RNA and RNase P holoenzyme under k_{cat}/K_M differ. For cleavage of pre-tRNA^{Asp} catalyzed by RNase P, the cleavage rate constant is faster than the substrate dissociation rate constant ($k_2>k_{-1}$) so that substrate association (k_1) is the main rate-contributing step. For the P RNA catalyzed reaction, the rate constant for substrate dissociation is faster than the cleavage rate constant ($k_2<k_{-1}$) so cleavage is mainly rate-limiting (k_1k_2/k_{-1}).

For the *B. subtilis* enzyme, addition of the protein component has only a modest effect on the value of the association rate constant (k_1) or the cleavage rate constant (k_2). The effect of the protein component on stabilizing the cleavage transition state may depend on the solution conditions and/or the enzyme source. A larger enhancement (>700-fold) of the single turnover rate constant for the *E. coli* RNase P holoenzyme compared to M1 RNA has been observed using different salt concentrations (50 mM MES, pH 5.75, 100 mM NaCl, 17.5 mM $MgCl_2$) and substrate (pre-$tRNA^{His595}$) (Sun et al. 2006). The largest effect of addition of the P protein subunit is to increase the affinity of the *B. subtilis* RNase P holoenzyme for pre-$tRNA^{Asp}$ by 10^4-fold compared the affinity of P RNA (Kurz et al. 1998). Similar enhancements (10^3-fold) in the affinity of the *E. coli* enzyme for pre-$tRNA^{Asp}$ have been reported to be due to the protein component (Buck et al. 2005). Since almost no differences are observed in the value of the microscopic association rate constant (k_1), the P protein enhances the substrate affinity of RNase P mainly by decreasing the value of the substrate dissociation rate constant, k_{-1}.

Interestingly, comparison of the kinetics of bacterial holoenzyme and the yeast nuclear RNase P (displayed in Table 6.1) demonstrate that these two enzymes have similar kinetic properties. For example, the k_{cat}/K_M values for both RNase P enzymes are comparable and much faster than the P RNA ribozyme. In addition, the k_{cat}/K_M value for the yeast nuclear RNase P, like the *B. subtilis* holoenzyme, becomes diffusion-limited (~ $1\times10^8 M^{-1} s^{-1}$) at 37°C (Ziehler et al. 2000; Xiao et al. 2005; Hsieh et al. 2009), indicating that pre-tRNA bound to yeast RNase P is more likely to be cleaved than to dissociate (or, $k_2>k_{-1}$ as in Scheme 6.2) (Fersht 1985; Hsieh et al. 2009).

In summary, the minimal kinetic mechanism of yeast RNase P, bacterial P RNA and bacterial holoenzyme can be described by Scheme 6.2. A more complete kinetic mechanism for P RNA alone has been derived and is displayed in Scheme 6.1, but should not be applied to either the holoenzyme or yeast enzyme. Comparison of the minimal kinetic parameters from the eukaryotic yeast enzyme and bacterial RNase P holoenzyme provide the exciting conclusion that the overall kinetic mechanism of RNase P may be conserved even though the architecture of RNase P varies significantly across phylogeny. This naturally leads to the question of the function of the additional proteins in eukaryotic and archael RNase P; this is the subject of on-going investigations.

6.4 Contributions of P Protein to RNase P Function

In addition to enhancing the rate constants for substrate binding and cleavage as discussed above, the *E. coli* and *B. subtilis* RNase P proteins decrease the metal ion concentration required for catalysis by RNase P, likely by stabilizing a metal-dependent conformational change important for the substrate affinity (Kurz and Fierke 2002; Sun et al. 2006; Sun and Harris 2007; Hsieh and Fierke 2009). However, there is limited information available about how the protein achieves these positive contributions to RNase P function, or about how specific regions of the protein contribute to particular steps in the kinetic mechanism. The currently identified and postulated roles of bacterial P protein in RNase P function are discussed below.

Although bacterial RNase P proteins are required in vivo and are structurally homologous, they have strikingly little sequence similarity (Jovanovic et al. 2002). Bacterial P proteins share a fold and structure similar to that of the *B. subtilis* protein, with an overall αβββαβα topology, and two probable RNA binding regions (Stams et al. 1998; Spitzfaden et al. 2000; Kazantsev et al. 2003). The structure of a protein subunit from archeal RNase P, a homolog of the eukaryotic RNase P protein Pop5p, is similar to that of the bacterial P protein, suggesting an analogous functional role across all three kingdoms of life (Walker and Engelke 2006; Wilson et al. 2006). There are two proposed RNA binding regions of bacterial P protein. The first region is the central cleft formed by the face of the central β-sheet and an α-helix. Time resolved fluorescence resonance energy transfer (trFRET), crosslinking and affinity cleavage data provide evidence that the central cleft interacts with pre-tRNA 5′ leaders to enhance the substrate recruitment and discrimination (Crary et al. 1998; Kurz et al. 1998; Niranjanakumari et al. 1998, 2007; Rueda et al. 2005). The second region of the protein identified as potentially important for RNA binding has an unusual left-handed βαβ cross-over connection between the parallel beta-strands of the central cleft. The alpha-helix portion of this region is termed the RNR motif because the first three residues are Arg-Asp-Arg, or R-N-R. Sequence alignment studies of the bacterial proteins demonstrate that there are only 14 residues with greater than 67% conservation; six of the 14 conserved residues are located in the RNR motif, making it the most highly conserved region of the bacterial P proteins (Jovanovic et al. 2002). However, no role for the RNR motif in the kinetic mechanism has yet been identified.

Mutations to the P protein have thus far not been extensively investigated. A limited number of residues, primarily in the RNR motif and central cleft (F22A, R62A, K66A, R67A, F73A, and W109A in *E. coli*) have weak to moderate effects on RNase P activity in vivo (Gopalan et al. 1997). In vitro studies of the contributions of these residues to catalysis are very limited. In general, single P protein mutants that have been analyzed alter the cleavage rate constant catalyzed by RNase P by less than ~two-fold at low pH and saturating magnesium concentrations (Gopalan et al. 1997; Niranjanakumari et al. 2007). These results are consistent with the modest enhancement of the cleavage rate constant catalyzed by the RNase P holoenzyme compared to P RNA (k_2 in Table 6.1). However, it is likely that mutations in the P protein could have larger effects on the affinity of pre-tRNA and/or important metals. Systematic analysis of the functional properties of RNase P reconstituted with P protein mutants is needed to fully understand how individual residues in the protein subunit contribute to the kinetic mechanism of RNase P.

6.5 Metal-Ion Association with Bacterial RNase P Holoenzyme

Divalent metal ions play a number of significant roles in RNase P function, ranging from nonspecific charge neutralization to enhancement of pre-tRNA binding and catalysis (Guerrier-Takada et al. 1986; Smith et al. 1992; Beebe et al. 1996; Brannvall and Kirsebom 1999; Kurz and Fierke 2002; Misra et al. 2003). A brief

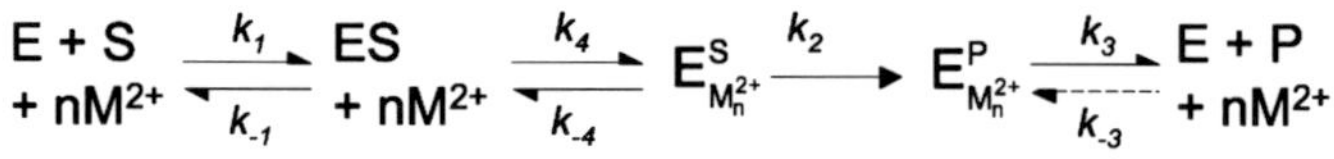

Scheme 6.3 Minimal kinetic mechanism of RNase P holoenzyme including the function of metal ions that enhance pre-tRNA affinity

overview of the role of divalent metal ions in the RNase P kinetic mechanism is given here. An expanded kinetic mechanism (Scheme 6.3) is also proposed which incorporates specific metal ion binding into the minimal mechanism displayed in Scheme 6.2.

Although monovalent cations are sufficient to support tight binding between bacterial RNase P RNA and tRNA (Smith et al. 1992; Kurz and Fierke 2002), pre-tRNA and tRNA affinity is further enhanced by addition of Mg(II) (Beebe et al. 1996). Both monovalent and divalent cations decrease the dissociation rate constant for the *B. subtilis* P RNA•tRNA^{Asp} complex (k_{-1}), suggesting that both nonspecific electrostatic shielding of RNA and specific Mg(II) binding sites stabilize the P RNA•tRNA interaction (Beebe et al. 1996). In addition, the concentration of Mg(II) required to activate catalysis is much lower in the holoenzyme than P RNA, such that the RNase P holoenzyme can be fully active at the physiological Mg(II) concentrations (Guerrier-Takada et al. 1983; Gardiner et al. 1985; Reich et al. 1988; Kurz and Fierke 2002). The number of divalent ions required for catalysis has been evaluated using several different approaches. Measurement of the number of divalent cations that interact with P RNA or RNase P holoenzyme indicates that ~150 divalent cations interact nonspecifically with both of these enzymes (Smith and Pace 1993; Beebe et al. 1996; Kurz and Fierke 2002). Hill analysis of activation of RNase P by divalent cations in the presence of cobalt hexamine, a exchange-inert mimic of hydrated magnesium, reveals that at least one class of metals that can form an inner sphere complex, such as Mg(II), Zn(II), or Mn(II), is required for pre-tRNA cleavage catalyzed by *B. subtilis* RNase P holoenzyme (Kurz and Fierke 2002). Additionally, Cd(II)-dependent rescue of cleavage catalyzed by thiophosphate-modified pre-tRNA suggests that optimal activity requires two different Cd(II) ions to bind to RNase P (Sun and Harris 2007). These studies also suggest that the binding of these catalytically pertinent (cocatalytic) divalent ions occurs concurrently with or subsequent to the formation of the RNase P•pre-tRNA "encounter" complex (ES) and before the cleavage step (k_2; Scheme 6.3). These observations are consistent with previous investigations demonstrating that Mg^{2+} enhances pre-tRNA and tRNA binding affinity for P RNA and RNase P holoenzyme due to the formation of metal-ion bound enzyme–substrate complexes (Beebe et al. 1996; Kurz and Fierke 2002).

Several potential binding sites for catalytic and/or cocatalytic metal ions have been proposed, including the P RNA helix P4, the nonbridging phosphodiester oxygen at the cleavage site, and the 2′ OH at the N(−1) nucleotide of the substrate (Hardt et al. 1995; Warnecke et al. 1999; Zahler et al. 2005). The preponderance of data suggests that sites in P4 and J18/2 are likely candidates in RNase P for binding the metal ion(s) required for catalysis; however, a detailed structure of these metal binding sites has not yet been completed. Moreover, the identification of other potential metal binding sites, and the enhancement of substrate binding and P RNA folding by divalent metal ions indicates that metal ions have multiple important functions in RNase P.

6.6 Isomerization in the Kinetic Mechanism of RNase P

There is an increasing body of kinetic evidence suggesting that a conformational change occurs after the formation of the initial RNase P-substrate complex and prior to 5′ leader cleavage. Pan and colleagues demonstrated that interactions between the *B. subtilis* P RNA and the substrate distal from the pre-tRNA cleavage site may affect the catalytic rate constant. This observation led to the proposal that an isomer of the enzyme–substrate complex positions nucleotide functional groups and Mg(II) ions at the pre-tRNA cleavage site (Loria and Pan 1998). Similar results were observed for *E. coli* RNase P (Brannvall et al. 2007). Furthermore, pre-tRNA binding to *E. coli* P RNA results in an unwinding of the acceptor stem of the bound tRNA, suggesting a conformational change of the complex consistent with that proposed by Pan and coworkers (Pomeranz Krummel et al. 2000). The contribution of the P protein to pre-tRNA and tRNA binding also suggests that a conformational change in the enzyme–substrate complex is enhanced by the P protein (Sun and Harris 2007). Recent transient kinetic studies by Hsieh have definitively demonstrated that a conformational change occurs after substrate binding (Hsieh and Fierke 2009).

Evidence for a conformational change prior to catalysis has also been found for yeast RNase P. In a recent study of yeast nuclear RNase P, curved Arrhenius plots for the steady-state kinetic parameters k_{cat} and (k_{cat}/K_M) were observed (Hsieh et al. 2009). These plots indicate large apparent Arrhenius activation energies (E_a) for both of these steady-state parameters at lower temperatures (~60 kcal/mol); this large activation energy disappears at high temperatures. The E_a for phosphodiester bond cleavage and diffusion-controlled substrate binding is much smaller (~13–20 kcal/mol) than the measured values (Uhlenbeck 1987) suggesting that the rate-limiting step for yeast RNase P includes conformational rearrangement. Together with the bacterial data, these results suggest at least one kinetically important intermediate between the initial collision complex (ES) and the cleavage step to form the product (EP). However, this conformational change in the kinetic pathway has not yet been directly observed but can only be indirectly inferred from the data analysis.

Together, the above observations suggest a simple model for the two proposed conformational states for the enzyme–substrate complex (Scheme 6.4): a low-affinity "open" complex forming as the result of the initial collision between the enzyme and the substrate, and a high affinity "closed" or "docked" conformation "ES*" state. A similar two-step binding mechanism has also been suggested for other ribozymes, including the hepatitis delta virus (HDV) ribozyme (Harris et al. 2002) and the *Tetrahymena* Group I intron (Hougland et al. 2006). Conformational change is a

$$E + S \underset{k_{-1}}{\overset{k_1}{\rightleftharpoons}} ES \underset{k_{-5}}{\overset{k_5}{\rightleftharpoons}} ES^* \xrightarrow{k_2} EP \underset{k_{-3}}{\overset{k_3}{\rightleftharpoons}} E + P$$

Scheme 6.4 Proposed two step binding mechanism for RNase P

common strategy employed by enzymes to increase the substrate affinity and to rearrange the active site for optimal catalysis (Narlikar and Herschlag 1997). Whether RNase P utilizes the same strategy remains an important question in this field.

6.7 Conclusions and Further Questions

Our understanding of the kinetic mechanism that RNase P uses to catalyze the cleavage of pre-tRNA has grown considerably in the past 15 years. Roles for the protein and metal components in the mechanism of RNase P have been assigned, leading to a better understanding of how this enzyme functions as a complex, not just a catalytic RNA. Furthermore, we have begun to unlock the mysteries of eukaryotic RNase P function and to observe that although the enzyme's structure has evolved, its mechanism has remained fundamentally the same.

Although a minimal kinetic mechanism with substrate binding, bond cleavage, and product release steps will suffice to explain most of the kinetic behavior of the RNase P (Scheme 6.2), there are still many questions regarding the RNase P mechanism. In particular, the location of the catalytically pertinent metal ions bound to the enzyme–substrate complex, and the details of the conformational change that occurs during catalysis (as proposed in Scheme 6.4) need to be addressed. Also, a mechanism that includes the contributions of all three components of RNase P (protein, metal and ribozyme) to catalysis and molecular recognitions is needed to fully comprehend the function of this picture. To answer these questions, new experimental approaches will be required to detect the transient (short-lived) kinetic intermediates.

Acknowledgments We would like to thank Nathan Zahler, Terry Watt, and James Hougland for their helpful discussions in the preparation of this manuscript. This project is supported National Institutes of Health (GM 55387 (CAF) and T32 GM08353 (KSK)).

References

Alifano P, Rivellini F, Piscitelli C, Arraiano CM, Bruni CB, Carlomagno MS (1994) Ribonuclease E provides substrates for ribonuclease P-dependent processing of a polycistronic mRNA. Genes Dev 8(24):3021–3031

Altman S (1989) Ribonuclease P: an enzyme with a catalytic RNA subunit. Adv Enzymol Relat Areas Mol Biol 62:1–36

Altman S, Wesolowski D, Guerrier-Takada C, Li Y (2005) RNase P cleaves transient structures in some riboswitches. Proc Natl Acad Sci U S A 102(32):11284–11289

Beebe JA, Fierke CA (1994) A kinetic mechanism for cleavage of precursor tRNA(Asp) catalyzed by the RNA component of *Bacillus subtilis* ribonuclease P. Biochemistry 33(34):10294–10304

Beebe JA, Kurz JC, Fierke CA (1996) Magnesium ions are required by *Bacillus subtilis* ribonuclease P RNA for both binding and cleaving precursor tRNAAsp. Biochemistry 35(32):10493–10505

Bothwell AL, Stark BC, Altman S (1976) Ribonuclease P substrate specificity: cleavage of a bacteriophage phi80-induced RNA. Proc Natl Acad Sci U S A 73(6):1912–1916

Brannvall M, Kirsebom LA (1999) Manganese ions induce miscleavage in the *Escherichia coli* RNase P RNA-catalyzed reaction. J Mol Biol 292(1):53–63

Brannvall M, Mattsson JG, Svard SG, Kirsebom LA (1998) RNase P RNA structure and cleavage reflect the primary structure of tRNA genes. J Mol Biol 283(4):771–783

Brannvall M, Pettersson BM, Kirsebom LA (2003) Importance of the +73/294 interaction in *Escherichia coli* RNase P RNA substrate complexes for cleavage and metal ion coordination. J Mol Biol 325(4):697–709

Brannvall M, Kikovska E, Wu S, Kirsebom LA (2007) Evidence for induced fit in bacterial RNase P RNA-mediated cleavage. J Mol Biol 372(5):1149–1164

Buck AH, Dalby AB, Poole AW, Kazantsev AV, Pace NR (2005) Protein activation of a ribozyme: the role of bacterial RNase P protein. EMBO J 24(19):3360–3368

Cassano AG, Anderson VE, Harris ME (2002) Evidence for direct attack by hydroxide in phosphodiester hydrolysis. J Am Chem Soc 124(37):10964–10965

Cassano AG, Anderson VE, Harris ME (2004a) Analysis of solvent nucleophile isotope effects: evidence for concerted mechanisms and nucleophilic activation by metal coordination in nonenzymatic and ribozyme-catalyzed phosphodiester hydrolysis. Biochemistry 43(32): 10547–10559

Cassano AG, Anderson VE, Harris ME (2004b) Understanding the transition states of phosphodiester bond cleavage: insights from heavy atom isotope effects. Biopolymers 73(1):110–129

Christian EL, Harris ME (1999) The track of the pre-tRNA 5′ leader in the ribonuclease P ribozyme–substrate complex. Biochemistry 38(39):12629–12638

Christian EL, McPheeters DS, Harris ME (1998) Identification of individual nucleotides in the bacterial ribonuclease P ribozyme adjacent to the pre-tRNA cleavage site by short-range photo-cross-linking. Biochemistry 37(50):17618–17628

Christian EL, Kaye NM, Harris ME (2000) Helix P4 is a divalent metal ion binding site in the conserved core of the ribonuclease P ribozyme. RNA 6(4):511–519

Christian EL, Smith KM, Perera N, Harris ME (2006) The P4 metal binding site in RNase P RNA affects active site metal affinity through substrate positioning. RNA 12(8):1463–1467

Crary SM, Niranjanakumari S, Fierke CA (1998) The protein component of *Bacillus subtilis* ribonuclease P increases catalytic efficiency by enhancing interactions with the 5′ leader sequence of pre-tRNAAsp. Biochemistry 37(26):9409–9416

Crary SM, Kurz JC, Fierke CA (2002) Specific phosphorothioate substitutions probe the active site of *Bacillus subtilis* ribonuclease P. RNA 8(7):933–947

DeRose VJ (2002) Two decades of RNA catalysis. Chem Biol 9(9):961–969

Doudna JA, Cech TR (2002) The chemical repertoire of natural ribozymes. Nature 418(6894): 222–228

Fersht A (1985) Enzyme structure and mechanism. W.H. Freeman, New York

Fierke CA, Hammes GG (1995) Transient kinetic approaches to enzyme mechanisms. Methods Enzymol 249:3–37

Forster AC, Altman S (1990) External guide sequences for an RNA enzyme. Science 249(4970): 783–786

Frank DN, Pace NR (1997) In vitro selection for altered divalent metal specificity in the RNase P RNA. Proc Natl Acad Sci U S A 94(26):14355–14360

Frank DN, Pace NR (1998) Ribonuclease P: unity and diversity in a tRNA processing ribozyme. Annu Rev Biochem 67:153–180

Gardiner KJ, Marsh TL, Pace NR (1985) Ion dependence of the *Bacillus subtilis* RNase P reaction. J Biol Chem 260(9):5415–5419

Getz MM, Andrews AJ, Fierke CA, Al-Hashimi HM (2007) Structural plasticity and Mg2+ binding properties of RNase P P4 from combined analysis of NMR residual dipolar couplings and motionally decoupled spin relaxation. RNA 13(2):251–266

Gimple O, Schon A (2001) In vitro and in vivo processing of cyanelle tmRNA by RNase P. Biol Chem 382(10):1421–1429

Gopalan V, Baxevanis AD, Landsman D, Altman S (1997) Analysis of the functional role of conserved residues in the protein subunit of ribonuclease P from *Escherichia coli*. J Mol Biol 267(4):818–829

Gossringer M, Kretschmer-Kazemi Far R, Hartmann RK (2006) Analysis of RNase P protein (rnpA) expression in *Bacillus subtilis* utilizing strains with suppressible rnpA expression. J Bacteriol 188(19):6816–6823

Guerrier-Takada C, Gardiner K, Marsh T, Pace N, Altman S (1983) The RNA moiety of ribonuclease P is the catalytic subunit of the enzyme. Cell 35(3 Pt 2):849–857

Guerrier-Takada C, Haydock K, Allen L, Altman S (1986) Metal ion requirements and other aspects of the reaction catalyzed by M1 RNA, the RNA subunit of ribonuclease P from *Escherichia coli*. Biochemistry 25(7):1509–1515

Hanna R, Doudna JA (2000) Metal ions in ribozyme folding and catalysis. Curr Opin Chem Biol 4(2):166–170

Hardt WD, Warnecke JM, Erdmann VA, Hartmann RK (1995) Rp-phosphorothioate modifications in RNase P RNA that interfere with tRNA binding. EMBO J 14(12):2935–2944

Harris ME, Christian EL (2003) Recent insights into the structure and function of the ribonucleoprotein enzyme ribonuclease P. Curr Opin Struct Biol 13(3):325–333

Harris ME, Pace NR (1995) Identification of phosphates involved in catalysis by the ribozyme RNase P RNA. RNA 1(2):210–218

Harris DA, Rueda D, Walter NG (2002) Local conformational changes in the catalytic core of the trans-acting hepatitis delta virus ribozyme accompany catalysis. Biochemistry 41(40):12051–12061

Hartmann RK, Heinrich J, Schlegl J, Schuster H (1995) Precursor of C4 antisense RNA of bacteriophages P1 and P7 is a substrate for RNase P of *Escherichia coli*. Proc Natl Acad Sci U S A 92(13):5822–5826

Heide C, Pfeiffer T, Nolan JM, Hartmann RK (1999) Guanosine 2-NH2 groups of *Escherichia coli* RNase P RNA involved in intramolecular tertiary contacts and direct interactions with tRNA. RNA 5(1):102–116

Heide C, Busch S, Feltens R, Hartmann RK (2001a) Distinct modes of mature and precursor tRNA binding to *Escherichia coli* RNase P RNA revealed by NAIM analyses. RNA 7(4):553–564

Heide C, Feltens R, Hartmann RK (2001b) Purine N7 groups that are crucial to the interaction of *Escherichia coli* rnase P RNA with tRNA. RNA 7(7):958–968

Hougland JL, Piccirilli JA, Forconi M, Lee J, Herschlag D (2006) How the group I intron works: a case study of RNA structure and function. In: Gesteland RF, Cech T, Atkins JF (eds) The RNA world. Cold Spring Harbor Laboratory Press, Cold Spring Harbor, NY

Hsieh J, Fierke CA (2009). Conformational change in the B. subtilis RNase P-pre-tRNA complex enhances substrate affinity and limits cleavage rate. RNA 15(8):1565–1577

Hsieh J, Andrews AJ, Fierke CA (2004) Roles of protein subunits in RNA–protein complexes: lessons from ribonuclease P. Biopolymers 73(1):79–89

Hsieh J, Walker S, Fierke CA, Engelke DR (2009) Pre-tRNA cleavage by the yeast nuclear RNase P holoenzyme is rate-limited by slow product release. RNA 15(2):224–234

Jovanovic M, Sanchez R, Altman S, Gopalan V (2002) Elucidation of structure–function relationships in the protein subunit of bacterial RNase P using a genetic complementation approach. Nucleic Acids Res 30(23):5065–5073

Kaye NM, Christian EL, Harris ME (2002) NAIM and site-specific functional group modification analysis of RNase P RNA: magnesium dependent structure within the conserved P1–P4 multihelix junction contributes to catalysis. Biochemistry 41(14):4533–4545

Kazantsev AV, Krivenko AA, Harrington DJ, Carter RJ, Holbrook SR, Adams PD, Pace NR (2003) High-resolution structure of RNase P protein from *Thermotoga maritima*. Proc Natl Acad Sci U S A 100(13):7497–7502
Kazantsev AV, Krivenko AA, Harrington DJ, Holbrook SR, Adams PD, Pace NR (2005) Crystal structure of a bacterial ribonuclease P RNA. Proc Natl Acad Sci U S A 102(38): 13392–13397
Kirsebom LA, Svard SG (1994) Base pairing between *Escherichia coli* RNase P RNA and its substrate. EMBO J 13(20):4870–4876
Komine Y, Kitabatake M, Yokogawa T, Nishikawa K, Inokuchi H (1994) A tRNA-like structure is present in 10Sa RNA, a small stable RNA from *Escherichia coli*. Proc Natl Acad Sci U S A 91(20):9223–9227
Kurz JC, Fierke CA (2000) Ribonuclease P: a ribonucleoprotein enzyme. Curr Opin Chem Biol 4(5):553–558
Kurz JC, Fierke CA (2002) The affinity of magnesium binding sites in the *Bacillus subtilis* RNase P x pre-tRNA complex is enhanced by the protein subunit. Biochemistry 41(30):9545–9558
Kurz JC, Niranjanakumari S, Fierke CA (1998) Protein component of *Bacillus subtilis* RNase P specifically enhances the affinity for precursor-tRNAAsp. Biochemistry 37(8):2393–2400
LaGrandeur TE, Darr SC, Haas ES, Pace NR (1993) Characterization of the RNase P RNA of *Sulfolobus acidocaldarius*. J Bacteriol 175(16):5043–5048
Loria A, Pan T (1996) Domain structure of the ribozyme from eubacterial ribonuclease P. RNA 2(6):551–563
Loria A, Pan T (1998) Recognition of the 5′ leader and the acceptor stem of a pre-tRNA substrate by the ribozyme from *Bacillus subtilis* RNase P. Biochemistry 37(28):10126–10133
Loria A, Pan T (2001) Modular construction for function of a ribonucleoprotein enzyme: the catalytic domain of *Bacillus subtilis* RNase P complexed with *B subtilis* RNase P protein. Nucleic Acids Res 29(9):1892–1897
Loria A, Niranjanakumari S, Fierke CA, Pan T (1998) Recognition of a pre-tRNA substrate by the *Bacillus subtilis* RNase P holoenzyme. Biochemistry 37(44):15466–15473
Marshall GR, Feng JA, Kuster DJ (2008) Back to the future: ribonuclease A. Biopolymers 90(3):259–277
Misra VK, Shiman R, Draper DE (2003) A thermodynamic framework for the magnesium-dependent folding of RNA. Biopolymers 69(1):118–136
Narlikar GJ, Herschlag D (1997) Mechanistic aspects of enzymatic catalysis: lessons from comparison of RNA and protein enzymes. Annu Rev Biochem 66:19–59
Nieuwlandt DT, Haas ES, Daniels CJ (1991) The RNA component of RNase P from the archaebacterium *Haloferax volcanii*. J Biol Chem 266(9):5689–5695
Niranjanakumari S, Stams T, Crary SM, Christianson DW, Fierke CA (1998) Protein component of the ribozyme ribonuclease P alters substrate recognition by directly contacting precursor tRNA. Proc Natl Acad Sci U S A 95(26):15212–15217
Niranjanakumari S, Day-Storms JJ, Ahmed M, Hsieh J, Zahler NH, Venters RA, Fierke CA (2007) Probing the architecture of the *B. subtilis* RNase P Holoenzyme active site by crosslinking and affinity cleavage. RNA 13:512–535
Pace NR, Brown JW (1995) Evolutionary perspective on the structure and function of ribonuclease P, a ribozyme. J Bacteriol 177(8):1919–1928
Pace NR, Smith D (1990) Ribonuclease P: function and variation. J Biol Chem 265(7):3587–3590
Pan T (1995) Higher order folding and domain analysis of the ribozyme from *Bacillus subtilis* ribonuclease P. Biochemistry 34(3):902–909
Peck-Miller KA, Altman S (1991) Kinetics of the processing of the precursor to 4.5 S RNA, a naturally occurring substrate for RNase P from *Escherichia coli*. J Mol Biol 221(1):1–5
Pomeranz Krummel DA, Kent O, MacMillan AM, Altman S (2000) Evidence for helical unwinding of an RNA substrate by the RNA enzyme RNase P: use of an interstrand disulfide crosslink in substrate. J Mol Biol 295(5):1113–1118
Radzicka A, Wolfenden R (1995) A proficient enzyme. Science 267(5194):90–93

Rasmussen TA, Nolan JM (2002) G350 of *Escherichia coli* RNase P RNA contributes to Mg2+ binding near the active site of the enzyme. Gene 294(1–2):177–185
Reich C, Olsen GJ, Pace B, Pace NR (1988) Role of the protein moiety of ribonuclease P, a ribonucleoprotein enzyme. Science 239(4836):178–181
Rox C, Feltens R, Pfeiffer T, Hartmann RK (2002) Potential contact sites between the protein and RNA subunit in the *Bacillus subtilis* RNase P holoenzyme. J Mol Biol 315(4):551–560
Rueda D, Hsieh J, Day-Storms JJ, Fierke CA, Walter NG (2005) The 5′ leader of precursor tRNA(Asp) bound to the *Bacillus subtilis* RNase P holoenzyme has an extended conformation. Biochemistry 44(49):16130–16139
Schmitz M, Tinoco I Jr (2000) Solution structure and metal-ion binding of the P4 element from bacterial RNase P RNA. RNA 6(9):1212–1225
Seif E, Altman S (2008) RNase P cleaves the adenine riboswitch and stabilizes pbuE mRNA in *Bacillus subtilis*. RNA 14(6):1237–1243
Serpersu EH, Shortle D, ASM (1986) Kinetic and magnetic resonance studies of effects of genetic substitution of a Ca2+-liganding amino acid in staphylococcal nuclease. Biochemistry 25(1): 68–77
Shan S, Kravchuk AV, Piccirilli JA, Herschlag D (2001) Defining the catalytic metal ion interactions in the Tetrahymena ribozyme reaction. Biochemistry 40(17):5161–5171
Smith D, Pace NR (1993) Multiple magnesium ions in the ribonuclease P reaction mechanism. Biochemistry 32(20):5273–5281
Smith D, Burgin AB, Haas ES, Pace NR (1992) Influence of metal ions on the ribonuclease P reaction. Distinguishing substrate binding from catalysis. J Biol Chem 267(4):2429–2436
Spitzfaden C, Nicholson N, Jones JJ, Guth S, Lehr R, Prescott CD, Hegg LA, Eggleston DS (2000) The structure of ribonuclease P protein from *Staphylococcus aureus* reveals a unique binding site for single-stranded RNA. J Mol Biol 295(1):105–115
Stams T, Niranjanakumari S, Fierke CA, Christianson DW (1998) Ribonuclease P protein structure: evolutionary origins in the translational apparatus. Science 280(5364):752–755
Sun L, Harris ME (2007) Evidence that binding of C5 protein to P RNA enhances ribozyme catalysis by influencing active site metal ion affinity. RNA 13(9):1505–1515
Sun L, Campbell FE, Zahler NH, Harris ME (2006) Evidence that substrate-specific effects of C5 protein lead to uniformity in binding and catalysis by RNase P. EMBO J 25(17):3998–4007
Svard SG, Kagardt U, Kirsebom LA (1996) Phylogenetic comparative mutational analysis of the base-pairing between RNase P RNA and its substrate. RNA 2(5):463–472
Tallsjo A, Kirsebom LA (1993) Product release is a rate-limiting step during cleavage by the catalytic RNA subunit of *Escherichia coli* RNase P. Nucleic Acids Res 21(1):51–57
Torres-Larios A, Swinger KK, Krasilnikov AS, Pan T, Mondragon A (2005) Crystal structure of the RNA component of bacterial ribonuclease P. Nature 437(7058):584–587
Tranguch AJ, Engelke DR (1993) Comparative structural analysis of nuclear RNase P RNAs from yeast. J Biol Chem 268(19):14045–14055
Uhlenbeck OC (1987) A small catalytic oligoribonucleotide. Nature 328(6131):596–600
Walker SC, Engelke DR (2006) Ribonuclease P: the evolution of an ancient RNA enzyme. Crit Rev Biochem Mol Biol 41(2):77–102
Warnecke JM, Held R, Busch S, Hartmann RK (1999) Role of metal ions in the hydrolysis reaction catalyzed by RNase P RNA from *Bacillus subtilis*. J Mol Biol 290(2):433–445
Waugh DS, Pace NR (1993) Gap-scan deletion analysis of *Bacillus subtilis* RNase P RNA. FASEB J 7(1):188–195
Westheimer FH (1987) Why nature chose phosphates. Science 235(4793):1173–1178
Wilson RC, Bohlen CJ, Foster MP, Bell CE (2006) Structure of Pfu Pop5, an archaeal RNase P protein. Proc Natl Acad Sci U S A 103(4):873–878
Xiao S, Day-Storms JJ, Srisawat C, Fierke CA, Engelke DR (2005) Characterization of conserved sequence elements in eukaryotic RNase P RNA reveals roles in holoenzyme assembly and tRNA processing. Rna 11(6):885–896
Zahler NH, Christian EL, Harris ME (2003) Recognition of the 5′ leader of pre-tRNA substrates by the active site of ribonuclease P. RNA 9(6):734–745

Zahler NH, Sun L, Christian EL, Harris ME (2005) The pre-tRNA nucleotide base and 2′-hydroxyl at N(-1) contribute to fidelity in tRNA processing by RNase P. J Mol Biol 345(5):969–985
Zahler NH, Koutmou KS, Kurz JC, Campbell FE, Harris ME, Fierke CA (2009) Protein-substrate contact leads to recognition of 5’ leaders by bacterial RNase P. (submitted)
Zaug AJ, Grosshans CA, Cech TR (1988) Sequence-specific endoribonuclease activity of the Tetrahymena ribozyme: enhanced cleavage of certain oligonucleotide substrates that form mismatched ribozyme–substrate complexes. Biochemistry 27(25):8924–8931
Ziehler WA, Day JJ, Fierke CA, Engelke DR (2000) Effects of 5′ leader and 3′ trailer structures on pre-tRNA processing by nuclear RNase P. Biochemistry 39(32):9909–9916

Chapter 7
Roles of Metal Ions in RNase P Catalysis

Leif A. Kirsebom

7.1 Introduction

The RNA processing enzyme RNase P is an endoribonuclease involved in the processing of tRNA precursors (Robertson et al. 1972). The catalytic activity is associated with the RNA moiety of RNase P and it has conclusively been demonstrated that the bacterial and some archaeal and eukaryal RNase P RNAs alone promote cleavage of a number of different substrates in vitro (Guerrier-Takada et al. 1983; Pannucci et al. 1999; Kikovska et al. 2007). However, in vivo the presence of the RNase P protein(s) are essential and dependent on origin the number of proteins vary: one in bacteria, at least four in archaea and 9–10 in eukaryotes (Gopalan and Altman 2006; Walker and Engelke 2006; Kirsebom 2007).

As for other RNAs metal(II)-ions such as Mg^{2+} play crucial roles for the function of RNase P and its catalytic RNA subunit. On average, there is one Mg^{2+}-ion bound per 3–4 nucleotides. In general, binding of metal(II)-ions affects the folding, RNA–RNA interactions, RNA–protein interactions and various catalytic processes such as cleavage of RNA, tRNA charging and codon–anticodon interaction (Gesteland et al. 2006). Other biological relevant metal(II)-ions, such as Ca^{2+} and Mn^{2+}, also bind to RNA and influence its function. Metal(II)-ions bound to an RNA can result in hydrolysis of the RNA giving products with 5′-hydroxyls and 2′;3′-cyclic phosphates while others produce 5′-phosphates and 3′-hydoroxyls at the ends of the cleavage products (Fig. 7.1). In the former case, the neighboring 2′OH is activated by a metal(II)-hydroxide ion, i.e., acts as a general base, or by the metal(II)-ion which then function as a Lewis acid. By contrast, when the nucleophilic attack comes from the opposite direction relative to the 2′OH immediately 5′ of the phosphorous, then the nucleophile can be a hydroxide ion coordinated to a

L.A. Kirsebom (✉)
Department of Cell and Molecular Biology, Biomedical Center, Uppsala University, Box 596, Uppsala, Sweden
e-mail: Leif.Kirsebom@icm.uu.se

F. Liu and S. Altman (eds.), *Ribonuclease P*, Protein Reviews 10,
DOI 10.1007/978-1-4419-1142-1_7, © Springer Science+Business Media, LLC 2010

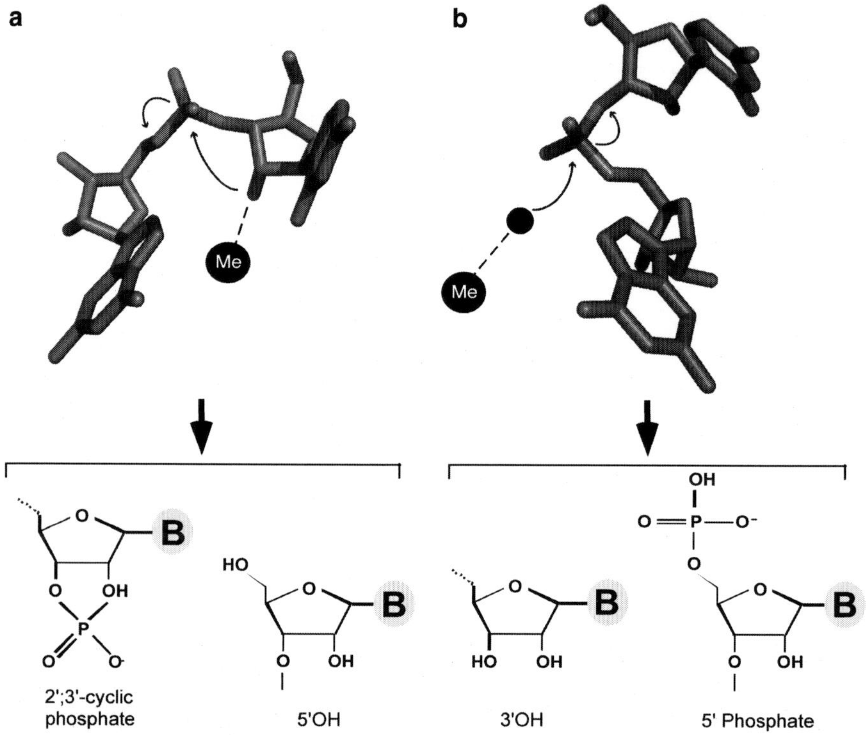

Fig. 7.1 The mechanisms of metal(II)-ion induced cleavage of RNA. In panel (**a**) the metal(II)-ion acts as a general base activating the 2′-hydroxyl on the ribose, which results in a nucleophilic attack on the neighboring phosphorous atom giving cleavage products with 5′OH and 2′;3′ cyclic phosphate at the ends. In panel (**b**), a metal(II)-ion coordinated hydroxyl ion acts as the nucleophile generating cleavage products with 5′ phosphate and 3′OH at the ends. Irrespective of mechanism the reaction proceed via a pentacoordinated transition state, S_N2 reaction. *Black sphere* marked with Me indicate metal(II)-ions while the small black sphere mark the activated H_2O nucleophile. (**b**) represents arbitrary bases

metal(II)-ion as is likely the case for RNase P which generates products with 5′-phosphates and 3′-hydroxyls at the ends. Irrespective of mechanism cleavage proceeds through a penta coordinated transition state via an S_N2 cleavage mechanism (see e.g., Steitz and Steitz 1993). Importantly, for RNase P the strategy has to be to prevent the metal(II)-ion induced cleavage and favor an attack on the phosphorous from the opposite side in relation to the 2′OH to ensure the products with correct ends. In this Chap. I will discuss our current knowledge of functionally important Mg^{2+}-ions in RNase P mediated cleavage of RNA. The emphasis will be on the RNA component of RNase P, cleavage site recognition and the catalytic center. However, first I will discuss some general aspects of how various metal(II)-ions have been instrumental to understand the RNase P mediated cleavage.

7.2 Identification of Metal(II)-Ion Binding Sites

7.2.1 Metal(II)-Binding to RNase P RNA

Metal(II)-ions can bind to RNA in two different modes, diffuse or nonspecific binding and site or specific binding. The former mode refers to nonspecific long-range electrostatic interactions between fully hexahydrated metal(II)-ions and the RNA. In the second mode dehydrated metal(II)-ions coordinate to specific ligands that is dictated by the folding of the RNA (Misra and Draper 1998). Considering that approximately 100 Mg^{2+} bind to the ≈400 residues long RNase P RNA (Beebe et al. 1996) an important question is to identify the Mg^{2+}-ions (or metal(II)-ion binding sites) that are important for function: (a) folding, (b) substrate interaction, (c) protein interaction and (d) chemistry of cleavage.

Certain metal(II)-ions induce degradation of RNA in a nonoxidative manner, and in some RNA molecules this process is exceptionally efficient and specific. The best-known example, yeast $tRNA^{Phe}$, undergoes specific fragmentation in the D-loop in the presence of Pb^{2+} (Werner et al. 1976; Sampson et al. 1987; Krzyzosiak et al. 1988) and Mg^{2+} (Wintermeyer and Zachau 1973; Marciniec et al. 1989). This is consistent with that Pb^{2+} and Mg^{2+} bind to an overlapping site 6–7 Å from the 2′OH of U16 in the D-loop that acts as the nucleophile (Brown et al. 1985; Jovine et al. 2000; Shi and Moore 2000). Hence, Pb^{2+}- (and Mg^{2+}) induced cleavage of RNA can provide information about metal(II)-binding to RNA. Subjecting bacterial RNase P RNA to metal(II)-ion induced cleavage results in specific cleavage patterns. Comparing the Mg^{2+}- and Pb^{2+}-induced cleavage patterns suggest that these two metal(II)-ions bind if not to the same sites at least to overlapping sites (Kazakov and Altman 1991; Ciesiolka et al. 1994; Tallsjö et al. 1993; Zito et al. 1993). This is in keeping with that an increasing concentration of Mg^{2+} suppresses Pb^{2+}-induced cleavage (Ciesiolka et al. 1994; Brännvall et al. 2001). As is shown in Fig. 7.2a, cleavage occurs at distinct positions referred to as sites Ia–V. In addition, week Pb^{2+}-induced cleavage has also been detected at positions 302-303 (site VI) and 326–335 (Hardt and Hartmann 1996; Brännvall et al. 2001; Fig. 7.2a). That site IIb is in close proximity to a metal(II)-ion is in keeping with the phosphorothioate modification interference studies which suggest that the Rp-oxygen of A130 and A132 coordinates to Mg^{2+} (Hardt et al. 1995). Importantly, not only increasing concentration of Mg^{2+} suppress Pb^{2+}-induced cleavage but also other metal(II)-ions such as Ca^{2+}, Mn^{2+} and Sr^{2+}, where the two former also promote the RNase P RNA mediated cleavage when present alone, while Sr^{2+} does not, at least not at neutral pH (Brännvall and Kirsebom 2001; Guerrier-Takada et al. 1986; Smith et al. 1992; Smith and Pace 1993; Kazakov and Altman 1991). Moreover, with respect to function, there are at least two categories of Mg^{2+}-ions (or sites) associated with the RNase P RNA and its substrate (see also below). One category that promotes folding of the active conformation and the other is involved in the chemistry of cleavage (Brännvall and Kirsebom 2001; Brännvall et al. 2001; Kurz and Fierke 2002). This is corroborated by data demonstrating the RNase P RNA mediated cleavage in the presence of Pb^{2+}

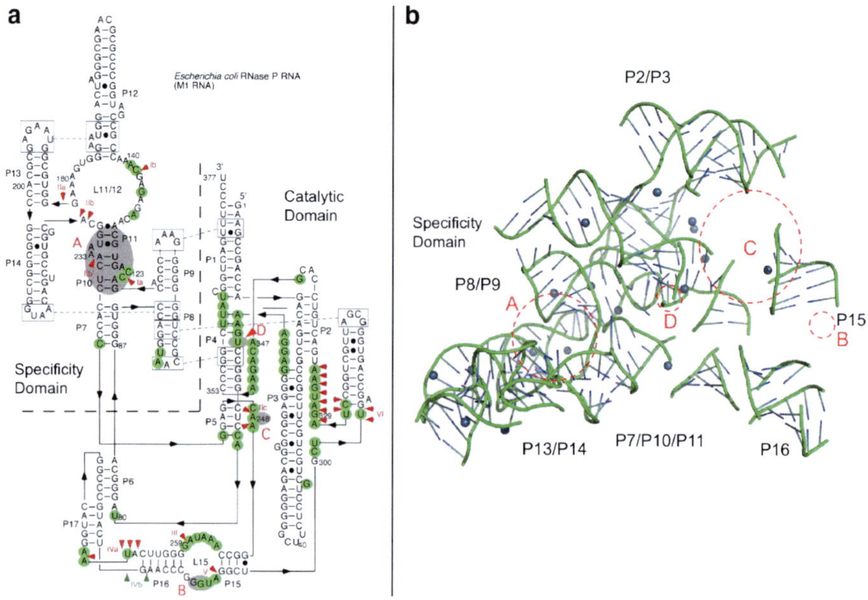

Fig. 7.2 Illustrations of the secondary structure and crystal structure of type A RNase P RNA. Panel (**a**). Secondary structure of *E. coli* RNase P RNA (M1 RNA) according to Massire et al. (1998). The *dashed line* marks the boundary between the specificity and catalytic domains while boxed residues and dashed lines indicate tertiary interactions. *Red arrow* heads (and roman numerals I–VI) mark metal(II)-ion (e.g., Mg^{2+} and Pb^{2+}; see e.g., Kazakov and Altman 1991; Ciesiolka et al. 1994; Zito et al. 1993; Tallsjö et al. 1993; Brännvall et al. 2001) induced cleavage sites and residues circled in *green* represents residues susceptible to cleavage in the presence of Tb^{3+} (Kaye et al. 2002a, b). Residues marked in *gray* indicate domains and residues (marked with (**A**)–(**D**)) that interact with or are in close contact with the substrate: (**A**) refers to the TBS-region, (**B**) the "RCCA–RNase P RNA interaction", (**C**) the "A_{248}/N_{-1} interaction" and (**D**) the "U_{69}/acceptor stem interaction". For details see the main text. Panel (**B**). The crystal structure of *Thermotoga maritima* RNase P RNA according to Torres-Larios et al. (2005) is illustrated. The *blue spheres* represent metal ions and the location of some of the helical regions (P2/P3, etc.) shown in panel (**A**) are indicated. The regions that interact or are in close contact with the substrate (**A**–**D**; see panel **A**) are marked with *red dashed circles* (see also Fig. 7.3b for regions/residues in the substrate that interact or are in close contact with RNase P RNA)

and Sr^{2+} (or $Co(NH_3)_6^{3+}$): while neither of these ions alone promote RNase P RNA mediated cleavage of the substrate with or without the protein subunit at physiological pH (Kikovska et al. 2005b; Kurz and Fierke 2002). The discussion above refers to in vitro studies on naked RNase P RNA. However, Pb^{2+}-induced cleavage studies of RNase P RNA in vivo and in the presence of C5 show that the intracellular environment and the RNase P protein has only minor effects on the cleavage pattern. This suggests that most of the metal(II)-ion binding sites are preserved in the holoenzyme complex and that the overall folding of the RNA is not significantly different from the fold it has in vitro in the presence of metal(II)-ions and without the protein (Lindell et al. 2005).

Additional metal(II)-ion binding sites in the P4 helix in RNase P RNA have been identified in nucleotide analog interference modification studies (Fig. 7.2; Hardt et al. 1995; Harris and Pace 1995; Christian et al. 2000; Kaye et al. 2002a, b; Kurz and Fierke 2002). Binding to metal(II)-ions to P4 is corroborated by Tb^{3+}-induced cleavage of RNase P RNA and by structural studies of a small model RNA representing the P4 helix by nuclear magnetic resonance spectroscopy (Schmitz and Tinoco 2000; Kaye et al. 2002a, b, Schmitz 2004; Getz et al. 2007). Studies demonstrating that substitutions and deletion of residues in the P4 helix results in increased activity in the presence of Ca^{2+} are also in keeping with metal(II)-binding in P4 (Frank and Pace 1997; Kaye et al. 2002a, b). Of notice is that the metal(II)-ion binding to the P4 helix does not appear to result in any significant structural changes (Crary et al. 2002). Recently, several metal ion binding sites have been mapped in the crystal structure of the catalytic domain of RNase P RNA (Kazantsev et al. 2009) however, these sites will not be further discussed in this chapter.

7.2.2 *Metal(II)-Ion Binding to the Substrate*

With respect to the substrate, the high-resolution crystal structures have identified ten specific Mg^{2+} binding sites in tRNA (Jovine et al. 2000; Shi and Moore 2000; Fig. 7.3b). In a tRNA precursor context, some of the metal(II)-ions bound to these sites might play a role, in particular in relation to the formation and stabilization of the a functional RNase P RNA–substrate complex. We do not yet have access to a crystal structure of a precursor tRNA. But metal(II)-ion induced cleavage studies of different precursor substrates and data using a phosphorothioate substitution approach suggest that there is a specific metal(II)-ion binding site(s) in the vicinity of the cleavage site in the substrate when bound to RNase P RNA and in its absence (see e.g., Kazakov and Altman 1991; Perreault and Altman 1992; Warnecke et al. 1996; 1999; Chen et al. 1997; Brännvall et al. 2001). The presence of a metal(II)-ion at this location has been corroborated by NMR studies of small RNAs representing the cleavage site, i.e., the acceptor-stem with a short 5′ leader (Zuleeg et al. 2001a, b).

7.2.3 *Summary*

Several specific metal(II)-ion binding sites have been identified both in RNase P RNA and in its substrate using different approaches. Moreover, studies using RNase P RNA derived from the other bacteria than *Escherichia coli* suggest that many of these sites are conserved among in particular bacterial RNase P RNA (Zito et al. 1993; Svärd et al. 1994). An important task in the future will be to get information about the metal(II)-ion binding sites in both archaeal and eukaryal RNase P RNA and whether the proteins affect the metal(II)-ion binding sites in these RNase P RNAs.

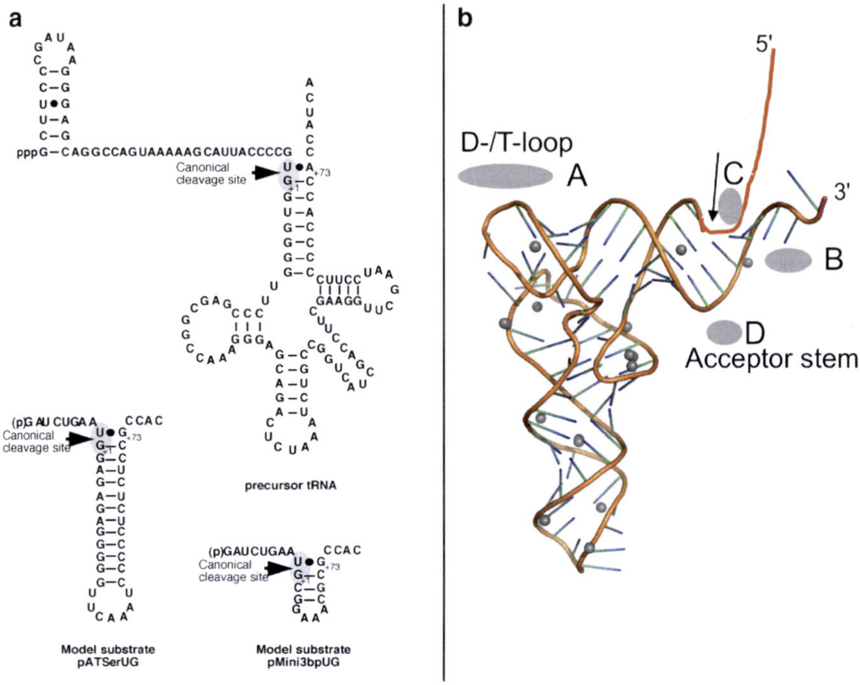

Fig. 7.3 Illustration of the structures of some RNase P substrates. Panel (**a**). The secondary structures of some commonly used substrates as indicated. The canonical cleavage sites between residues −1 and +1 (shaded in *gray*) are marked with *arrows*. The discriminator base, i.e., the residue immediately 5′ of the 3′ terminal CCA-motifs is referred to as residue +73 and the corresponding position in the two other substrates is marked with +73. Panel (**b**). A three dimensional structural representation of a model tRNA precursor using the crystal structure of yeast tRNA^{Phe} according to Jovine et al. (2000) and Shi and Moore (2000). The 5′ leader is shown in *red* and the arrow mark the canonical RNase P cleavage site. The *gray spheres* represent metal(II)-ions that have been mapped and the *gray areas* marked (**A**)–(**D**) refers to region/residues demonstrated to interact or to be in close contact with RNase P RNA (see also Fig. 7.2a): (**A**) interact with the TBS-region, (**B**) 'the RCCA–RNase P RNA interaction", (**C**) the "A_{248}/N_{-1} interaction" and (**D**) the "U_{69}/acceptor stem interaction". For details see the main text

7.3 Different Metal(II)-Ions and RNase P RNA

As for other large RNA molecules (Murphy and Cech 1993; Wang et al. 1994) RNase P RNA can be separated into different domains, the specificity (S-) and catalytic (C-) domains (Fig. 7.2; Pan 1995; see also Guerrier-Takada and Altman 1992). The folding of the two domains as well as the full-size RNase P RNA depend on binding of metal(II)-ions both to specific sites and to nonspecific sites. Folding of the active conformation of RNase P RNA is a cooperative process that is completed at 5–6 mM Mg^{2+} (Loria and Pan 1996; Zarrinkar et al. 1996; Fang et al. 1999; Kent et al. 2000). Here, Mg^{2+} is likely to coordinate both to the neighboring and to distal nonbridging oxygens as well as to the other chemical groups in the RNA. In this context, it is noted that the folding transition of RNase P RNA requires the binding of at least 2–3 additional Mg^{2+} (Loria and Pan 1996; Zarrinkar et al. 1996; Fang et al. 1999).

It was early realized that RNase P is active not only in the presence of Mg^{2+} but also in the presence of the other metal(II)-ions, e.g., Mn^{2+}, Ca^{2+} and Zn^{2+} (Gardiner et al. 1985; Guerrier-Takada et al. 1986). In line with this is that comparison of the Pb^{2+}-induced cleavage patterns in the presence of different metal(II)-ions suggest that the overall conformation of RNase P RNA is very similar in the presence of Mg^{2+}, Mn^{2+}, Ca^{2+}, Sr^{2+} or Ba^{2+} while it is different in the presence of, e.g., Cd^{2+}. Although Mg^{2+} and Ca^{2+} bind with approximately the same affinity to RNA (Brännvall et al. 2001) cleavage in the presence of Ca^{2+} as the only divalent metal ion is significantly reduced (Kazakov and Altman 1991; Smith et al. 1992; Smith and Pace 1993; Kufel and Kirsebom 1994; Warnecke et al. 1999; Brännvall and Kirsebom 2001). Sr^{2+} and Ba^{2+} on the other hand bind with less affinity to the RNA when compared with Mg^{2+} and Ca^{2+} (Brännvall et al. 2001) and do not promote cleavage when present alone at physiological pH (Brännvall and Kirsebom 2001), but do so in the presence of 10% of ethanol at pH 9.5 (Kazakov and Altman 1991). Moreover, dependent on the structural architecture at the cleavage site cleavage can occur at alternative sites. For example, substrates with a C_{-1}/G_{+73} pair result in a low frequency of cleavage 5′ of C_{-1}, i.e., between residues at −2 and −1 in the Mg^{2+} alone cleavage reaction (G_{+73} corresponds to the tRNA discriminator base; Crothers et al. 1972). This is referred to as miscleavage since correct cleavage occurs between the −1 and the +1 positions (Fig. 7.3). Compared to the Mg^{2+} alone reaction addition of Ca^{2+}, Sr^{2+} and Ba^{2+} can increase the frequency of correct cleavage of substrates with a C_{-1}/G_{+73} pair (Brännvall and Kirsebom 2001).

By contrast Mn^{2+} has a higher affinity to RNA when compared with Mg^{2+} (Brännvall et al. 2001). However, like addition of Ca^{2+}, adding Mn^{2+} to the cleavage reaction reduces the rate. Its presence can also increase the frequency of miscleavage both for reconstituted RNase P holoenzyme as well as for RNase P RNA alone (Brännvall and Kirsebom 1999; 2001; Brännvall et al. 2003). This miscleavage is as discussed above dependent on the structural architecture at and in the vicinity of the cleavage site. An increased miscleavage is also observed as a result of adding other transition metal(II)-ions (Ni^{2+}, Zn^{2+}, Cd^{2+}, Cu^{2+} or Co^{2+}) to the reaction (Brännvall and Kirsebom 2001; Warnecke et al. 1996). Here it is noted that the Pb^{2+}-induced cleavage patterns in the presence of Mg^{2+} when compared with that observed for example with Zn^{2+} is different. Hence, it appears that binding of Zn^{2+} to RNase P RNA gives an alteration in the overall conformation relative to that in the presence of Mg^{2+}. This might rationalize why Zn^{2+} results in weak substrate binding and promotes RNase P RNA mediated cleavage poorly (Cuzic and Hartmann 2005). On the other hand Cd^{2+}, which appears to result in a more substantial change in overall conformation compared to Zn^{2+}, promotes substrate binding better than Zn^{2+} while the rate is higher in the presence of Zn^{2+}. In this context, note that upon formation of the RNase P RNA–substrate complex involve binding of at least two metal(II)-ions and one that stabilize the transition state (Beebe et al. 1996).

Together these data clearly indicate that the nature of the metal(II)-ion not only affect the folding of RNase P RNA but also substrate binding and rate of cleavage in a differential manner. This is in keeping with that there are at least two categories of metal(II)-ions (or binding sites) in RNase P RNA mediated cleavage. One that promotes correct folding and facilitate/stabilize substrate binding. The other has a more direct role in the chemistry of cleavage. Hence, certain metal(II)-ions can fulfill the former role but does not have the potential to (or to a much lesser extent) participate in the chemistry of cleavage and vice versa. This is consistent with the finding that Pb^{2+} and $Co(NH_3)_6^{3+}$ when present alone do not promote cleavage but together they do promote cleavage at the correct position resulting in products with the correct ends (5′-phosphate and 3′-hydroxyl; Kikovska et al. 2005a, b).

7.4 Metal(II)-Ions and the RNase P Protein

Very little is known about the functions of the archaeal and eukaryal RNase P protein subunits. Consequently we do not have any information about how these proteins influence metal(II)-ion binding to RNase P RNA and its RNA substrate. However, studies of the bacterial RNase P protein, C5, demonstrate that addition of the protein lowers the Mg^{2+} requirement and increase the overall cleavage efficiency relative to the RNA-alone reaction (Guerrier-Takada et al. 1983; Gardiner et al. 1985; Reich et al. 1988; Tallsjö and Kirsebom 1992; with respect to archaeal RNase P see Tsai et al. 2006). Moreover, using *Bacillus subtilis* RNase P RNA data suggest that metal(II)-ions cooperatively increase the affinity between RNase P RNA and tRNA (and likely also precursor tRNA) three orders of magnitude (Beebe et al. 1996). Addition of C5, which binds to the 5′ leader of the tRNA precursor substrate (Crary et al. 1998; Loria et al. 1998; Niranjanakumari et al. 1998; see also Jovanovic et al. 2002) not only increase the binding affinity of the substrate but also in increased binding affinities of metal(II)-ions (Mg^{2+}) bound to the RNase P substrate complex. This rationalizes one way how the C5 protein facilitates catalysis (Kurz and Fierke 2002; see also Day-Storms et al. 2004. Another is that C5 prevents rebinding of the 5′ matured tRNA product. In addition, C5 broadening substrate specificity and influence the cleavage site recognition process. Here, the latter might be related to that both C5 and metal(II)-ions bind the 5′ leader of the precursor substrate.

7.5 Metal(II)-Ions, Substrate Interaction and Cleavage

Early data suggested that RNase P recognized the characteristic tRNA fold in a tRNA precursor (Chang and Smith 1973; Kirsebom 2007 and references therein). Subsequent experiments demonstrated that an RNA hairpin-loop representing the amino acid acceptor-stem, T-stem, and T-loop is sufficient to result in efficient cleavage by the RNase P RNA alone and in the presence of the C5 protein (see e.g., McClain et al. 1987; Perreault and Altman 1992, 1993; Schlegl et al. 1992; Pan and Jackaka 1996; Kufel and Kirsebom 1998). The smallest substrate that has been shown to act as a substrate to date in the RNA alone reaction in trans is an RNA hairpin loop composed of three base pairs, a GAAA-tetra loop and a 3′ CCAC tail (Brännvall et al. 2007). However, it has been shown that the bacterial RNase P holoenzyme can cleave the single stranded RNA oligonucleotides (Hansen et al. 2001). This is consistent with the earlier data demonstrating that the cleavage site can be single stranded (Svärd and Kirsebom 1992).

Biochemical and genetic studies have provided evidence that the tRNA D-/T-loop region of a tRNA precursor, the 3′ CCA tail including the discriminator base at position +73, the residue at the position immediately 5′ of the cleavage site and the acceptor stem five base pair downstream of the cleavage site interact with RNase P RNA. These interactions have been referred to as the TSL-/TBS interaction, the RCCA–RNase P RNA interaction (interacting residues underlined), the A_{248}/N_{-1} interaction and the U_{69}-acceptor stem interaction, respectively (Kirsebom 2007). For all these interactions, the data are consistent with that metal(II)-ions play a role and that changes of residues at and in the vicinity of these interactions influence metal(II)-ion binding in the RNase P RNA alone and/or in the RNase P RNA substrate complex. In addition as discussed above, the C5 protein binds to the 5′ leader of the substrate and affect the affinity of metal(II)-ion binding. Below, I will briefly discuss the different interactions and metal(II)-ion binding.

7.5.1 The TSL-/TBS Interaction

Changing the structural topography of the TBS-region in RNase P RNA influence the positioning of metal(II)-ion binding in its vicinity (Tallsjö et al. 1993; Mattsson et al. 1994; Brännvall et al. 2007; Fig. 7.2a). In addition, the Mg^{2+} requirement increases in the RNase P RNA alone reaction for model hairpin-loop substrates with an intact T-loop that can interact with the TBS-region compared to cleavage of tRNA precursor substrates. Moreover, changing the structural topography of the T-loop by substitution the T-loop with a GAAA-tetra loop increase the Mg^{2+} requirement further (Brännvall et al. 2007). On the basis of these findings, it is conceivable that the metal(II)-ion(s) play an important role for the establishment of a productive TSL-/TBS interaction. In this context, it is important to note that metal(II)-ions also bind in the vicinity of the D-/T-loop in the substrate (Jovine et al. 2000; Shi and Moore 2000; Fig. 7.3b) and these might also play a role for proper binding of the substrate. Importantly, in the RNase P RNA alone reaction a productive TSL-/TBS interaction has been suggested to result in a conformational change comprising the positioning of chemical groups and Mg^{2+} at and in the vicinity of the site of cleavage (Brännvall et al. 2007).

7.5.2 The RCCA–RNase P RNA Interaction

Establishment of the RNase P RNA substrate complex involves pairing between the discriminator base and the two cytidines at the 3′ termini of the substrate and a conserved 5′GGU-motif in the P15-loop of RNase P RNA (Kirsebom and Svärd 1994; Figs. 7.2 and 7.4). Metal(II)-induced cleavage studies have identified two metal(II)-ions in the P15-loop referred to as site III and V (Kazakov and Altman 1991; Ciesiolka et al. 1994; Zito et al. 1993; Tallsjö et al. 1993; Kufel and Kirsebom 1996; see above). In addition from nuclear magnetic spectroscopy studies, it appears that there is a third metal(II)-ion positioned close to the residue 254 (Glemarec et al. 1996; Fig. 7.2a). Among these metal(II)-ions genetic and biochemical studies suggest that the metal(II)-ion bound in the vicinity of U294 (site V; Fig. 7.2a) play an important role during catalysis. This metal(II)-ion is repositioned

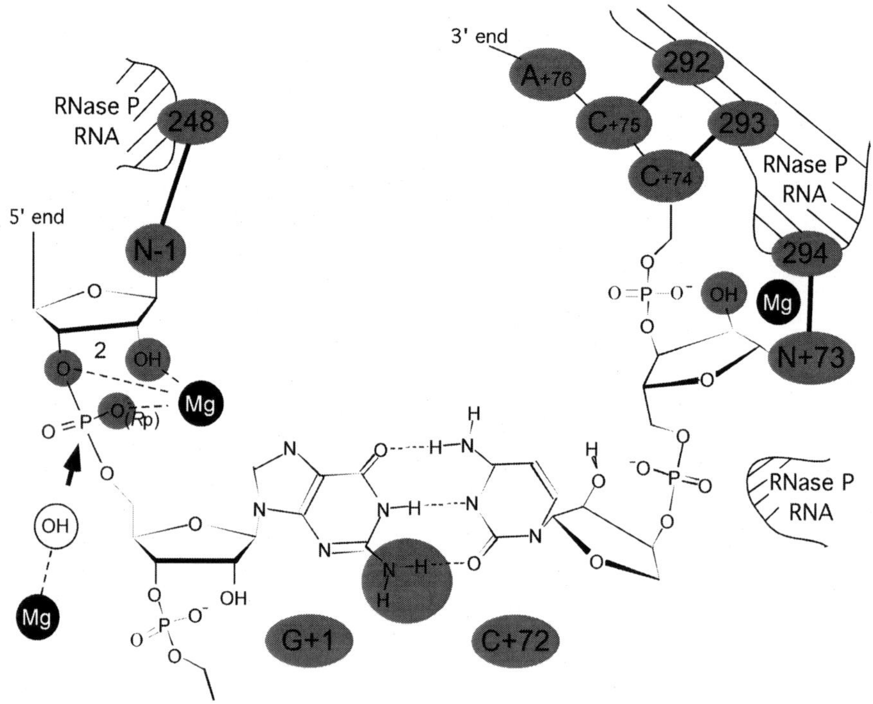

Fig. 7.4 Model representing the canonical RNase P cleavage site. Important chemical groups and RNase P RNA substrate interactions, the "RCCA–RNase P RNA" (including the "+73/294 interaction") and the "A_{248}/N_{-1}" interactions, are highlighted in *gray*. *Black spheres* refers to Mg(II)-ions suggested to play a role in catalysis are shown, for details see the main text. The Mg-hydroxyl ion represents the nucleophile (*arrow*) that attacks the phosphorous atom at the cleavage site. In addition available data suggest that the exocyclic amine of G_{+1} (when present) acts as an inner or outer sphere ligand for Mg^{2+} or to interact with RNase P RNA that affect Mg^{2+} binding near the cleavage site (Kikovska et al. 2006)

as a result of formation of the RCCA–RNase P RNA interaction and it is suggested to stabilize this interaction (Kirsebom 2007 and references therein). In addition, it has been discussed that the metal(II)-ion bound at this location influence the positioning of metal(II)-ion(s) at and in the vicinity of the cleavage site (Brännvall et al. 2004). The mechanism for how this is accomplished is not clear apart from that we know that the structural topography of the interaction between the discriminator base and 294 not only affects the metal(II)-ion binding in its vicinity but also influence the charge distribution at the cleavage site (Brännvall et al. 2003; 2004; Kikovska et al. 2005a, b).

7.5.3 The A_{248}/N_{-1} Interaction

The residue immediately 5′ of the scissile bond is suggested to interact with the conserved residue A248 (Zahler et al. 2003). Addition of lead(II)-ions result in cleavage of RNase P RNA 5′ and 3′ of A248 (site IIc; Fig. 7.2a). Moreover, biochemical and structural data suggest that there is a metal(II)-ion positioned in the substrate close to the cleavage site (Perreault and Altman 1992; 1993; Smith and Pace 1993; Brännvall et al. 2001; 2004; Persson et al. 2003; Zuleeg et al. 2001a, b). Cleavage of substrates with C at position −1 and G at +73 require higher concentrations of Mg^{2+} in the RNase P RNA alone reaction when compared with substrates with U at −1 (Brännvall et al. 2007). These data raise the possibility that metal(II)-ion binding in the vicinity of the cleavage site play a role in the formation of the A_{248}/N_{-1} interaction. In fact, it has been suggested that the true substrate for RNase P is a substrate with Mg^{2+} coordinated in the vicinity of the cleavage site (Perreault and Altman 1993). Moreover, given that the identity of N_{-1} varies it is conceivable that metal(II)-ion binding close to the cleavage site and/or in the vicinity of A248 in RNase P RNA has a role to ensure that a productive A_{248}/N_{-1} interaction being formed.

7.5.4 The U_{69}-Acceptor Stem Interaction

The well conserved P4 helix consititute a metal(II)-ion binding site and cross-linking studies have provided evidence that the universally conserved bulge (uridine at position 69; Fig. 7.2a) in P4 interacts with the amino acid acceptor-stem of the substrate. The site of interaction is suggested to be located five base pairs 3′ of the cleavage site in the substrate. Hence, it was concluded that the metal(II)-ion binding pocket in P4 does not directly participate in the binding of catalytic metal(II)-ion binding. Rather the data are consistent with that the P4 metal(II)-ion binding site influences active site metal(II)-ion affinity through substrate positioning (Christian et al. 2006).

7.6 Orchestration of the Cleavage Site and Cleavage

The chemistry of cleavage in RNase P mediated cleavage is strictly dependent on the presence of metal(II)-ions and it is generally believed that cleavage proceed through an inline S_N2 reaction mechanism (Steitz and Steitz 1993). At least two metal(II)-ions participate where one is involved in generating the nucleophile, activated H_2O coordinated to a metal(II)-ion. In Table 7.1, a number of metal(II)-ions that have been demonstrated to promote cleavage are listed. However, note that certain divalent metal ions such as Pb^{2+} do only promote the RNase P RNA mediated cleavage when another ion is also present for example Sr^{2+} and $Co(NH_3)_6^{3+}$. The second metal(II)-ion bound at the site of cleavage is suggested to stabilize the developing oxyanion in the transition state.

As discussed above, several contacts between RNase P and its substrate have been identified. The data suggest that these contacts ensure cleavage at the correct position generating cleavage products with 3′-hydroxyls and 5′-phosphates at their ends (Robertson et al. 1972; Guerrier-Takada et al. 1983). Metal(II)-ions can act as a Lewis acid or general base and activate 2′-hydroxyls in an RNA or H_2O. In the former case, this results in cleavage of the RNA with 2′;3′-cyclic phosphates and 5′-hydroxyls as end products. To ensure cleavage products with correct ends in RNase P mediated the nucleophilic attack has to come from the opposite side of the phosphorous atom relative to the 2′OH at the −1 position (Fig. 7.1). Hence, the alignment of the scissile bond in the RNase P substrate complex has to prevent a nucleophilic attack by the neighboring −1 2′OH and favor an attack from the other side. Biochemical studies combined with chemical genetic approaches have generated

Table 7.1 Some properties of metal(II)-ions relevant to RNase P activity

Metal(II)-ion	Rel. affinity to RNA	Coord. number	Ionic radius (Å)	pK_a of aqua ion	Substrat. binding	Promote catalysis
Mg^{2+}	1	6	0.72	11.4	Yes	Yes
Ca^{2+}	1	8	1.12	12.7	Yes	Yes
Sr^{2+}	0.4	6	1.18	13.2	Yes	No[a]
Ba^{2+}	Nd	6	1.35	13.8	Nd	No[a]
Mn^{2+}	2.5	6	0.67	10.6	Yes	Yes
Zn^{2+}	3 (0.3)[b]	6	0.74	9(9.6)[c]	Yes	Yes
Cd^{2+}	Nd	6	0.95	10.1(11.7)[c]	Yes	Yes
Pb^{2+}	Nd	6	1.19	7.8	Nd	Yes[d]
$Co(NH_3)_6^{3+}$	0.1	6	0.61		Yes	No
Co^{2+}	Nd	6	0.75	9.65	Yes[e]	Yes[e]

Data taken from Feig and Uhlenbeck (1999)

[a]Sr^{2+} and Ba^{2+} promotes catalysis in the presence of 10% ethanol at pH 9.5 (Kazakov and Altman 1991) but not at physiological pH

[b]Dependent on RNA

[c]Different number given dependent on reference (see Feig and Uhlenbeck 1999)

[d]Pb^{2+} promote catalysis in the presence of an additional metal(II)-ion or $Co(NH_3)_6^{3+}$, [5]cleavage activity only observed for the *B. subtilis* holoenzyme (Kurz and Fierke 2002)

[e]Cleavage activity only observed for the *B. subtilis* holoenzyme (Kurz and Fierke 2002)

a picture of which chemical groups play a role in this process, i.e., the positioning of metal(II)-ions at and in the vicinity of the cleavage site. These groups are: the 2′-hydroxyls at positions −2 and −1 in the 5′ leader of the substrate, and the 2′-hydroxyls and N7 (when present) at position +73 in the substrate, the (pro)-Rp-oxygen of the phosphorous atom to be attacked, the bridging 3′-oxygen (3′ leaving group) and the exocyclic amine of the guanosine (when present) immediately 3′ of the scissile bond. Note that the majority of tRNA precursors irrespective of origin carry a guanosine at this position, position +1 (Sprinzl and Vassilenko, 2005; Fig. 7.3). There are also experimental evidence that the 2′OH of the residue at the +1 position in the substrate and the (pro)-Sp-oxygen at the cleavage site influence catalysis (Perreault and Altman 1993; Kikovska et al. 2006; Warnecke et al. 1996; 1999; Chen et al. 1997; Cuzic et al. 2008). The underlying reasons to these observations are at present unclear and will not be further discussed here. Below the 2′OH at −1, the exocyclic amine of G_{+1} and the +73/294 interaction (which is part of the RCCA–RNase P RNA interaction, see above) in relation to metal(II)-ion will be discussed in more detail.

7.6.1 The 2′OH at the −1 and −2 Positions in the Substrate

The 2′OH at −1 is not essential for cleavage but its presence is important for ground state binding, rate of cleavage, cleavage site recognition and metal ion binding in its vicinity. This is evident from experiments where the 2′OH at −1 has been substituted with, e.g., 2′H or $2'NH_2$ which results in changes in these steps in RNase P mediated cleavage (Forster and Altman 1990; Perreault and Altman 1992; 1993; Smith et al. 1992; Gaur and Krupp 1993; Kleineidam et al. 1993; Smith and Pace 1993; Loria and Pan 1997; Persson et al. 2003; Brännvall et al. 2004; Brännvall and Kirsebom 2005; Zahler et al. 2005; Kikovska et al. 2006). More specifically, cleavage studies at different pH in the presence of different metal ions using substrates with various modifications (2′H, $2'NH_2$ and 2′F) suggested that in the RNase P RNA substrate complex the 2′OH at the −1 position acts as an outer (or inner) sphere ligand for Mg^{2+} (Persson et al. 2003; Brännvall et al. 2004; see also Zahler et al. 2005). Note that on the basis of that the 2′OH acts as a ligand for Mg^{2+} this should not result in a nucleophilic attack generating hydrolysis of the "−1/+1" phosphodiester bond. Here one possibility is that Mg^{2+} coordinating the 2′OH also uses the 3′ bridging- and the (*pro*)-Rp-oxygens as ligands (outer or inner sphere). The participation of the (*pro*)-Rp-oxygen as a metal(II)-ion coordinater in RNase P RNA mediated cleavage has clearly been documented for both bacterial type A and type B RNase P RNA (Warnecke et al. 1996; 1999; Chen et al. 1997) as well as the for eukaryal RNase P (Pfeiffer et al. 2000; Thomas et al. 2000). However, attempts to demonstrate that the 3′-oxyanion coordinates Mg^{2+} in the transition state in RNase P RNA mediated cleavage has so far failed. But this does not exclude this possibility since these trials have been based on replacing the bridging oxygen with sulfur and this might result in exclusion of Mg^{2+} at this position (Warnecke et al. 2000).

Nevertheless, this structural positioning of Mg^{2+} has been observed in the crystal structure of a catalytically active Group I intron intermediate (Stahley and Strobel 2005). Moreover, on the basis of biochemical and structural studies of the Group II ribozyme it has been suggested that one Mg^{2+} coordinate the 2′OH, the 3′ bridging oxygen and the nonbridging oxygens in the transition state (Gordon et al. 2000; Toor et al. 2008). To resolve if one metal(II)-ion indeed use the 2′OH, the 3′ bridging oxygen and the (pro)-Rp-oxygen as ligands in RNase P mediated cleavage has to wait until we have access to the crystal structure of RNase P in complex with its substrate. The outlined scenario described above is however not unlikely.

On the basis of studies of substrates where the 2′OH at the −2 position it appears that this 2′OH influence substrate binding and is part of a functionally important Mg^{2+} binding site (Perreault and Altman 1992; Loria and Pan 1998). This is consistent with the idea that the true RNase P substrate is precursor with Mg^{2+} coordinated in the vicinity of the cleavage site at the junction between single- and double-stranded regions (Perreault and Altman 1993).

7.6.2 The Exocyclic Amine of G_{+1} and the 2′OH and N7 of Base +73 in the Substrate

The guanosine (G_{+1}) at the +1 position in the substrate (and the G_{+1}/C_{+72} base pair) has a role in the cleavage site recognition process, substrate binding and cleavage under physiological pH conditions as well as under conditions where chemistry of cleavage is suggested to be rate limiting (Kirsebom 2007; Kirsebom and Trobro 2009 and references therein). In fact, it was suggested that G_{+1} in a tRNA precursor acts as a guiding nucleotide (Perreault and Altman 1992; Svärd and Kirsebom 1992). A detailed study provided evidence that the exocyclic amine of the guanosine at +1 contributes to cleavage site recognition, ground state binding and rate of cleavage. In addition, its absence changes the charge distribution at and in the vicinity of the cleavage site and conceivably metal(II)-ion binding (Kikovska et al. 2006). In this study, we discussed the possibility that the exocyclic amine interacts with RNase P RNA and/or affects the Mg^{2+} positioning at and in the vicinity of the cleavage site. Interestingly, in an RNA A-helix mimicking the structure of the RNase P cleavage site the exocyclic amine of G_{+1} is positioned close to the 2′OH and N7 of G_{+73}, the residue that interacts with U_{294} in RNase P RNA, the +73/294 interaction (Fig. 7.4). The 2′OH, N7 of G_{+73} and the structural architecture of the +73/294 interaction appears to influence the positioning of metal ions in its vicinity. Metal(II)-ion bound to this site has been suggested to stabilize this interaction (Brännvall and Kirsebom 1999; 2001; 2005; Brännvall et al. 2004; Kikovska et al. 2005b). On the basis of this it is plausible that the exocyclic amine of G_{+1} is involved in positioning of this metal ion perhaps by acting as an outer sphere ligand (see also Fig. 7.4). In this context note that replacement of G_{+1} in the substrate with

U or C, but not A, promotes lead(II)-induced hydrolysis of the scissile bond indicating the importance of the structural topography of the +1/+72 pair for metal(II)-ion binding (Kikovska et al. 2005b).

7.6.3 *Summary*

Orchestration of chemical groups at the cleavage site and positioning of Mg^{2+} are underlying events that ultimately generates a nucleophilic attack on a specific phosphorous atom generating 3′-hydroxyls and 5′-phosphates as cleavage products (Figs. 7.1 and 7.4). The scissile bond is correctly positioned for an efficient in line attack by an activated H_2O coordinated to Mg^{2+} after binding of the substrate. The TSL-/TBS interaction (applies to in particular tRNA precursor substrates) results in a conformational change that likely influences the positioning of the phosphorous target and Mg^{2+} at and in the vicinity of the cleavage site (see above; Brännvall et al. 2007). The RCCA–RNase P RNA interaction anchors the substrate and ensures that the cleavage site is exposed. Here the Mg^{2+} positioned in the vicinity of the +73/294 interaction is suggested to stabilize the interaction and to influence catalysis, i.e., rate of cleavage perhaps by affecting the positioning of Mg^{2+} at the cleavage site (Brännvall et al. 2002, 2003, 2004; Brännvall and Kirsebom 2001, 2005; Kikovska et al. 2005b). Establishment of the RCCA–RNase P RNA interaction makes N_{-1} available to interact with A_{248}. On the basis that the Mg^{2+} requirement changes when the U_{-1}/G_{+73} pair in the substrate is replaced with C_{-1}/G_{+73} formation of the A248/N-1 interaction appears to involve Mg^{2+} (Kikovska et al. 2005a; 2006; Brännvall et al. 2007). Both the RCCA–RNase P RNA and the A248/N-1 interactions result in structural changes at the cleavage site that affect positioning of the scissile bond and Mg^{2+}. In this scenario, it has been proposed that the Mg^{2+} coordinating the 2′OH at −1 also coordinates to the (*pro*)-Rp-oxygen and the leaving 3′OH. This positioning of Mg^{2+} at the cleavage site would prevent a nucleophilic attack by the 2′OH at the −1 position and facilitate the nucleophilic attack on the phosphorous target. In addition it would stabilize the developing 3′-oxyanion on the leaving group in the transition state (Kirsebom and Trobro 2009; Fig. 7.4). Moreover, the data are consistent with that there is a functionally important Mg^{2+} positioned in the vicinity of the cleavage site also before the RNase P substrate complex is formed (Kazakov and Altman 1991; Perreault and Altman 1992; 1993; Brännvall and Kirsebom 1999; Kikovska et al. 2005a; 2005b; 2006; Zuleeg et al. 2001a, b). Inspecting structures mimicking the cleavage site and on the basis that hexahydrated Mg^{2+} is positioned close to the cleavage site (≈7–8 Å) and to the phosphorous centers of +1, −1 and −2 we have raised the possibility that this substrate bound hexahydrated Mg^{2+} activates the nucleophile (metal(II)-hydroxide ion) (Kikovska et al. 2005b see also Cassano et al. 2004). Consequently this would invoke at least two Mg^{2+} directly involved in RNase P RNA mediated cleavage of RNA. Hence, the function of RNase P RNA is to prevent the nonspecific metal(II)-ion induced hydrolysis of the RNA substrate and to ensure a nucleophilic attack such that the cleavage products with correct ends, 5′-phosphate and 3′OH, are generated.

7.7 RNase P RNA, Antibiotics and "Metal Mimics"

Many small ligands interact and interfere with RNA function. This is witnessed by the fact that many clinically relevant antibiotics target ribosomal RNA. In general, to function as a suitable drug target the targeted RNA has to be unique for the infectious agent, e.g., bacterial pathogen. Alternatively, the structural differences comparing the targeted RNA in the infectious agent with that of the host homologue are large enough to result in higher tolerance towards the drug for the host RNA. Given that RNase P and its catalytic RNA subunit is essential for bacterial growth, that bacterial RNase P RNA is structurally different when compared with the human RNase P RNA and that the composition of bacterial and human RNase P differs makes RNase P a suitable drug target. In fact, it has been demonstrated that various aminoglycoside derivatives as well as other compounds interact with RNase P and inhibits its function (Mikkelsen et al. 1999; Tekos et al. 2000; Eubank et al. 2002; Kawamoto et al. 2008). More specifically, aminoglycosides have been shown to bind to the catalytic RNA subunit. Aminoglycosides are secondary metabolites that are produced and secreted by the producer to ensure a growth advantage in relation to its neighbors (Davies 1994; Davies and Wright 1997; Zembower et al. 1998, for further information about classification and biosynthesis of aminoglycosides and other antibiotics, see Walsh 2003). Structural and biochemical studies of RNA in complex with different aminoglycoside derivatives (for reviews see for example Davies et al. 1993; Kotra et al. 2000; Walter et al. 1999; Hermann 2003; Vicens and Westhof 2003; Yonath and Bashan 2004; Kirsebom et al. 2006) have revealed that aminoglycosides often bind in the deep groove of the RNA where hexahydrated Mg^{2+}-ions are known to bind. Electrostatic interactions, hydrogen bond formation directly between RNA residues/backbone and amino/hydroxyl groups on the aminoglycoside and water-mediated interactions are crucial to achieve high affinity. More relevant to the present context aminoglycosides can, when it binds to an RNA, result in displacement of metal(II)-ions (Hermann and Westhof 1998; Mikkelsen et al. 2001 see also Carter et al. 2000 and Summers et al. 2002). Hence, it has been suggested that aminoglycosides can be considered "metal mimics". On the basis of this, it is likely that the binding of an aminoglycoside to RNase P RNA interferes with its function by displacing functionally important metal(II)-ion(s). For example, metal(II)-ions bound to the P15-loop (Mikkelsen et al. 1999). Hence, characterization of metal(II)-ions in, e.g., RNase P RNA and use these sites to search for small ligands with new scaffolds that displace metal(II)-ions is a clear possibility. These new scaffolds can be used in the process to develop new antibiotics.

Taken together, an increased knowledge of the way metal ions and the other small ligands such as aminoglycosides interact with RNA and carry out their functions is fundamental and necessary in order to understand the mechanism of action of RNase P RNA and other RNA molecules that exist in a cell. This knowledge can subsequently be exploited to identify small ligands that bind a given RNA specifically and interferes with its function. Needless to say these ligands can be used as leads to develop novel drugs/antibiotics.

Acknowledgments I thank my colleagues over the years for a pleasant and stimulating working atmosphere, Dr. S. Dasgupta for comments on the manuscript, and Drs N.E. Mikkelsen and S. Trobro for help with the figures 1, 2 and 3. This work was supported by the Strategic Research Foundation, and the Swedish Research Council.

References

Beebe JA, Kurz JC, Fierke CA (1996) Magnesium ions are required by *Bacillus subtilis* ribonuclease P RNA for both binding and cleaving precursor $tRNA^{Asp}$. Biochemistry 35:10493–10505

Brännvall M, Kirsebom LA (1999) Manganese ions induce miscleavage in the *Escherichia coli* RNase P RNA-catalyzed reaction. J Mol Biol 292:53–63

Brännvall M, Kirsebom LA (2001) Metal ion cooperativity in ribozyme cleavage of RNA. Proc Natl Acad Sci U S A 98:12943–12947

Brännvall M, Kirsebom LA (2005) Complexity in the orchestration of the chemical groups near different cleavage sites in RNase P RNA mediated cleavage. J Mol Biol 351:251–257

Brännvall M, Mikkelesen NE, Kirsebom LA (2001) Monitoring the structure of *Escherichia coli* RNase P RNA in the presence of various metal ion. Nucl Acids Res 29:1426–1432

Brännvall M, Pettersson BMF, Kirsebom LA (2002) The residue immediately upstream of the RNase P cleavage site is a positive determinant. Biochemie 84:693–703

Brännvall M, Pettersson BMF, Kirsebom LA (2003) Importance of the +73/294 interaction in *Escherichia coli* RNase P RNA substrate complexes for cleavage and metal ion coordination. J Mol Biol 325:697–709

Brännvall M, Kikovska E, Kirsebom LA (2004) Cross talk in RNase P RNA mediated cleavage. Nucl Acids Res 32:5418–5429

Brännvall M, Kikovska E, Wu S, Kirsebom LA (2007) Evidence for induced fit in bacterial RNase P RNA-mediated cleavage. J Mol Biol 372:1149–1164

Brown RS, Dewan JC, Klug A (1985) Crystallographic and biochemical investigation of the lead(II)-catalyzed hydrolysis of yeast phenylalanine tRNA. Biochemistry 24:4785–4801

Carter AP, Clemons WM, Brodersen DE, Morgan-Warren RJ, Wimberly BT, Ramakrishnan V (2000) Functional insights from the structure of the 30 S ribosomal subunit and its interaction with antibiotics. Nature 407:340–348

Cassano AG, Anderson VE, Harris ME (2004) Analysis of solvent nucleophile isotope effects: evidence for concerted mechanisms and nucleophilic activation by metal coordination in nonenzymatic and ribozyme-catalyzed phosphodiester hydrolysis. Biochemistry 43: 10547–10559

Chang SE, Smith JD (1973) Structural studies on a tyrosine tRNA precursor. Nat New Biol 246:165–168

Chen Y, Li X, Gegenheimer P (1997) Ribonuclease P catalysis requires Mg2+ coordinated to the pro-Rp oxygen of the scissile bond. Biochemistry 36:2425–2438

Christian EL, Kaye NM, Harris ME (2000) Helix P4 is a divalent metal ion binding site in the conserved core of the ribonuclease P ribozyme. RNA 6:511–519

Christian EL, Smith KMJ, Perera N, Harris ME (2006) The P4 metal binding site in RNase P RNA affects active site metal affinity through substrate positioning. RNA 12:1463–1467

Ciesiolka J, Hardt WD, Schlegl J, Erdmann VA, Hartmann RK (1994) Lead-ion-induced cleavage of RNase P RNA. Eur J Biochem 219:49–56

Crary SM, Niranjanakumari S, Fierke CA (1998) The protein component of *Bacillus subtilis* ribonuclease P increases catalytic efficiency by enhancing interactions with the 5′ leader sequence of pre-tRNAAsp. Biochemistry 37:9409–9416

Crary SM, Kurz JC, Fierke CA (2002) Specific phosphorothioate substitutions probe the active site of *Bacillus subtilis* ribonuclease P. RNA 8:933–947

Crothers DM, Seno T, Söll DG (1972) Is there a discriminator site in transfer RNA? Proc Natl Acad Sci U S A 69:3063–3067

Cuzic S, Hartmann RK (2005) Studies on *Escherichia coli* RNase P RNA with Zn2+ as the catalytic cofactor. Nucl Acids Res 33:2464–2474

Cuzic S, Heidemann KA, Wöhnert J, Hartmann RK (2008) Escherichia coli RNase P RNA: substrate ribose at G+1, but not nucleotide -1/+73 base pairing, affect the transition state for cleavage chemistry. J Mol Biol 379:1–8

Davies J (1994) New pathogens and old resistance genes. Microbiologica 10:9–12

Davies J, Wright GD (1997) Bacterial resistance to aminoglycoside antibiotics. Trends Microbiol 5:234–240

Davies J, von Ahsen U, Schroeder R (1993) Antibiotics and the RNA world: a role for low-molecular-weight effectors in biochemical evolution? In: Gesteland RF, Atkins JF (eds) The RNA world. Cold Spring Harbor Laboratory Press, Cold Spring Harbor, NY, pp 185–204

Day-Storms JJ, Niranjanakumari S, Fierke CA (2004) Ionic interactions between P RNA and P protein in *Bacillus subtilis* RNase P characterized using a magnetocapture-based assay. RNA 10:1595–1608

Eubank TD, Biswas R, Jovanovic M, Litovchick A, Lapidot A, Gopalan V (2002) Inhibition of bacterial RNase P by aminoglycoside–arginine conjugates. FEBS Lett 511:107–112

Fang X, Pan T, Sosnick TR (1999) Mg2+-dependent folding of a large ribozyme without kinetic traps. Nat Struct Biol 6:1091–1095

Feig AL, Uhlenbeck OC (1999) The role of metal ions in RNA biochemistry. In: Gesteland R, Cech T, Atkins J (eds) RNA world II. Cold Spring Harbor Laboratory Press, Cold Spring Harbor, NY, pp 287–319

Forster AC, Altman S (1990) External guide sequences for an RNA enzyme. Science 249:783–786

Frank DN, Pace NR (1997) In vitro selection for altered divalent metal specificity in the RNase P RNA. Proc Natl Acad Sci U S A 94:14355–14360

Gardiner KJ, Marsh TL, Pace NR (1985) Ion dependence of the *Bacillus subtilis* RNase P reaction. J Biol Chem 260:5415–5419

Gaur RK, Krupp G (1993) Modification interference approach to detect ribose moieties important for the optimal activity of a ribozyme. Nucl Acids Res 21:21–26

Gesteland RF, Cech TR, Atkins JF (2006) The RNA world, 3rd edn. Cold Spring Harbor Press, Cold Spring Harbor, NY

Getz MM, Andrews AJ, Fierke CA, Al-Hashimi HM (2007) Structural plasticity and Mg2+ binding properties of RNase P P4 from combined analysis of NMR residual dipolar couplings and motionally decoupled spin relaxation. RNA 13:251–266

Glemarec C, Kufel J, Földesi A, Maltseva T, Sandström A, Kirsebom LA, Chattopadhyaya J (1996) The NMR structure of 31mer RNA domain of *Escherichia coli* RNase P RNA using its non-uniformly deuterium labelled counterpart [the 'NMR-window' concept]. Nucl Acids Res 24:2022–2035

Gopalan V, Altman S (2006) Ribonuclease P: structure and catalysis. In: Gesteland RF, Cech TR, Atkins JF (eds) The RNA world, 3rd ed. Cold Spring Harbor Press, Cold Spring Harbor, NY, available only online http://rna.cshl.edu/

Gordon PE, Sontheimer EJ, Piccirilli JA (2000) Kinetic characterization of the second step of group II intron splicing: Role of metal ions and the cleavage site 2′-OH in catalysis. Biochemistry 39:12939–12952

Guerrier-Takada C, Altman S (1992) Reconstitution of enzymatic activity from fragments of M1 RNA. Proc Natl Acad Sci U S A 89:1266–1270

Guerrier-Takada C, Gardiner K, Marsh T, Pace N, Altman S (1983) The RNA moiety of riboNucleicease P is the catalytic subunit of the enzyme. Cell 35:849–857

Guerrier-Takada C, Haydock K, Allen L, Altman S (1986) Metal ion requirements and other aspects of the reaction catalyzed by M1 RNA, the RNA subunit of ribonuclease P from *Escherichia coli*. Biochemistry 25:1509–1515

Hansen A, Pfeiffer T, Zuleeg T, Limmer S, Ciesiolka J, Feltens R, Hartmann RK (2001) Exploring the minimal substrate requirements for trans-cleavage by RNase P holoenzymes from *Escherichia coli* and *Bacillus subtilis*. Mol Microbiol 41:131–143

Hardt W-D, Hartmann RK (1996) Mutational analysis of the joining regions flanking helix P18 in *E. coli* RNase P RNA. J Mol Biol 259:422–433

Hardt W-D, Warnecke JM, Erdmann VA, Hartmann RK (1995) Rp-phosphorothioate modifications in RNase P RNA that interfere with tRNA binding. EMBO J 14:2935–2944

Harris ME, Pace NR (1995) Identification of phosphates involved in catalysis by the ribozyme RNase P RNA. RNA 1:210–218

Hermann T (2003) Chemical and functional diversity of small molecule ligands for RNA. Biopolymers 70:4–18

Hermann T, Westhof E (1998) Aminoglycoside binding to the hammerhead ribozyme: a general model for the interaction of cationic antibiotics with RNA. J Mol Biol 276:903–912

Jovanovic M, Sanchez R, Altman S, Gopalan V (2002) Elucidation of structure–function relationships in the protein subunit of bacterial RNase P using a genetic complementation approach. Nucl Acids Res 30:5065–5073

Jovine L, Djordjevic S, Rhodes D (2000) The crystal structure of yeast phenylalanine tRNA at 2.0 Å resolution: Cleavage by Mg^{2+} in 15-year old crystals. J Mol Biol 301:401–414

Kawamoto SA, Sudhahar CG, Hatfield CL, Sun J, Behrman EJ, Gopalan V (2008) Studies on the mechanism of inhibition of bacterial ribonuclease P by aminoglycoside derivatives. Nucl Acids Res 36:697–704

Kaye NM, Zahler NH, Christian EL, Harris ME (2002a) Conservation of helical structure contributes to functional metal ion interactions in the catalytic domain of ribonuclease P RNA. J Mol Biol 324:429–442

Kaye NM, Christian EL, Harris ME (2002b) NAIM and site-specific functional group modification analysis of RNase P RNA: magnesium dependent structure within the conserved P1–P4 multihelix junction contributes to catalysis. Biochemistry 41:4533–4545

Kazakov S, Altman S (1991) Site-specific cleavage by metal ion cofactors and inhibitors of M1 RNA, the catalytic subunit of RNase P from *Escherichia coli*. Proc Natl Acad Sci U S A 88:9193–9197

Kazantsev AV, Krivenko AA, Pace NR (2009) Mapping metal-binding sites in the catalytic domain of bacterial RNase P RNA. RNA 15:266–276

Kent O, Chaulk SG, MacMillan AM (2000) Kinetic analysis of the M1 RNA folding pathway. J Mol Biol 304:699–705

Kikovska E, Brännvall M, Kufel J, Kirsebom LA (2005a) Substrate discrimination in RNase P RNA-mediated cleavage: importance of the structural environment of the RNase P cleavage site. Nucl Acids Res 33:2012–2021

Kikovska E, Mikkelsen N-E, Kirsebom LA (2005b) The naturally trans-acting ribozyme RNase P RNA has leadzyme properties. Nucl Acids Res 33:6920–6930

Kikovska E, Brännvall M, Kirsebom LA (2006) The exocyclic amine at the RNase P cleavage site contributes to substrate binding and catalysis. J Mol Biol 359:572–584

Kikovska E, Svärd SG, Kirsebom LA (2007) Eukaryotic RNase P RNA mediates cleavage in the absence of protein. Proc Natl Acad Sci U S A 104:2062–2067

Kirsebom LA, Virtanen A, Mikkelsen NE (2006) Aminoglycoside interaction with RNA and nucleases. In: Barciszewski J, Brosius J, Erdmann VA (eds) Handbok Exp Pharmacol 173 RNA towards medicine. Wiley Press, pp 73–96

Kirsebom LA (2007) RNase P RNA mediated cleavage: substrate recognition and catalysis. Biochimie 89:1183–1194

Kirsebom LA, Svärd SG (1994) Base pairing between *Escherichia coli* RNase P RNA and its substrate. EMBO J 13:4870–4876

Kirsebom LA, Trobro S (2009) RNase P RNA-mediated cleavage. IUBMB Life 61:189–200

Kleineidam RG, Pitulle C, Sproat B, Krupp G (1993) Efficient cleavage of pre-tRNAs by *E. coli* RNase P RNA requires the 2′-hydroxyl of the ribose at the cleavage site. Nucl Acids Res 21:1097–1101

Kotra LP, Haddad J, Mobashery S (2000) Aminoglycosides: Perspectives on mechanisms of action and resistance and strategies to counter resistance. Antimicrob Agents Chemother 44:3249–3256

Krzyzosiak WJ, Marciniec T, Wiewiórowski M, Romby P, Ebel J, Giegé R (1988) Characterization of the lead(II)-induced cleavages in tRNAs in solution and effect of the Y-base removal in yeast tRNA^{Phe}. Biochemistry 27:5771–5777

Kufel J, Kirsebom LA (1994) Cleavage site selection by M1 RNA the catalytic subunit of *Escherichia coli* RNase P, is influenced by pH. J Mol Biol 244:511–521

Kufel J, Kirsebom LA (1996) Residues in *Escherichia coli* RNase P RNA important for cleavage site selection and divalent metal ion binding. J Mol Biol 263:685–698

Kufel J, Kirsebom LA (1998) The P15-loop of Escherichia coli RNase P RNA is an autonomous divalent metal ion binding domain. RNA 4:777–788

Kurz JC, Fierke CA (2002) The affinity of magnesium binding sites in the *Bacillus subtilis* RNase P-pre-tRNA complex is enhanced by the protein subunit. Biochemistry 41:9545–9558

Lindell M, Brännvall M, Wagner EGH, Kirsebom LA (2005) Lead(II) cleavage analysis of RNase P RNA in vivo. RNA 11:1348–1354

Loria A, Pan T (1996) Domain structure of the ribozyme from eubacterial ribonuclease P. RNA 2:551–563

Loria A, Pan T (1997) Recognition of the T stem-loop of a pre-tRNA substrate by the ribozyme from *Bacillus subtilis* RNase P. Biochemistry 36:6317–6325

Loria A, Pan T (1998) Recognition of the 5′ leader and the acceptor stem of a pre-tRNA substrate by the ribozyme from *Bacillus subtilis* RNase P. Biochemistry 37:10126–10133

Loria A, Niranjanakumari S, Fierke CA, Pan T (1998) Recognition of a pre-tRNA substrate by the *Bacillus subtilis* RNase P holoenzyme. Biochemistry 37:15466–15473

Mans RM, Guerrier-Takada C, Altman S, Pleij CW (1990) Interaction of RNase P from *Escherichia coli* with pseudoknotted structures in viral RNAs. Nucl Acids Res 18:3479–3487

Marciniec T, Ciesiołka J, Wrzesiński J, Wiewiórowski M, Krzyzosiak WJ (1989) Specificity and mechanism of the cleavages induced in yeast tRNAPhe by magnesium ions. Acta Biochim Polon 36:115–122

Massire C, Jaeger L, Westhof E (1998) Derivation of the thres-dimensional architecture of bacterial ribonuclease P RNAs from comparative sequence analysis. J Mol Biol 279:773–793

Mattsson JG, Svärd SG, Kirsebom LA (1994) Characterization of the *Borrelia burgdorferi* RNase P RNA gene reveals a novel tertiary interaction. J Mol Biol 241:1–6

McClain WH, Guerrier-Takada C, Altman S (1987) Model substrates for an RNA enzyme. Science 238:527–530

Mikkelsen N-E, Brännvall M, Virtanen A, Kirsebom LA (1999) Inhibition of RNase P RNA cleavage by aminoglycosides. Proc Natl Acad Sci U S A 96:6155–6160

Mikkelsen N-E, Johansson K, Virtanen A, Kirsebom LA (2001) Aminoglycoside binding displaces a divalent metal ion in a tRNA–neomycin B complex. Nat Struct Biol 8:510–514

Misra VK, Draper DE (1998) On the role of magnesium ions in RNA stability. Biopolymers 48:113–135

Murphy FL, Cech TR (1993) An independently folding domain of RNA tertiary structure within the Tetrahymena ribozyme. Biochemistry 32:5291–5300

Niranjanakumari S, Stams T, Crary SM, Christianson DW, Fierke CA (1998) Protein component of the ribozyme ribonuclease P alters substrate recognition by directly contacting precursor tRNA. Proc Natl Acad Sci U S A 95:15212–15217

Paisley T, Van Tuyle GC (1994) The processing of wild type and mutant forms of rat nuclear pre-tRNALys by homologous RNase P. Nucl Acids Res 22:3347–3353

Pan T (1995) Higher order folding and domain analysis of the ribozyme from *Bacillus subtilis* ribonuclease P. Biochemistry 34:902–909

Pan T, Jakacka M (1996) Multiple substrate binding sites in the ribozyme from *Bacillus subtilis* RNase P. EMBO J 15:2249–2255

Pannucci JA, Haas ES, Hall TA, Brown JW (1999) RNase P RNAs from some archaea are catalytically active. Proc Natl Acad Sci U S A 96:7803–7808

Perreault J-P, Altman S (1992) Important 2′-hydroxyl groups in model substrates for M1 RNA, the catalytic RNA subunit of RNase P from *Escherichia coli*. J Mol Biol 226:399–409

Perreault J-P, Altman S (1993) Pathway of activation by magnesium ions of substrates for the catalytic subunit of RNase P from *Escherichia coli*. J Mol Biol 230:750–756

Persson T, Cuzic S, Hartmann RK (2003) Catalysis by RNase P RNA: unique features and unprecedented active site plasticity. J Biol Chem 278:43394–43401

Pfeiffer T, Tekos A, Warnecke JM, Drainas D, Engelke DR, Séraphin B, Hartmann RK (2000) Effects of phosphorothioate modifications on precursor tRNA processing by eukaryotic RNase P enzymes. J Mol Biol 298:559–565

Reich C, Olsen GJ, Pace B, Pace NR (1988) Role of the protein moiety of ribonuclease P, a ribonucleoprotein enzyme. Science 239:178–181

Robertson HD, Altman S, Smith JD (1972) Purification and properties of a specific *Escherichia coli* ribonuclease which cleaves a tyrosine transfer ribonucleic acid precursor. J Biol Chem 247:5243–5251

Sampson JR, Sullivan FX, Behlen AB, DiRenzo AB, Uhlenbeck OC (1987) Characterization of two RNA-catalyzed RNA cleavage reactions. Cold Spring Harbor Symp Quant Biol 52:267–275

Schlegl J, Fürste JP, Bald R, Erdmann VA, Hartmann RK (1992) Cleavage efficiences of model substrates for ribonuclease P from *Escherichia coli* and *Thermus thermophilus*. Nucl Acids Res 20:5963–5970

Schmitz M (2004) Change of RNase P RNA function by single base mutation correlates with perturbation of metal ion binding in P4 as determined by NMR spectroscopy. Nucl Acids Res 32:6358–6366

Schmitz M, Tinoco I Jr (2000) Solution structure and metal-ion binding of the P4 element from bacterial RNase P RNA. RNA 6:1212–1225

Shi H, Moore PB (2000) The crystal structure of yeast phenylalanine tRNA at 1.93 Å resolution: a classic structure revisited. RNA 6:1091–1105

Smith D, Pace NR (1993) Multiple magnesium ions in the ribonuclease P reaction mechanism. Biochemistry 32:5273–5281

Smith D, Burgin AB, Haas ES, Pace NR (1992) Influence of metal ions on the ribonuclease P reaction. Distinguishing substrate binding from catalysis. J Biol Chem 267:2429–2436

Sprinzl M, Vassilenko KS (2005) Compilation of tRNA sequences and sequences of tRNA genes. Nucl Acids Res 33:D139–D140

Stahley MR, Strobel SA (2005) Structural evidence for two-metal-ion mechanism of group I intron splicing. Science 309:1587–1590

Steitz TA, Steitz JA (1993) A general two-metal-ion mechanism for catalytic RNA. Proc Natl Acad Sci U S A 90:6498–6502

Summers JS, Shimko J, Freedman FL, Badger CT, Sturgess M (2002) Displacement of Mn^{2+} from RNA by K^{+}, Mg^{2+}, neomycin B, and an arginine-rich peptide: indirect detection of nucleic acid/ligand interactions using phosphorus relaxation enhancement. J Am Chem Soc 124:14934–149339

Svärd SG, Kirsebom LA (1992) Several regions of a tRNA precursor determine the *Escherichia coli* RNase P cleavage site. J Mol Biol 227:1019–1031

Svärd SG, Mattsson JG, Johansson KE, Kirsebom LA (1994) Cloning and characterization of the RNase P RNA genes from two porcine mycoplasmas. Mol Microbiol 11:849–859

Tallsjö A, Kirsebom LA (1992) Product release is a rate-limiting step during cleavage by the catalytic RNA subunit of *Escherichia coli* RNase P. Nucl Acids Res 21:51–57

Tallsjö A, Svärd SG, Kufel J, Kirsebom LA (1993) A novel tertiary interaction in M1 RNA, the catalytic subunit of *Escherichia coli* RNase P. Nucl Acids Res 21:3927–3933

Tekos A, Tsagla A, Stathopoulos C, Drainas D (2000) Inhibition of eukaryotic ribonuclease P activity by aminoglycosides: kinetic studies. FEBS Lett 485:71–75

Thomas BC, Chamberlain J, Engelke DR, Gegenheimer P (2000) Evidence for an RNA-based catalytic mechanism in eukaryotic nuclear ribonuclease P. RNA 6:554–562

Toor N, Rajashankar K, Keating KS, Pyle AM (2008) Structural basis for exon recognition by a group II intron. Nat Struct Mol Biol 15:1221–1222

Torres-Larios A, Swinger KK, Krasilnikov AS, Pan T, Mondragón A (2005) Crystal structure of the RNA component of bacterial ribonuclease P. Nature 437:584–587

Tsai H-Y, Pulukkunat DK, Woznick WK, Gopalan V (2006) Functional reconstitution and characterization of *Pyrococcus furiosus* RNase P. Proc Natl Acad Sci U S A 103:16147–16152

Vicens Q, Westhof E (2003) RNA as a drug target: the case of aminoglycosides. ChemBioChem 4:1018–1023

Walker SC, Engelke DR (2006) Ribonuclease P: The evolution of an ancient RNA enzyme. Crit Rev Biochem Mol Biol 41:77–102

Walsh C (2003) Antibiotics: actions, origins, resistance. ASM Press, Washington DC, USA

Walter F, Vicens Q, Westhof E (1999) Aminoglycoside–RNA interactions. Curr Opin Chem Biol 3:694–704

Wang YH, Murphy FL, Cech TR, Griffith JD (1994) Visualization of a tertiary structural domain of the Tetrahymena group I intron by electron microscopy. J Mol Biol 236:64–71

Warnecke JM, Fürste JP, Hardt W-D, Erdmann VA, Hartmann RK (1996) Ribonuclease P (RNase P) RNA is converted to a Cd2+-ribozyme by a single Rp-phosphorothioate modification in the precursor tRNA at the RNase P cleavage site. Proc Natl Acad Sci U S A 93:8924–8928

Warnecke JM, Held R, Busch S, Hartmann RK (1999) Role of metal ions in the hydrolysis reaction catalyzed by RNase P RNA from *Bacillus subtilis*. J Mol Biol 290:433–445

Warnecke JM, Sontheimer EJ, Piccirilli JA, Hartmann RK (2000) Active site constraints in the hydrolysis reaction catalyzed by bacterial RNase P: analysis of precursor tRNAs with a single 3′-S-phosphorothiolate internucleotide linkage. Nucl Acids Res 28:720–727

Werner C, Krebs B, Keith G, Dirheimer G (1976) Specific cleavages of pure tRNAs by plumbous ions. Biochem Biophys Acta 432:161–175

Wintermeyer W, Zachau HG (1973) Mg^{2+}-catalyzed specific splitting of tRNA. Biochim Biophys Acta 299:82–90

Yonath A, Bashan A (2004) Ribosomal crystallography: Initiation, peptide bond formation, and amino acid polymerization are hampered by antibiotics. Annu Rev Microbiol 58:233–251

Zahler NH, Christian EL, Harris ME (2003) Recognition of the 5′ leader of pre-tRNA substrates by the active site of ribonuclease P. RNA 9:734–745

Zahler NH, Sun L, Christian EL, Harris ME (2005) The pre-tRNA nucleotide base and 2′-hydroxyl at N(-1) contribute to fidelity in tRNA processing by RNase P. J Mol Biol 345:969–985

Zarrinkar PP, Wang J, Williamson JR (1996) Slow folding kinetics of RNase P RNA. RNA 2:564–573

Zembower TR, Noskin GA, Postelnick MJ, Nguyen C, Peterson LR (1998) The utility of aminoglycosides in an era of emerging drug resistance. Int J Antimicrob Agents 10:95–105

Zito K, Hüttenhofer A, Pace NR (1993) Lead-catalyzed cleavage of ribonuclease P RNA as a probe for integrity of tertiary structure. Nucl Acids Res 21:5916–5920

Zuleeg T, Hartmann RK, Kreutzer R, Limmer S (2001a) NMR spectroscopic evidence for Mn(2+) (Mg(2+)) binding to a precursor-tRNA microhelix near the potential RNase P cleavage site. J Mol Biol 305:181–189

Zuleeg T, Hansen A, Pfeiffer T, Schubel H, Kreutzer R, Hartmann RK, Limmer S (2001b) Correlation between processing efficiency for ribonuclease P minimal substrates and conformation of the nucleotide −1 at the cleavage position. Biochemistry 40:3363–3369

Chapter 8
Challenges in RNase P Substrate Recognition: Considering the Biological Context

Michael Harris and Lindsay Yandek

8.1 Introduction

As demonstrated by the detailed descriptions of RNase P structure and function presented in the chapters of this volume, our understanding of substrate recognition and catalytic mechanism of this essential enzyme is highly advanced. The kinetic mechanisms for the processing of model substrates by the P RNA ribozyme and by the reconstituted holoenzyme have been described (Chapter by Fierke et al.). Comparative analyses of the differences in these mechanisms have provided insight into the unique partnership between the two subunits. The role of essential metal ions has been investigated, and in some cases binding sites for functional metal ions have been identified (Chapter by Kirsebom et al.). New three dimensional structures now give us high-resolution models of the RNA and protein subunits that provide contexts in which to further define structure-function relationships (Chapters by Mondragon and Westhof). Despite these important advances, it can be argued that we still lack a comprehensive understanding of how the functional and biophysical properties of the enzyme are integrated into its specific biological function. However, these emerging data on folding and function now put us in a position to develop such an integrated perspective together as a field. Considering the problem from different points of view, by different researchers, will be required to achieve this goal. In this chapter, we consider selected aspects of the biochemistry and biophysics of RNase P in the context of its biological role in tRNA processing. This exercise points out several potential gaps in our understanding of RNase P reaction kinetics and substrate binding specificity that may serve as points of departure for achieving a comprehensive understanding of the function of this essential enzyme in biology.

Biosynthesis of the translational machinery takes up most of the resources in rapidly dividing cells (Bremer and Dennis 2008). The most ubiquitous and important role

M. Harris (✉) and L. Yandek
RNA Center and Department of Biochemistry, Case Western Reserve University School of Medicine, Cleveland, OH 44106, USA
e-mail: michael.e.harris@case.edu

F. Liu and S. Altman (eds.), *Ribonuclease P*, Protein Reviews 10,
DOI 10.1007/978-1-4419-1142-1_8, © Springer Science+Business Media, LLC 2010

played by RNase P is catalysis of the essential step of 5′-end maturation during tRNA biosynthesis (Kirsebom 2007; Deutscher 2009). The ability to recognize multiple substrates is inherent to RNase P, and in addition to all pre-tRNAs it is well known that it also processes pre-5S, pre-tmRNA and other substrates. Given the ubiquitous occurrence of P RNA subunit genes in all three Kingdoms, an ancestral RNase P enzyme was present in the last common ancestor, most likely due to its essential role in the tRNA biosynthesis (Evans et al. 2006). In considering the relationships that may be drawn between the biophysical properties of RNase P and its biological role, we will primarily consider its function in pre-tRNA processing. Yet, it is recognized that there are likely to be important and interesting idiosyncratic features to the kinetics and thermodynamics of processing of non-tRNA substrates. Nonetheless, focusing on the specific role of RNase P in tRNA processing allows the properties of the enzyme to be more related directly to one, basic aspect of organism "fitness"- the relative steady state abundance of tRNAs. First, we will review the potential contribution of RNase P processing to the rate of tRNA biosynthesis and the equilibrium distribution of different tRNA species, and subsequently discuss the known kinetic and thermodynamic properties of the enzyme in this context.

8.2 Contribution of RNase P Processing to the Overall Rate of tRNA Biosynthesis

The production of stable RNAs, of which tRNA is a major component, is obviously essential to cell growth and consumes the majority of metabolic resources in rapidly dividing cells. In *Escherichia coli* cells, tRNA and rRNA biosynthesis are tightly regulated by the stringent response, which senses the accumulation of uncharged tRNA and negatively regulates the initiation of transcription of tRNA and rRNA operons (Potrykus and Cashel 2008). *E. coli* has 79 tRNA genes coding for 46 different amino acid acceptor species (Komine et al. 1990). However, the distribution of these tRNA species is not uniform. The tRNAs that are present at higher concentrations are those that recognize the preferred codons of the genes encoding the highly expressed proteins of rapidly growing bacteria (Ikemura 1981; Ikemura 1985). The overall abundance in rapidly dividing cells appears to be largely set by gene copy number. This correspondence of codon usage and tRNA abundance is believed to function to maximize the translation efficiency and therefore growth rates (Kurland 1993).

tRNAs can be encoded in polycistronic genes along with rRNA and mRNA, as well as in single tRNA genes. Precursors to individual tRNAs that are part of a rRNA primary transcripts are thought to be released during the course of rRNA maturation by endonculeolytic cleavage by RNase III or RNase E. Separation of individual tRNA precursors from transcripts containing mRNAs or other tRNAs is accomplished primarily by RNase E (Li and Deutscher 2002; Ow and Kushner 2002). Cleavage by RNase P serves to mature the 5′ terminus of all tRNA molecules by endonucleolytic phosphodiester bond hydorlysis. In *E.coli,* all tRNAs have their 3′-CCA sequence encoded in the gene and 3′ maturation is carried out by exoribonucleases, primarily RNase T and RNase PH (Li and Deutscher 1996). These multiple exoribonucleases act in processing the 3′ end of a particular tRNA precursor in a stochastic manner depending on secondary structure and sequence. Whether 5′ end processing precedes 3′ end processing has not been systematically examined and may depend on the sequence length and structure of the individual pre-tRNA. The otherwise mature tRNAs are also modified internally and these modifications can be common to all tRNAs like pseudouridine in the TpsiC loop, or idiosyncratic to an individual tRNA. The involvement of many common enzymes during tRNA biosynthesis indicates that tRNA maturation is a complex process with multiple possible points of regulation.

Nonetheless, the observation that tRNA abundance is largely dictated by gene copy number indicates that to a first approximation the overall *rates* (υ_{synth} mole/time) of tRNA biosynthesis are greater for the more abundant tRNAs proportional to the number of genes that encode them. Furthermore, maintaining constant relative ratios of different tRNAs is important since changes in these ratios results in misincorporation of amino acids by the ribosome (Szaflarski et al. 2008). Consequently, the overall *rate constants* (k_{synth}) for the irreversible first order synthesis of each tRNA are assumed to be roughly equal (Fig. 8.1). That is, the overall rate for biosynthesis of any individual tRNA is expressed as $\upsilon_{synth} = k_{synth}$[gene]. Although models in which k_{synth} is different for different tRNAs, or in which there are different numbers of steps for different tRNAs are possible, it is useful to consider tRNA biosynthesis

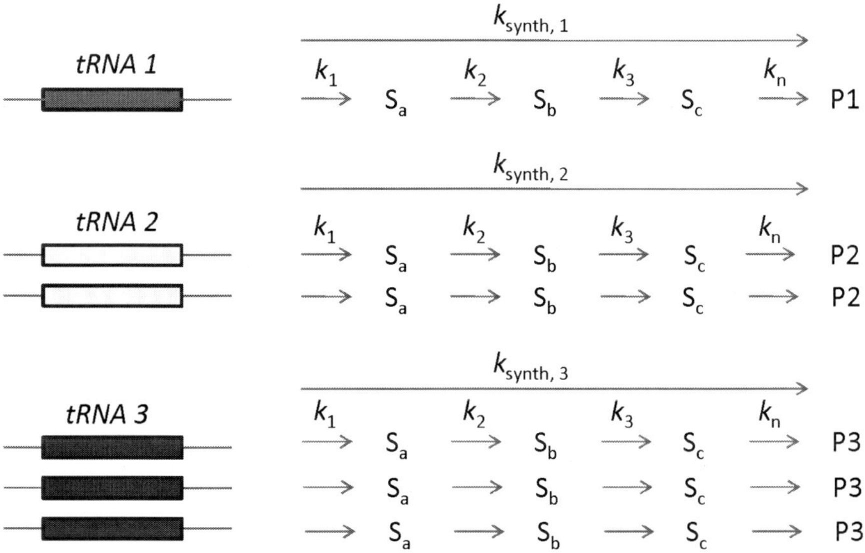

Fig. 8.1 tRNA biosynthesis as a series of irreversible pseudo first order reactions. Different tRNA genes are designated tRNA1, tRNA2, and tRNA3. The overall rate constant for biosynthesis of the product tRNA Pn is $k_{synth,n}$. The rate constants for individual steps are indicated above each reaction *arrow*

process to be a series of irreversible first order reactions with the same number of steps for each tRNA. The overall rate constant for biosynthesis of a particular tRNA, k_{synth}, then depends the individual rate constants for each step in the following way:

$$k_{synth} = \frac{k_1 k_2 k_3}{k_1 k_2 + k_2 k_3 + k_1 k_3}$$

where k_1, k_2, k_3 represent the individual rate constants for the conversion of S_a to S_b, S_b to S_c and S_c to P, respectively (Cleland and Cook 2007). A greater abundance of a particular tRNA may also arise due to a higher k_{synth}, however, this different will not change the basic relationship between k_{synth} and the rate constants of the individual steps. Here, the process represented by k_1 would be transcription to generate pre-tRNA and k_2 and k_3 represent additional steps in tRNA maturation including cleavage to yield the mature tRNA 5′ end by RNase P. Additional steps expand series accordingly:

$$k_{synth} = \frac{k_1 k_2 k_3 k_n}{k_1 k_2 k_3 + k_1 k_2 k_n + k_1 k_3 k_n + k_2 k_3 k_n}$$

where k_n is the rate constant for the nth step. Given the requirement that k_{synth} for different tRNAs must be essentially equal, it is instructive to consider the ways in which k_1, k_2, and k_3 may vary, and yet maintain this requirement. If each step proceeds with the same rate constant, then the overall rate constant for production of the product (P) will be approximately $k_1/3$ (or k_n/n), and at steady state the total population of intermediates for a specific tRNA partitions equally between S_a, S_b and S_c. A decrease in any one of the rate constants will result in a proportional decrease in the k_{synth}; however the magnitude of the decrease in k_{synth} will be attenuated proportional to the total number of steps. By inspection of the equation above, it is evident that the overall rate constants for the biosynthesis of different tRNAs can be maintained as equal, even if the rate constant for the RNase P catalyzed step varies for different substrates, so long as one or more of the remaining rate constants co-vary proportionally. That is, the rate constant for RNase P processing (k_2, say) may be much faster (for example 10-fold) for one substrate when compared with another, but if the rate constant for synthesis of pre-tRNA (k_1), *or* a downstream step (k_3) is proportionally slower (by tenfold, or $0.1 \times k_2$), then the same k_{synth} is obtained. Indeed, an equivalent k_{synth} is also obtained for different substrates if the rate constant for P processing is slow and co-rate limiting with that of another step for both, as long as the remaining step is proportionately faster, regardless of whether that step comes before or after the RNase P processing step.

Two boundary conditions can be considered for how RNase P processing may fit into this simplified mechanism for tRNA biosynthesis. In one case, the rate constant for RNase P-catalyzed 5′ end maturation is the same for all pre-tRNA substrates, while in the second case, the rate constant for the RNase P processing step may be different for different substrates. If the RNase P processing rate constant is equivalent for all pre-tRNAs, then the remaining rate constants may nonetheless vary; however, to maintain an equivalent k_{synth} they must co-vary proportionally. It also follows that if RNase P processing is not rate limiting, there can be large changes in its rate constant for processing different substrates, yet the overall rate constant will be little affected. For example, if the rate constant for RNase P processing is 100,000 times faster than the other remaining steps, then a 1,000-fold decrease in its value for a different substrate would only decrease k_{synth} by less than a percent.

The turnover of mature tRNA does not appear to be a major mechanism for the modulation of tRNA abundance (Dittmar et al. 2004; Deutscher 2009) and so does not significantly impact the simplistic treatment, above. Stable RNAs are generally only turned over under stress conditions or when they are incorrectly folded. This latter process of quality control in *E. coli* is very similar to that observed for mRNAs, that is, polyadenylation of the fragments and degradation by 3′-to-5′ exoribonucleases, principally RNase R and PNPase (Deutscher 2009). However, global analysis of tRNA distribution in *B. subtilis* using microarrays confirmed that transcription contributes significantly to the distribution of the different tRNAs (Dittmar et al. 2004). Aside from transcription, the final tRNA distribution was derived from a balance between processing and precursor degradation activities. Interestingly, these data indicated that the turnover of pre-tRNAs may be an important mode of regulation under conditions of changing growth rate.

The relative abundance of each tRNA within the same operon is typically constant in dividing cells, but can be nonequal. This distribution becomes more irregular for mutant cells in which the levels of RNase P are reduced and this result is particularly obvious for operons containing larger numbers of tRNA genes (Dittmar et al. 2004). This observation suggests that transcription is sufficient to account for the relative tRNA distribution for the wild-type cells, but not when RNase P processing is rate limiting. Pan and colleagues proposed that the uneven relative level of tRNAs within the same operon may be explained by two models. The first model suggests the rate constant for RNase P processing is different for tRNAs within the same operon, and is significantly greater than rate constants for degradation. Lower concentrations of enzyme result in lower apparent rate constants for processing and the unprocessed tRNAs are immediately targeted for precursor degradation. Under these conditions, the distribution of tRNAs will be more significantly influenced by the differences in RNase P processing. The second model suggests that degradation of the mature tRNA is different for tRNAs within the same operon. As discussed above, this possibility is less likely since mature tRNA degradation is slow and non-specific.

In sum, consideration of the role of RNase P in the overall process of tRNA biosynthesis yields the following points of interest. First, there is little evidence that RNase P processing plays a major role in establishing the relative levels of different tRNAs. Rather, transcription initiation appears to be the main determinant of abundance with pre-tRNA turnover also playing a role. Differences in the rate constant for the RNase P catalyzed step may be different for different tRNAs, however, depending on the relative number, and magnitude of the rate constants for the other steps in the pathway, the effect on the overall rate for tRNA biosynthesis may be small or undetectable. If there are major differences in the magnitude of rate constants for RNase P cleavage of different substrates, then an overall rate constant for tRNA biosynthesis may be maintained so long as there are equivalent, offsetting differences in the rate constants for the other steps. An implication of the simple kinetic model presented above is that the order of steps may be different for different substrates; yet, an equivalent k_{synth} may still be maintained. Thus, the role for RNase P appears to be to recognize as rapidly and as specifically as possible pre-tRNA and hand it off to the next enzyme in the maturation pathway. If there are differences in the rate constants for RNase P processing, either it is not rate limiting or there are offsetting differences in the rate constants for other processing steps. However, the simplest model is one in which the processing of different pre-tRNAs by RNase P is essentially uniform.

8.3 The Power (and limitations) of the "Reductionist" Perspective on RNase P Substrate Recognition

As comprehensively discussed in the other chapters in this volume and in several recent reviews (Hsieh et al. 2004; Kirsebom 2007; Smith et al. 2007; Kirsebom and Trobro 2009), our understanding of the structural determinants in pre-tRNA that are directly contacted by residues in RNase P is very well developed. Briefly, the current model based on extensive biochemical analyses is that RNase P recognizes the stacked acceptor stem and TpsiC stems of the tRNA as well as nucleotides in the 5′ leader sequence. The P RNA subunit contacts 2′-hydroxyl groups in the TpsiC stem as well as individual nucleotides at the base of the acceptor stem that flank the site of phosphodiester hydrolysis. The specific sequence determinants adjacent to the cleavage site include the conserved nucleotides R(73)C(74)C(75) at the 3′ end, the conserved G(1)-(C72) basepair and a uridine residue at N(-1) immediately 5′ to the site of phosphodiester bond hydrolysis. Nucleotides that are further 5′ distal to the cleavage site at N(-4) to N(-7) in the leader sequence are contacted by the RNase P protein subunit.

For the most part, this perspective comes from considering data from experiments in which the the RNA alone reaction was the functional read-out. While P RNA can fold and function at near physiological concentrations of Mg^{2+} ions an at moderate 0.1 M ionic strength (Baird et al. 2007), its affinity for pre-tRNA is low under these conditions (Beebe et al. 1996; Buck et al. 2005a; Sun et al. 2006). However, early studies of the enzyme reported a requirement for very high 1–2 M ionic strength for P RNA alone activity (Gardiner et al. 1985) which increases the substrate binding affinity. Subsequently, most structure-function studies of substrate recognition have been performed under these convenient, but decidedly non-physiological conditions. In some labs, the required high ioinic strength was obtained by elevating the monovalent ion concentration (Smith et al. 1992; Smith and Pace 1993), and in others by elevating the divalent ion concentration (Kirsebom and Altman 1989), with uncertain differences in the results. Most studies of RNase P substrate recognition have focused on dissecting the determinants for high binding affinity and fast cleavage rate by manipulating the structure of one, or a few model substrates (Loria and Pan 1998; Zahler et al. 2003; Kikovska et al. 2006). Often the substrates are short duplexes or stem loops, cut off on the reasonable simplifying assumption that they bind in the same fashion as pre-tRNA (Hansen et al. 2001; Brannvall et al. 2007). Often the substrate used has been from a different organism altogether than the source of the RNase P enzyme (Christian et al. 1998; Loria and Pan 1999). In a blatant example of molecular profiling, many of us are guilty of assuming that pre-tRNAs are all alike.

However, there are multiple lines of evidence that the information gained in analyzing the RNA alone reaction at high salt reflects, at some level, the in vivo function of the ribonuleoprotein holoenzyme. First, the structure of the P RNA as revealed by chemical probing is essentially the same in the presence and absence of the protein subunit, thus, it is anticipated that its functional properties will be globally similar (Christian et al. 2002; Buck et al. 2005b). The nucleotides identified as important for binding in pre-tRNA and in the catalytic core of P RNA are highly, or universally conserved in phylogeny (Frank and Pace 1998; Kazantsev and Pace 2006). Disruption of at least a few of the interactions between P RNA and tRNA, most notably the interaction with the R(73)C(74)C(75) substrate sequence, have been shown to disrupt the tRNA processing and cell growth in vivo (Wegscheid and Hartmann 2007). In several cases, the holoenzyme has been found to be sensitive to the same substrate mutations that disrupt the ribozyme reaction (Zahler et al. 2003). However, the magnitude of the effects are often attenuated relative to the effects of the same mutations on the RNA alone reaction.

Thus, the problem of understanding RNase P substrate recognition has been approached from a largely "reductionist" perspective. That is, understanding has been gained by experimentally examining one or a few substrates in detail by structure function analyses. It is undeniable that this strategy has been highly successful and the experimental data that support the current models of enzyme-substrate interactions in the Michaelis-Menten complex are very strong. A model that emerges from these data is that the efficient cleavage of pre-tRNA substrates depends on the several determinants described above. More specifically, it has been argued that, dependent on the number of determinants in a given substrate, RNase P processing will be affected to a varying extent (Kikovska et al. 2006; Brannvall et al. 2007). In strict terms, this model holds that there is a direct thermodynamic relationship between the presence of a particular functional group or interaction and its contribution to the free energy of binding or of transition state stabilization. This perspective further implies that the sequence variation among pre-tRNA substrates in vivo will be reflected at some level in the rate constant for the RNase P catalyzed step of tRNA maturation. Indeed, it has been proposed that this model of progressive increase in binding affinity and cleavage rate with increasing numbers of specificity determinants provides a rationale as to why RNase P processes such a large variety of substrates with different efficiencies (Brannvall et al. 2007).

A quantitative variation on this model, which takes into account the changes in the magnitudes of the effects of substrate modifications when they are made in combination, holds that the binding mechanism is not simply two-state, and that binding thermodynamics is governed by threshold effects. That is, to be efficiently processed a substrate must contain any subset of the possible interactions with the enzyme, as long as their thermodynamic contribution is greater than a certain threshold. Indeed, such a two step binding mechanism explains the observed linkage between the rate constant for catalysis and binding contacts distant from the cleavage site. The basic model takes the same form as the two step binding model which underlies threshold effects in the Tetrahymena L21Sca ribozyme (e.g. (Shan and Herschlag 2000; Karbstein et al. 2004). In its simple form, the binding mechanism is considered to involve the formation of an initial enzyme substrate complex (ES) with equilibrium constant K_a that isomerizes to form a catalytically competent complex ES', governed by equilibrium constant K_c, that undergoes reaction with rate constant k_c.

$$E+S \underset{}{\overset{K_a}{\rightleftarrows}} ES \underset{}{\overset{K_c}{\rightleftarrows}} ES' \overset{k_c}{\longrightarrow} E+P$$

A general expression for the observed rate constant would be:

$$k_{obs} = \frac{k_c}{1 + \frac{1}{K_c} + \frac{K_d}{[S]}\left(\frac{1}{K_c}\right)}$$

At saturating substrate concentration ($[S] >> 1/K_a$) the observed rate constant will be approximated by:

$$k_{obs} = \frac{k_c}{1 + \frac{1}{K_c}}$$

In this model the interactions between the enzyme and the substrate are partitioned between the ES and ES' complexes. The modifications to substrate structure that disrupt ribozyme-substrate interactions that primarily affect catalysis, such as changes to functional groups at the cleavage site, would be likely to contribute to formation of ES'. In the RNA alone reaction, it is assumed that the ES to ES' equilibrium is unfavorable, and that the reaction is sensitive to substrate modifications that lower K_c.

The single turnover rate constant for the RNA alone reaction is dependent on reaction pH, and this sensitivity has been noted as evidence that the chemical step is rate limiting. Substrate modification that reduce the rate constant for catalysis typically do not alter this pH dependence. As noted previously by Pan and colleagues, this framework assumes that both binding and conformational change steps are at equilibrium, and thus, the chemical step remains rate limiting (Loria and Pan 1998; Zahler et al. 2005). However, the k_{obs} is found to decrease due to substrate mutations that decrease the fraction of enzyme in the ES' state.

If the docking equilibrium is unfavorable and the ES state is favored (that is the equilibrium constant is low), then k_{obs} will approach zero, despite the fact that the substrate is fully bound. If the conformational change is highly favorable and the equilibrium constant is very large then k_{obs} will approximate k_c. The important feature is that k_{obs} will vary hyperbolically with changes in K_c. That is, if K_c is already very large, then a significant change in its magnitude will have little effect on the observed rate constant for the reaction. For example, a decrease in the magnitude of K_c 1,000 to 100 reduces the observed rate constant by only 0.1%, while a similar tenfold decrease of K_c from 10 to 1 results in a 50% decrease in k_{obs}.

An important feature is that when K_c is large changes in its magnitude have little effect on catalysis, but the thermodynamic contribution of these interactions to K_c will "show up" in equilibrium binding. In the above model, the fraction of bound substrate f_{bound} is given by:

$$f_{bound} = \frac{ES + ES'}{E + ES + ES'}$$

which by virtue of the definition of the equilibrium constants given above is:

$$f_{bound} = \frac{1 + K_c}{1 + K_c + \dfrac{1}{K_a[S]}}$$

From this it follows that when docking is highly unfavorable then K_c will be small and the fraction of bound enzyme will vary with substrate concentration according to:

$$f_{bound} = \frac{1}{1 + \dfrac{1}{K_a[S]}}$$

And the equilibrium association constant for the initial complex will govern binding affinity. In this case, mutations that affect K_c will not significantly alter the dependence of complex formation on substrate concentration. However, these changes will affect the observed catalytic rate constant as indicated above. On the other hand, when the conformation change step is favorable and K_c is large then the fraction of enzyme bound will be approximated by:

$$f_{bound} = \frac{1}{1 + \dfrac{K_c}{K_a[S]}}$$

And thus, substrate modifications that affect K_c will be reflected in the dependence of f_{bound} on substrate concentration.

Several lines of evidence support such a two step binding mechanism. Pan and colleagues suggested such a docking step to explain how substrate modifications at the correct RNase P processing site can affect both the rate constants for correct cleavage and for mis-cleavage (Loria and Pan 1998). The quantitative model is essentially the same as the one outlined above. A conformational change involving interdomain motion was proposed. Additionally, the Kirsebom laboratory proposed an "induced fit" mechanism for RNase P substrate recognition base on the observation that the modification of enzyme-substrate contacts distal to the cleavage site can also reduce the rate constant for catalysis (Brannvall et al. 2007). While no specific reaction mechanism or quantitative expression for the changes in the observed catalytic constants was reported, it is assumed that the concept of "induced fit" conforms to the simple quantitative equilibrium model described above. A good example of the context dependence of the effects of substrate modification that support a two step binding mechanism is the effect of modifications to the nucleobase and ribose moieties of N(-1) the residue immediately upstream of the cleavage site in the 5′ leader sequence (Zahler et al. 2003; Zahler et al. 2005). Disruption of the adjacent 2′-hydroxyl contact has a greater effect on catalysis when pairing between the ribozyme and N(-1) is also disrupted. Importantly, the effects of simultaneously disrupting these contacts on binding are also non-additive in that both make only small reductions in binding affinity when the other modification is present.

Thus, the "reductionist" perspective on RNase P substrate recognition has provided important insights into the function of the enzyme. Furthermore, such strict structure-function dissections of substrate specificity have yielded specific quantitative models of the binding mechanism that successfully explain most of the observations that have been made regarding the magnitude of the effects on binding and catalysis. However, it is not clear how this perspective interfaces with the constraints on RNase P's role in tRNA biosynthesis outlined in the previous section. Does the equilibrium for the conformational change influence the rate constant for catalysis for substrate recognition in vivo? If so, then the rate constant for RNase P processing may be important for influencing the steady state distribution of tRNAs. How does the two step binding mechanism alter our expectations regarding the multiple turnover reaction and enzyme specificity? Is a conformational change an important component of the mechanism for ensuring cleavage at the correct site, or is it simply a reflection of the propensity of RNA to form multiple stable conformational states? One approach to answer these questions is to move beyond the "reductionist" perspective and focus on the properties of the holoenzyme and the biological context in which RNase P operates and attempt to develop an integrated perspective.

8.4 Facing up to the Biological Context

The motivation to develop a more integrated perspective on how the kinetic and the modynamic properties of RNase P are tuned to fulfill its biological role raises several challenges. The first challenge, as outlined above, is to understand how the structural variation in the naturally occurring population of pre-tRNAs in a cell affects the rate of tRNA processing by RNase P and potentially the relative levels of different tRNAs in vivo. The second challenge comes from the realization that the RNase P substrates exist in competition with each other. Understanding the quantitative effects of this competition on multiple turnover reaction rates should shed light on the fundamental properties of RNase P function. Furthermore, it is important to recognize that in addition to competition between cognate pre-tRNA substrates, there is the potential for competition between these authentic substrates and the large concentration of non-cognate RNA substrates in the cell. In this section, we discuss each of these issues in light of recently published work, and attempt to identify those avenues of investigation which may be most useful in gaining an integrated model of RNase P in vitro biochemistry and in vivo function.

Even for artless enzymologists not trained in comparative sequence analysis, simple inspection of the tRNA genes in *E. coli* reveals there is a lot of structural diversity. From a compilation of *E. coli* pre-tRNA sequences adjacent to the site of RNase P cleavage, inspection reveals that the majority contains the full complement of consensus elements at the cleavage site that have been identified as important for RNase P recognition (Sun et al. 2006). Nonetheless, a significant number (ca. 30%) lack a consensus U necessary for interaction at the N(-1) position, but nonetheless retain a G(1)–C(72) or other consensus Watson–Crick base pair at the cleavage site. About 20% have Watson–Crick base pairs at the N(1)–N(72) position other than the consensus G–C, but the majority of these retain a consensus U at N(-1). Several E. coli pre-tRNAs have an even greater degree of variation in these consensus recognition elements and would be predicted, based on the "reductionist" model for RNAse P recognition, to be very poor substrates for the enzyme. These substrates include a pre-$tRNA^{HIS}$, which retains a closing G–C pair and U(-1) contacts, but has an extra base pair in its acceptor stem. Additionally, the pre-tRNA for initiator tRNAf-met and certain pre-tRNALEU substrates have nonconsensus C–A or G–U pairs, respectively, at the cleavage site and also lack a U at N(-1). Thus, pre-tRNAs fall into two broad catagories: consensus substrates containing all of the recognized RNase P recognition elements and non-consensus pre-tRNAs that lack one or more of these features. Note that both classes can be considered cognate substrates, that is they are substrates for RNase P in vivo, and not non-cognate substrates (like rRNA or mRNA) where RNase P cleavage would be detrimental to cell function.

To better understand how such variation affects the processing by P RNA and the RNase P holoenzyme, we examined the binding and cleavage examples of pre-tRNAs that show the greatest deviation from consensus structure as well as examples of consensus substrates (Sun et al. 2006). As expected from the reductionist structure-function studies of the RNA alone reaction, the non-consensus substrates bind with lower affinity and undergo cleavage with a much lower rate constant than consensus substrates. Most interestingly, the results revealed that the affinities of both consensus and non-consensus substrates for the RNase P holoenzyme are essentially uniform. Comparative analyses of pre-tRNA and tRNA binding to the RNase P holoenzyme and P RNA alone reveal differential contributions of the protein subunit to 5′-leader and tRNA affinity. Structure swapping experiments in which leader sequences were exchanged between different substrates showed that uniform binding results from variations in the energetic contribution of the 5′-leader, which serve to compensate for weaker tRNA interactions. Furthermore, kinetic analyses reveal uniformity in the rates of substrate cleavage that result from dramatic (4,900-fold) contributions of the protein subunit to catalysis for some non-consensus pre-tRNAs.

Together, these data suggest that an important biological function of RNase P protein is to offset differences in pre-tRNA structure such that binding and catalysis are uniform. Such a mechanism is similar to previously described instances of thermodynamic compensation by structural variation for tRNA binding to EF-Tu and the ribosome (e.g. (LaRiviere et al. 2001; Fahlman et al. 2004). The uniformity in kinetics and thermodynamic clearly extends to substrates that differ significantly from the consensus derived from structure function studies of individual substrates. Such a result has an important implication concerning the biological role of RNase P introduced in the first section in that it predicts that the multiple turnover rate constant for the RNase P catalyzed step will be essentially identical for different pre-tRNAs. Thus, it is unlikely that this step will contribute significantly to the overall distribution of tRNA species. The model that emerges is that the biological role of RNase P is to apply no discrimination between different structurally variant pre-tRNAs, and to move the intermediates along as rapidly as possible to the next processing step. Such a model further underscores that the importance of regulation at the level of transcription, but does not preclude a contribution to the regulation at the level of pre-tRNA turnover (Dittmar et al. 2004).

An unanticipated development was the observation that C5 makes significant contributions to catalysis for the non-consensus substrates that were examined (Sun et al. 2006). The robust catalytic activity of *E. coli* P RNA with tight binding substrates and the ability of the *B. subtilis* protein to affect catalysis similar to C5 argue that the protein does not contribute functional groups to active site formation or interact directly with the substrate phosphate. Instead, the large protein effects on both binding and catalysis suggest that the influence on the catalytic rate constant is likely to be indirect, involving stabilization of the catalytic enzyme-substrate complex rather than affecting the intrinsic rate of catalysis (k_c in the two-step mechanism outlined, above).

The two step binding scheme provides a potential mechanism for the contribution of *E. coli* C5 to RNase P holoenzyme function with non-consensus substrates without evoking a direct participation in active site interactions with the reactive phosphate of pre-tRNA. It is assumed that the variation from consensus pre-tRNA structure results in an unfavorable conformational change in the RNA alone reaction. That is, K_c for such substrates is low and they bind primarily in the ES rather than the ES' complex. In the RNase P holoenzyme reaction, tight, direct contacts between the protein subunit and leader, as well as the interactions of tRNA with P RNA would act to overcome strain introduced by the juxtaposition of RNA functional groups in poor geometric and electrostatic environments for substrates lacking consensus elements. Thus, the main effect of the protein would be to contribute to shifting the equilibrium into the ES' state. The proposed mechanism for protein effects on catalysis (stabilization of E–S*) is consistent with the observation that binding and catalysis effects are correlated because measurements of each involve a K_c component (see above). If the docked ES' and undocked ES states are in equilibrium, then the observed rate (k_{obs}) is a function of both k_{chem} and K_c (Loria and

Pan 1998; Zahler et al. 2003; Zahler et al. 2005) (see above). The observed binding affinity will depend on both K_d for pre-tRNA binding and K_c, and the protein can have effects on both, resulting in a large shift to the docked state in the presence of the protein.

In addition to variation in the sequence of tRNA and the 5′ leader sequence there are also differences in the context. Of the 87 tRNA genes in *E. coli* only 18 are located in rRNA operons, while the remainder are transcribed as individual genes (Komine et al. 1990). The influence of this difference in pre-tRNA context, if any, is not known. It has been assumed that rRNA processing steps and trimming to generate the tRNA 3′ end precede cleavage by RNase P. However, there is no functional property of the RNase P enzyme that we know of that would preclude its acting on large precursor transcripts. If 3′ processing precedes 5′ processing, then the presence of flanking sequences in a large precursor will not be present to interfere with or modulate subsequent RNase P processing. However, whether 5′ end processing truly precedes 3′ end processing has not been systematically examined, and may depend on the sequence length and structure of the individual pre-tRNA. Results from conditional expression of RNase P in the cyanobacterium *Synechocystis* indicate a preferred order in the action of RNase P when it processes a dimeric tRNA precursor. Additionally, in RNase P depleted cells precursors were detected that were 3′-processed, indicating that 3′ processing can occur before 5′ processing by RNase P (Tous et al. 2001). Regardless, the presence of 3–5 additional nucleotides on the 5′ end of pre-tRNA does not appear to have any significant effect on the RNA alone reaction (LaGrandeur et al. 1994). Thus, the observation that RNase P is not sensitive to extra 3′ sequences essentially means that it can act on substrates that either have, or have not already undergone 3′ end maturation.

However, the sensitivity the RNase P holoenzyme to 3′ flanking sequences has not been systematically nor quantitatively established. Indeed, the assumption that 3′ flanking sequences have little or no information content is not a safe one. Pairing between nucleotides proximal to the 5′ and 3′ ends of the tRNA sequence clearly can perturb RNase P recognition. Pettersson and Kirsebom have demonstrated that pairing between N(−1) and G(73) can result in cleavage at an adjacent phosphodieser bond 5′ to the canonical cleavage site in vitro and in vivo (Pettersson and Kirsebom 2008). Disruption of C−1/G+73 restores processing at the correct RNase P cleavage site. The results provided are important in demonstrating that the local structure can be crucial for RNase P processing. In addition, they highlight the potential for there to be additional instances in which the overall context of the flanking sequences that a particular tRNA is embedded in can influence its expression.

In contrast to variation in sequence and structure at the 3′ end of substrates, more information is available for the impact of variation in 5′ leader sequence and length on RNase P processing. If 3′ end processing can precede processing by RNase P, then the length of leader sequences for poly-cistronic substrates can vary or ambiguous. The 3′ processing of an upstream tRNA in a poly-cistronic transcript defines the 5′ leader length for the downstream substrate. However, for the numerous pre-tRNAs in *E. coli* that are encoded by individual transcription units, the variation in the length of 5′ leaders can be more directly examined. Based on computational analyses the average length of the 5′ leaders in tRNA precursors in *E. coli* is 17–18 residues long. Kirsebom and colleagues also showed using an in vivo assay based on tRNA nonsense suppression investigated the effect of the 5′ leader length variation on tRNA processing and bacterial growth (Fredrik Pettersson et al. 2005). The data indicate that there are growth advantages to have a leader sequence, however, they demonstrated that at least for one example the expression of tRNA gene constructs that lacked sequences encoding the 5′ leader resulted in a functional tRNA. Thus, the occurrence of an RNase P processing step is not obligatory for cells to function. However, it is possible that more subtle effects may be important for optimal cell growth, or that individual tRNA species may have idiosyncratic differences regarding the requirement for leader length and sequence.

An additional facet to the biological context of RNase P function is that it must process all pre-tRNAs in the cell. Obviously it must do so in an environment where all of the substrates are present simultaneously. In this context, the rate of formation of tRNA from pre-tRNA S_1 can be expressed in general as a function of the concentration of the free enzyme ($[E]_0$), the substrate concentration ($[S_1]$) and k_{cat}/K_m according to:

$$v_1 = (k_{cat}/K_m)_1 [E]_0[S_1]$$

Therefore, the ratio of the rates of processing for two different substrates (S_1 and S_2), and an important definition of specificity (Herschlag 1988), will be:

$$\frac{v_1}{v_2} = \frac{(k_{cat}/K_m)_1 [E]_0[S_1]}{(k_{cat}/K_m)_2 [E]_0[S_2]} = \frac{(k_{cat}/K_m)_1 [S_1]}{(k_{cat}/K_m)_2 [S_2]}$$

An important point is that the dependence on enzyme concentration cancels out and the discrimination at any given concentration of the two competing pre-tRNA substrates will be determined by the ratio of the values of k_{cat}/K_m for the two substrates in question(Herschlag 1988). This perspective indicates that if k_{cat}/K_m is similar for different pre-tRNAs, then the rates of processing will be solely determined by their relative concentration. Also, it is clear that the relevant biochemical parameter is (of course) k_{cat}/K_m, and that the k_{cat}/K_m for different substrates may be monitored in complex reaction mixtures where there are multiple substrates in competition where the k_{cat}/K_m of at least one of the substrates is known.

In this example, specificity is defined as the ratio of the rates for the two different substrates (v_1/v_2). This concept applies to two different cognate substrates (regardless of whether they are canonical or non-canonical); however, it also applies to the discrimination between cognate and non-cognate substrates. Regarding specificity mis-cleavage of pre-tRNA substrates has been studied extensively and has provided important information and an experimental basis for dissecting ribozyme-substrate interactions. However, the kind of specificity implied by comparing pre-tRNA and non-pre-tRNA substrates is a different kind of specificity with a different biological role. Mis-cleavage of a pre-tRNA substrate to generate tRNA with extra nucleotides, or a deficit in nucleotides at the 5′, but that is otherwise normal would likely be recognized by the cellular RNA turnover machinery and degraded. However, cleavage of non-cognate substrates by a ribonuclease is obviously counterproductive to normal cell function.

It would appear obvious that the cleavage of non-cognate substrates is not a problem at the concentration of RNase P in the cell is reasonably high (see above). However, this conclusion says nothing about the kinetic or thermodynamic mechanisms by which RNase P avoids cleavage of non-cognate substrates. Interestingly, at high (micromolar) enzyme concentrations, RNase P will cleave RNA hairpins and duplexes with 5′ overhangs, typically within GU sequences(Hansen et al. 2001). Cleavage can also occur even single stranded oligonucleotides with some limited sequence specificity. At present, the overall kinetic and thermodynamic features of the processing of such non-cognate substrates have not been examined. Accomplishing this goal would be important in defining the constraints on the biological function of RNaseP. An important questions to ask in this regard seem to be: What are the limits to how fast RNase P will cleave non-cognate substrates in vitro?

8.5 Summary and Perspective

It is clear that RNase P will continue to be a useful experimental system in which to explore the structure function relationships that underlie its catalytic function. These essential avenues of investigation will certainly continue to include, among other things, RNA-metal ion interactions for both folding and catalysis and RNA-protein interactions. With respect to RNA-protein interactions it is clear that both the RNA and protein subsunits undergo some conformational transitions upon assembly which have profound consequences for its function. While extensive progress has been made in defining the optimal elements of RNase P recognition, developing a more complete and quantitative understanding of how the enzyme accommodates variations amongst different substrates will be in interesting challenge. As evidenced by several of the exciting studies described above in which RNase P processing was considered in vivo, we have much to learn about the interplay between substrate structure and specificity, both on a local (cleavage site recognition), regional (competition between canonical and non-canonical substrates) and global scale (competition between cognate and non-cognate substrates). As shown by the difficult advances and dramatic achievements outlined elsewhere in this volume, it can be expected that developing in integrated perspective that combines detailed descriptions of three dimensional structure, reaction kinetics, and thermodynamics with the biologically imposed constraints on its function is not unrealistic.

References

Baird NJ, Fang XW et al (2007) Folding of a universal ribozyme: the ribonuclease P RNA. Q Rev Biophys 40(2):113–161

Beebe JA, Kurz JC et al (1996) Magnesium ions are required by Bacillus subtilis ribonuclease P RNA for both binding and cleaving precursor tRNAAsp. Biochemistry 35(32):10493–10505

Brannvall M, Kikovska E et al (2007) Evidence for induced fit in bacterial RNase P RNA-mediated cleavage. J Mol Biol 372(5):1149–1164

Bremer H, Dennis P (2008) Feedback control of ribosome function in *Escherichia coli*. Biochimie 90(3):493–499

Buck AH, Dalby AB et al (2005a) Protein activation of a ribozyme: the role of bacterial RNase P protein. EMBO J 24(19):3360–3368

Buck AH, Kazantsev AV et al (2005b) Structural perspective on the activation of RNAse P RNA by protein. Nat Struct Mol Biol 12(11):958–964

Christian EL, McPheeters DS et al (1998) Identification of individual nucleotides in the bacterial ribonuclease P ribozyme adjacent to the pre-tRNA cleavage site by short-range photo-cross-linking. Biochemistry 37(50):17618–17628

Christian EL, Zahler NH et al (2002) Analysis of substrate recognition by the ribonucleoprotein endonuclease RNase P. Methods 28(3):307–322

Cleland WW, Cook PF (2007) Enzyme Kinetics and Mechanism. Garland Publishers, London and New York

Deutscher MP (2009) Maturation and degradation of ribosomal RNA in bacteria. Prog Mol Biol Transl Sci 85:369–391

Dittmar KA, Mobley EM et al (2004) Exploring the regulation of tRNA distribution on the genomic scale. J Mol Biol 337(1):31–47

Evans D, Marquez SM et al (2006) RNase P: interface of the RNA and protein worlds. Trends Biochem Sci 31(6):333–341

Fahlman RP, Dale T et al (2004) Uniform binding of aminoacylated transfer RNAs to the ribosomal A and P sites. Mol Cell 16(5):799–805

Frank DN, Pace NR (1998) Ribonuclease P: unity and diversity in a tRNA processing ribozyme. Annu Rev Biochem 67:153–180

Fredrik Pettersson BM, Ardell DH et al (2005) The length of the 5′ leader of Escherichia coli tRNA precursors influences bacterial growth. J Mol Biol 351(1):9–15

Gardiner KJ, Marsh TL et al (1985) Ion dependence of the Bacillus subtilis RNase P reaction. J Biol Chem 260(9):5415–5419

Hansen A, Pfeiffer T et al (2001) Exploring the minimal substrate requirements for trans-cleavage by RNase P holoenzymes from *Escherichia coli* and *Bacillus subtilis*. Mol Microbiol 41(1):131–143

Herschlag D (1988) The role of induced fit and conformationla changes of enzymes in specificity and catalysis. Bioorg Chem 16:62–96

Hsieh J, Andrews AJ et al (2004) Roles of protein subunits in RNA-protein complexes: lessons from ribonuclease P. Biopolymers 73(1):79–89

Ikemura T (1981) Correlation between the abundance of Escherichia coli transfer RNAs and the occurrence of the respective codons in its protein genes: a proposal for a synonymous codon choice that is optimal for the E. coli translational system. J Mol Biol 151(3):389–409

Ikemura T (1985) Codon usage and tRNA content in unicellular and multicellular organisms. Mol Biol Evol 2(1):13–34

Karbstein K, Tang KH et al (2004) A base triple in the Tetrahymena group I core affects the reaction equilibrium via a threshold effect. RNA 10(11):1730–1739

Kazantsev AV, Pace NR (2006) Bacterial RNase P: a new view of an ancient enzyme. Nat Rev Microbiol 4(10):729–740

Kikovska E, Brannvall M et al (2006) The exocyclic amine at the RNase P cleavage site contributes to substrate binding and catalysis. J Mol Biol 359(3):572–584

Kirsebom LA (2007). RNase P RNA mediated cleavage: Substrate recognition and catalysis. Biochimie

Kirsebom LA, Altman S (1989) Reaction in vitro of some mutants of RNase P with wild-type and temperature-sensitive substrates. J Mol Biol 207(4):837–840

Kirsebom LA, Trobro S (2009) RNase P RNA-mediated cleavage. IUBMB Life 61(3):189–200

Komine Y, Adachi T et al (1990) Genomic organization and physical mapping of the transfer RNA genes in Escherichia coli K12. J Mol Biol 212(4):579–598

Kurland CG (1993) Major codon preference: theme and variations. Biochem Soc Trans 21(4):841–846

LaGrandeur TE, Huttenhofer A et al (1994) Phylogenetic comparative chemical footprint analysis of the interaction between ribonuclease P RNA and tRNA. EMBO J 13(17):3945–3952

LaRiviere FJ, Wolfson AD et al (2001) Uniform binding of aminoacyl-tRNAs to elongation factor Tu by thermodynamic compensation. Science 294(5540):165–168

Li Z, Deutscher MP (1996) Maturation pathways for *E. coli* tRNA precursors: a random multienzyme process in vivo. Cell 86(3):503–512

Li Z, Deutscher MP (2002) RNase E plays an essential role in the maturation of Escherichia coli tRNA precursors. RNA 8(1):97–109

Loria A, Pan T (1998) Recognition of the 5′ leader and the acceptor stem of a pre-tRNA substrate by the ribozyme from Bacillus subtilis RNase P. Biochemistry 37(28):10126–10133

Loria A, Pan T (1999) The cleavage step of ribonuclease P catalysis is determined by ribozyme-substrate interactions both distal and proximal to the cleavage site. Biochemistry 38(27): 8612–8620

Ow MC, Kushner SR (2002) Initiation of tRNA maturation by RNase E is essential for cell viability in E. coli. Genes Dev 16(9):1102–1115

Pettersson BM, Kirsebom LA (2008) The presence of a C-1/G+73 pair in a tRNA precursor influences processing and expression in vivo. J Mol Biol 381(5):1089–1097

Potrykus K, Cashel M (2008) (p)ppGpp still magical? Annu Rev Microbiol 62:35–51

Shan SO, Herschlag D (2000) An unconventional origin of metal-ion rescue and inhibition in the Tetrahymena group I ribozyme reaction. RNA 6(6):795–813

Smith D, Pace NR (1993) Multiple magnesium ions in the ribonuclease P reaction mechanism. Biochemistry 32(20):5273–5281

Smith D, Burgin AB et al (1992) Influence of metal ions on the ribonuclease P reaction Distinguishing substrate binding from catalysis. J Biol Chem 267(4):2429–2436

Smith JK, Hsieh J et al (2007) Importance of RNA-protein interactions in bacterial ribonuclease P structure and catalysis. Biopolymers 87(5–6):329–338

Sun L, Campbell FE et al (2006) Evidence that substrate-specific effects of C5 protein lead to uniformity in binding and catalysis by RNase P. EMBO J 25(17):3998–4007

Szaflarski W, Vesper O et al (2008) New features of the ribosome and ribosomal inhibitors: non-enzymatic recycling, misreading and back-translocation. J Mol Biol 380(1):193–205

Tous C, Vega-Palas MA et al (2001) Conditional expression of RNase P in the cyanobacterium Synechocystis sp. PCC6803 allows detection of precursor RNAs. Insight in the in vivo maturation pathway of transfer and other stable RNAs. J Biol Chem 276(31):29059–29066

Wegscheid B, Hartmann RK (2007) In vivo and in vitro investigation of bacterial type B RNase P interaction with tRNA 3′-CCA. Nucleic Acids Res 35(6):2060–2073

Zahler NH, Christian EL et al (2003) Recognition of the 5′ leader of pre-tRNA substrates by the active site of ribonuclease P. RNA 9(6):734–745

Zahler NH, Sun L, et al (2005) The pre-tRNA nucleotide base and 2′-hydroxyl at N(-1) contribute to fidelity in tRNA processing by RNase P. J Mol Biol 345(5): 969–985. Epub 2004 Dec 8

Chapter 9
Archaeal RNase P: A Mosaic of Its Bacterial and Eukaryal Relatives

Lien B. Lai, I-Ming Cho, Wen-Yi Chen, and Venkat Gopalan

9.1 Introduction

The literature on RNase P is dominated by studies on the bacterial prototype given its historical significance in establishing the catalytic potential of RNAs (Guerrier-Takada et al. 1983). However, a small but growing number of reports on the non-bacterial variants indicate that RNase P continues to be a treasure trove of surprises and all its guises merit further study. With the notable exceptions of a few organellar versions, RNase P is a ribonucleoprotein (RNP) in all organisms (Gopalan and Altman 2006; Walker and Engelke 2006; Holzmann et al. 2008; Marquez et al. 2008; also see Chap. 11). The activity of RNase P RNAs (RPRs) *decreases* from bacteria to archaea to eukarya and exhibits an inverse correlation with the protein:RNA mass ratio which *increases* from bacteria to archaea to eukarya (10 to 50 to 70% of total RNP, respectively). RNase P is therefore an appealing model to address how structural and functional attributes of RPRs might have been reassigned to the RNase P protein (RPP) cofactors that were recruited during evolution.

With respect to macromolecular recognition and biocatalysis, RNase P exemplifies a rare instance in that there are striking differences among (1) the subunit compositions of its natural variants in the three domains of life and (2) its numerous substrates which vary both in sequence and structure. Additional catalytic challenges are likely to arise when the RNase P-catalyzed reaction needs to be

L.B. Lai, I.-M. Cho, W.-Y. Chen, and V. Gopalan (✉)
Department of Biochemistry, The Center for RNA Biology, The Ohio State University, Columbus, OH 43210, USA
e-mail: gopalan.5@osu.edu

I.-M. Cho and V. Gopalan
Department of Molecular Genetics, The Ohio State University, Columbus, OH 43210, USA

W.-Y. Chen and V. Gopalan
The Molecular Cellular and Developmental Biology Program,
The Ohio State University, Columbus, OH 43210, USA

F. Liu and S. Altman (eds.), *Ribonuclease P*, Protein Reviews 10,
DOI 10.1007/978-1-4419-1142-1_9,

performed in extreme physical or chemical habitats. Therefore, it is necessary to study divergent RNase P representatives including those from extremophiles to obtain a comprehensive picture of how, despite plurality in the nature of the enzyme (E) and its substrate (S), rearrangements in the ES complex permit a common biological function.

Characterizing and comparing variants that differ in subunit makeup is likely to yield insights into the functional alterations that accompanied the gradual transformation of an ancient catalytic RNA to a protein-rich RNP. An essential first step in this regard is to reconstitute and study the different holoenzymes in vitro. Such experiments on the one-RNA, one-protein bacterial RNase P have revealed that although the RPR is catalytically active in vitro, the single protein cofactor exerts pleiotropic effects on RPR structure and catalysis thus accounting for its essentiality in vivo. Archaeal RNase P has the promise of being an excellent surrogate for the experimentally intractable eukaryal relative based on three main reasons (Tsai et al. 2006). First, due to the high sequence conservation between archaeal and eukaryal RPPs (Hall and Brown 2002), insights on the function of RPPs in archaeal RNase P are likely to be applicable to eukaryal RNase P, whose reconstitution has proven difficult. Second, with relatively fewer (predicted) and smaller proteins, archaeal RNase P is a simpler RNP complex that might constitute the minimal catalytic RNP core in eukaryal RNase P. Third, archaeal RPPs are strongly basic and thermostable, appealing features for their purification, structural studies and functional reconstitution. Some of these expectations have already been borne out. In the following sections, we discuss our current knowledge of structure–function relationships in the archaeal RPR and RPPs, functional coordination among the subunits, and conclude by enumerating research directions on archaeal RNase P that should be profitable.

9.2 Isolation and Characterization of Native Archaeal RNase P Holoenzymes

Native RNase P has been partially purified and characterized from four archaea: three euryarchaea [*Haloferax volcanii* (*Hvo*), *Methanothermobacter thermautotrophicus* (*Mth*), and *Methanocaldococcus jannaschii* (*Mja*)] and a crenarchaeon [*Sulfolobus acidocaldarius* (*Sac*)] (Lawrence et al. 1987; Darr et al. 1990; Andrews et al. 2001). Following the seminal findings on bacterial RPRs, the early studies focused on establishing the presence of an essential RNA subunit in *Hvo* and *Sac* RNase P. Identification of the *Hvo* and *Sac* RPRs, a first for archaea, was achieved by a painstaking approach of enzyme purification using chromatographic approaches followed by use of precipitated RNAs as probes to find the encoding genomic sequences (Nieuwlandt et al. 1991; LaGrandeur et al. 1993). Although the sequence and secondary structure models of *Hvo* and *Sac* RPRs revealed their shared lineage with bacterial relatives, neither RPR (without RPPs) showed any precursor tRNA (ptRNA)-processing activity when tested under conditions similar to those optimal for bacterial RPRs. These observations already hinted at a greater dependence of archaeal RPR on protein cofactors.

To assess the protein content in archaeal RNase P, buoyant density was measured on Cs_2SO_4 gradients and found to be 1.61, 1.42, 1.39, and 1.27 g/mL for *Hvo, Mth, Mja,* and *Sac*, respectively (Hamilton et al. 1971; Lawrence et al. 1987; Darr et al. 1990; Andrews et al. 2001). From these values, the corresponding RPP:RPR mass ratios were estimated as 10, 50, 50, and 70%, with the first and last ratios being those calculated from the masses of known subunits in bacterial and eukaryal RNase P, respectively. This wide range suggests that using buoyant density to estimate RPP content could serve only as a rough yardstick because modifications of proteins or co-purification of lipids could influence the buoyant density of an RNP (LaGrandeur et al. 1993).

In addition to providing a glimpse of the diversity among archaeal RNase P that remained to be unraveled, these native RNase P studies also revealed activity requirements and at least one limitation with regard to in vitro characterization of RNase P from thermophiles (Lawrence et al. 1987; Darr et al. 1990; Andrews et al. 2001). With *Hvo*, *Mja* and *Sac* RNase P, the optimal $[NH_4^+]$ is ≤50 mM; with *Mth*, it is 800 mM. The optimal $[Mg^{2+}]$ was 5, 7.5, 30, and 60 mM for *Mth*, *Sac*, *Mja* and *Hvo* RNase P, respectively. The optimal assay temperature was not determined for the mesophile *Hvo*, but was reported to be 77° C for *Sac* and ≥80° C for *Mja* and *Mth*. In all cases, RNase P from the thermophilic archaea could not be assayed at temperatures above 80° C due to thermal instability of the in vitro transcribed ptRNA substrates. The melting temperature of a tRNA from the hyperthermophile *Pyrodictium occultum* (*Poc*) was determined as 80° C for the in vitro transcribed version compared to ≥100° C for that purified in its native form from *Poc*; the latter was shown to contain a variety of modified nucleosides, which might underlie its thermostability (Ushida et al. 1996).

Although the purification of RNase P from the four archaea had led to the identification of their RPRs in some cases, none of these preparations was sufficiently homogeneous to permit elucidation of the RPP composition. Toward this goal, we have initiated collaboration with Tom Santangelo and John Reeve (OSU) to introduce by homologous recombination an affinity-tagged variant of an RNase P protein subunit in *Thermococcus kodakaraensis* (*Tko*). Such archaeal genetic manipulation has become available only recently and our preliminary results suggest that purification of native *Tko* RNase P to homogeneity and subsequent identification of all RPPs should be feasible (unpublished results).

9.3 Archaeal RNase P RNA (RPR)

9.3.1 Identification

The search for an RPR sequence in the rapidly growing databases of genomic sequences, although not straightforward, has been facilitated by computational methods. While BLAST searches are fruitful if a sequence from a closely related species is available as a query, customized search algorithms are generally more successful. For example, Li and Altman (2004) reported a two-step procedure for finding and validating an RPR sequence from a sequenced microbial genome. After first retrieving 600-nt sequences having two universally conserved segments, GAGGAANNUCNNNNC in the conserved region (CR) I and AGNNNNAU... [10–60 nt]...ACANAANNNNGNNUA in CR IV and V, these candidate sequences are then examined for other notable RPR features using the pattern-profiling program RNAMotif. Using this computational method, they found a single RPR sequence in 51 of 52 bacterial and six of eight archaeal genomes, where an RPR sequence was not previously annotated.

The three genomes, where Li and Altman (2004) could not identify an RPR, were from the bacterium *Aquifex aeolicus* and the archaea *Nanoarchaeum equitans (Neq)* and *Pyrobaculum aerophilum (Pae)*. Moreover, failure to detect a 5′ ptRNA-processing activity in crude lysates of *Neq* and *Pae* led Randau et al. (2008) to propose that in *Neq* the pressure of genome reduction might have influenced strict promoter placement and transcription of leaderless tRNAs, thus obviating the need for a 5′ ptRNA-processing enzyme. However, tRNA sequencing data are required to establish unambiguously that all *Neq* tRNAs are indeed transcribed leaderless. It is possible that as *Neq* is an obligate parasitic archaeon, found physically attached throughout its life cycle to an *Ignicoccus* host, its parasitic lifestyle might have relaxed the dependence on a self-encoded ptRNA-processing enzyme. In contrast to Li and Altman (2004) and Randau et al. (2008), we recently discovered a radically smaller form of RPR in *Pae* and four other *Pyrobaculum* species using comparative genomics and improved computational methods. We also confirmed ptRNA-processing activity of this RPR in vitro and demonstrated *bona fide* RNase P activity in partially purified cell extracts from *Pae* and *P. calidifontis (Pca)*, suggesting that an RPR-dependent RNase P is present in these free-living archaea (Lai et al. 2009; see Sect. 5).

9.3.2 RPR General Features and Structure

The archaeal RPR is evolutionarily and structurally related to those from bacteria and eukarya (Fig. 9.1; Brown 1999; Hall and Brown 2001). When compared to bacterial RPRs, both eukaryal and archaeal RPRs lack P13, P14 and P18, but share 13 conserved nucleotides most of which are clustered around the active site (Brown 1999; Walker and Engelke 2006; Gopalan 2007; Marquez et al. 2008). Additional information on the phylogenetic covariation analyses that have yielded insights on the structure of archaeal RPRs is provided elsewhere (see Chap. 2).

Bacterial RPRs have two independently folding domains which cooperate to generate a functional tertiary fold: the specificity (S) domain interacts with the T stem-loop of the ptRNA and positions it for cleavage by the catalytic (C) domain (Loria and Pan 1996). Although thematic parallels have not been fully established, the C domain in *Pyrococcus furiosus* (*Pfu*) and *Mja* RPRs harbors all the determinants

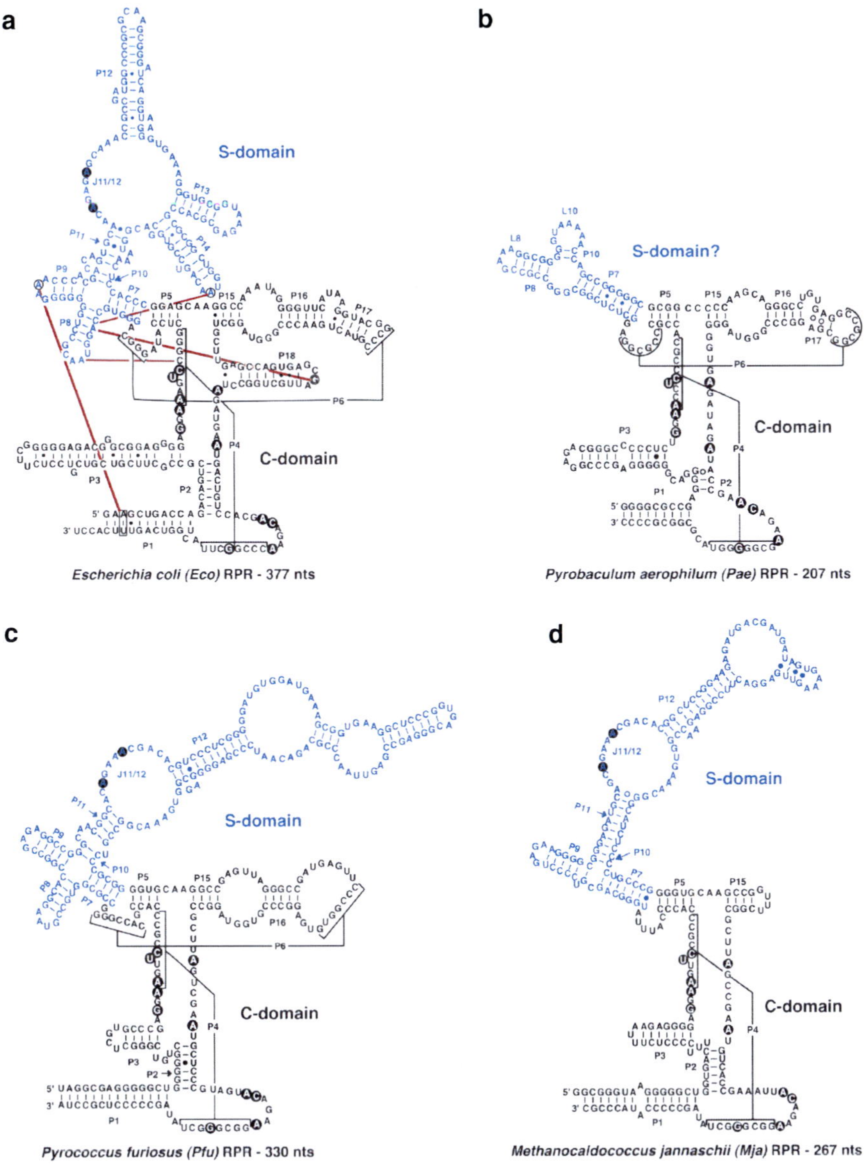

Fig. 9.1 Secondary structures of RPRs from bacteria (**a,** type A), crenarchaea (**b**) and euryarchaea (**c,** type A; **d,** type M). Paired regions are labeled as P1, P2, etc. consecutively from 5′ to 3′. Universally conserved nucleotides are highlighted in *black circles*. The C and S domains are indicated in *black* and *blue* color, respectively; the S domain assignments for archaeal RPRs require further experimental validation. *Red lines* indicate the tertiary contacts in bacterial RPR (**a**), which were predicted initially by phylogenetic covariation analyses and subsequently validated by high-resolution structures. Parts of this figure were modified from versions available in the RNase P database (Brown 1999)

for catalysis, as it could perform ptRNA cleavage without the S domain when cognate RPPs are present or when the ptRNA is conjugated to the RPR (Tsai et al. 2006; Pulukkunat and Gopalan 2008).

Most characterized archaeal RNase P enzymes come from the kingdom Euryarchaeota and have been classified into types A and M based on the secondary structure of the RPR (Harris et al. 2001). The majority are type A RPRs, so-called based on their resemblance to the bacterial type A RPRs (Fig. 9.1c). Type M RPRs lack P6, P8, and P16-17, and possess a smaller, atypical L15 (Fig. 9.1d); to date, only RPRs from *Methanococcales* and *Archaeoglobales* make up this type. All RPRs from the kingdom Crenarcheaota present in the RNase P database (http://www.mbio.ncsu.edu/RNaseP/home.html) are similar to type A, except for a shorter P15 and a smaller L15 that lacks the GGU sequence vital for anchoring the 3′ RCCA of ptRNAs (Kirsebom and Svard 1994; Oh and Pace 1994). In fact, these structural differences have functional consequences (see below) and appear to be the basis for discrimination of type A and M RPRs in terms of inhibition by aminoglycoside-arginine conjugates (Kawamoto et al. 2008).

9.3.3 RPR-Alone Activity In Vitro

Whether archaeal RPRs are catalytic in the absence of RPPs was unclear initially. Cell lysates of various archaea were able to process ptRNAs, but only a few retain activity after deproteinization by extraction with organic solvents (Pannucci et al. 1999). In some cases, the lysate activity was resistant to detergent in assay buffer, suggesting that the ptRNA-processing activity was dependent on RNA. Although the C domain of archaeal RPRs is highly similar to that of bacterial RPRs, it took a decade since *Hvo* (the first archaeal) RPR was identified before Pannucci et al. (1999) successfully demonstrated RPP-independent activity for it and some other archaeal RPRs. The finding that the main requirement for activity was an extremely high ionic strength (4 M NH_4OAc, 300 mM $MgCl_2$) suggests that these RPRs probably depend on RPPs for structural stability. Mn^{2+} (but not Ca^{2+}, Zn^{2+} or Cu^{2+}) could replace Mg^{2+} in promoting catalysis at an even lower concentration, but Mg^{2+} is preferred as Mn^{2+} increases nonspecific cleavage of the substrate, product, and likely the catalytic RNA moieties (Pannucci et al. 1999).

All archaeal RPRs lack P13, P14, and P18, elements revealed by the crystal structures of bacterial RPRs to play a pivotal role in bringing together the S and C domains (Kazantsev et al. 2005; Torres-Larios et al. 2005), whose proximity in turn underlies efficient and accurate cleavage of ptRNAs. As noted in the pioneering studies by Pannucci et al. (1999), high concentrations of monovalent and divalent cations occasionally mitigate the absence of protein cofactors, which might promote intramolecular braces that bridge the S and C domains. However, not all archaeal RPRs can be "forced" to cleave ptRNAs even at high concentrations of Mg^{2+} and NH_4^+ (Pannucci et al. 1999). Most notable of these intractable RPRs are the type M

RPRs that lack P8, a typical L15, and P16-P17 region (which includes half of the P6 pseudoknot). As these structural elements have been demonstrated to participate in substrate binding in bacterial RPRs (Harris and Christian 2003; Kirsebom 2007), absence of such elements might render the archaeal RPRs defective in binding and positioning ptRNAs for cleavage.

A comparison of the secondary structures of bacterial and archaeal RPRs reveals that the most divergent regions are those in the S domain (Fig. 9.1). This difference is most likely the basis for the lower substrate affinity in archaeal RPRs. Compared to the nanomolar range for the *Escherichia coli* (*Eco*) RPR, the K_m was determined as 30 μM for the *Pfu* RPR (Tsai et al. 2003) and estimated as >40 μM for the *Mth* RPR (Pannucci et al. 1999). Li et al. (2009) found that under identical assay conditions, the single-turnover rate at pH 6 for the *Eco* and *Mth* RPRs are 5 and ~0.03 min^{-1}, respectively. Replacing the S domain of *Mth* RPR with that from *Eco* RPR generated a chimera which is 34-fold more active than the *Mth* RPR, attesting to the poor ability of the archaeal S domain to engage in productive interactions with ptRNA. This poor substrate affinity is particularly true for the archaeal type M RPRs which are not active in vitro without the RPPs (Pannucci et al. 1999). We recently reported that the substrate-binding defect of the *Mja* (type M) RPR could be alleviated by two different RNA engineering strategies: either replacing its S domain with that from the *Eco* RPR or by directly tethering a ptRNA to the *Mja* RPR (Pulukkunat and Gopalan 2008). Both RNA constructs support efficient cleavage unaided by RPPs.

The above findings on archaeal RPRs, together with the demonstration that the human RPR can process different ptRNAs and model substrates (Kikovska et al. 2007), suggest that all RPRs possess the crucial structural elements required for generating the active site. In addition, as all characterized RPPs, either individually or in combination, cannot promote ptRNA processing without their RPRs, the RPR must be the catalytic moiety in all RNase P holoenzymes that function as an RNP. That the active site rests with the RPR during evolution is also supported by mutagenesis experiments in which the universally conserved bulge-helix motif in P4 was altered in bacterial, archaeal and eukaryal RPRs. The identity and position of the bulged U in this motif was demonstrated to be vital for maintenance of a unique helical geometry, which in turn is essential in *Eco* RPR for substrate positioning and binding of catalytically important Mg^{2+} ions (Kaye et al. 2002; Christian et al. 2006; Kikovska et al. 2007). Similarly, we found that removing (ΔU) or increasing the size of the bulge (+U) resulted in decreased activity in *Mth* RPR (unpublished results). *Pyrococcus horikoshii* (*Pho*) RNase P holoenzymes assembled using RPRs with identity/geometry alterations in the P4 helix also displayed lower activity compared to the wild type (Terada et al. 2007). Both *Mth* and some *Pho* RPR mutants could be rescued at high [Mg^{2+}], akin to the scenario with *Eco* RPR mutants. In the human RPR, deletion of the bulged U and the two flanking nucleotides eliminated activity (Kikovska et al. 2007). Therefore, some aspects of catalysis including positioning of substrate and catalytic Mg^{2+} ions are likely to be evolutionarily conserved in the various RNP forms of RNase P.

9.4 Archaeal RNase P Proteins (RPPs)

9.4.1 Identification

A common ancestry for the RPRs in all three domains of life is indicated by sequence and structure similarities, especially in regions expected to form the catalytic core (Fig. 9.1). However, the four currently known archaeal RPPs share homology only with those associated with eukaryal RNase P (Hall and Brown 2002), reflecting the shared archaeal and eukaryal origin deduced from the 16S rRNA-based phylogeny. Genetic and biochemical studies established the association of yeast and human nuclear RNase P activity with at least nine RPPs, of which seven are homologs (Eder et al. 1997; Chamberlain et al. 1998; Jarrous 2002; Walker and Engelke 2006). The archaeal RPPs, which were first computationally identified, are called POP5, RPP30, RPP21, and RPP29 and correspond to the human RPP homologs with the same names. Western analysis and immunoprecipitation subsequently confirmed that partially purified *Mth* RNase P activity co-elutes with these four RPPs (Hall and Brown 2002). The archaeal RPPs, particularly RPP29 and RPP30, are smaller than eukaryal homologs. The larger RPPs might be required in eukarya for nuclear localization (Jarrous et al. 1999) and for interacting with additional RPPs not present in archaea (Jiang and Altman 2001; Houser-Scott et al. 2002).

Recently, two additional archaeal proteins with some homology to human RPPs were examined as RPP candidates. The first is the ribosomal protein L7Ae that is homologous to human RPP38. Addition of *Pho* L7Ae to the in vitro reconstituted *Pho* RNase P (RPR+4 RPPs) elevated the maximal-activity temperature from 55 to 70°C, a temperature similar to that of partially purified native *Pho* RNase P assayed in vitro; the V_{max} of the *Pho* holoenzyme (assayed at 55°C) also increased five-fold upon addition of L7Ae (assayed at 65°C) (Fukuhara et al. 2006). Despite these interesting observations hinting at L7Ae being an RPP, there is currently no evidence for its direct association with *Pho* RNase P in vivo; however, if proven, L7Ae would present an interesting example of a single protein associated with different RNPs acting in tRNA processing (RNase P), RNA modification (snoRNPs) and translation (ribosomes). The second candidate is the archaeal chromatin protein Alba, whose family members include human RPP20 and RPP25 (Aravind et al. 2003). However, *Pho* Alba did not affect the cleavage rate or temperature optimum of the in vitro reconstituted *Pho* RNase P activity (Hada et al. 2008). Moreover, *Mth* Alba did not co-purify with *Mth* RNase P and polyclonal antisera against it failed to immunoprecipitate *Mth* RNase P activity (Ellis et al. 2007). Hence, based on currently available data, only POP5, RPP30, RPP21, and RPP29 can be considered as *bona fide* subunits of archaeal RNase P.

9.4.2 High-Resolution Structures of RPPs

Despite being identified after their eukaryal relatives, the structures of all four archaeal RPPs have been solved by NMR and/or X-ray crystallography in a relatively short time.

The high solubility, thermostability and simplified domain structure of the archaeal RPPs aided the rapid pace. Because yeast two-hybrid data on *Mth* and *Pho* RPPs (Hall and Brown 2004; Kifusa et al. 2005) and biochemical reconstitution studies with *Pfu* RNase P (Tsai et al. 2006) revealed that the four RPPs function as pairs, with POP5 pairing with RPP30 and RPP21 with RPP29, the structures of these binary RPP complexes have also been determined (Fig. 9.2). Although some RPPs were solved by both NMR spectroscopy and X-ray crystallography and some differences were reported between the solution and crystal structures (in part due to differences in the size and source of the RPP), we focus on the common core structure determined in both instances.

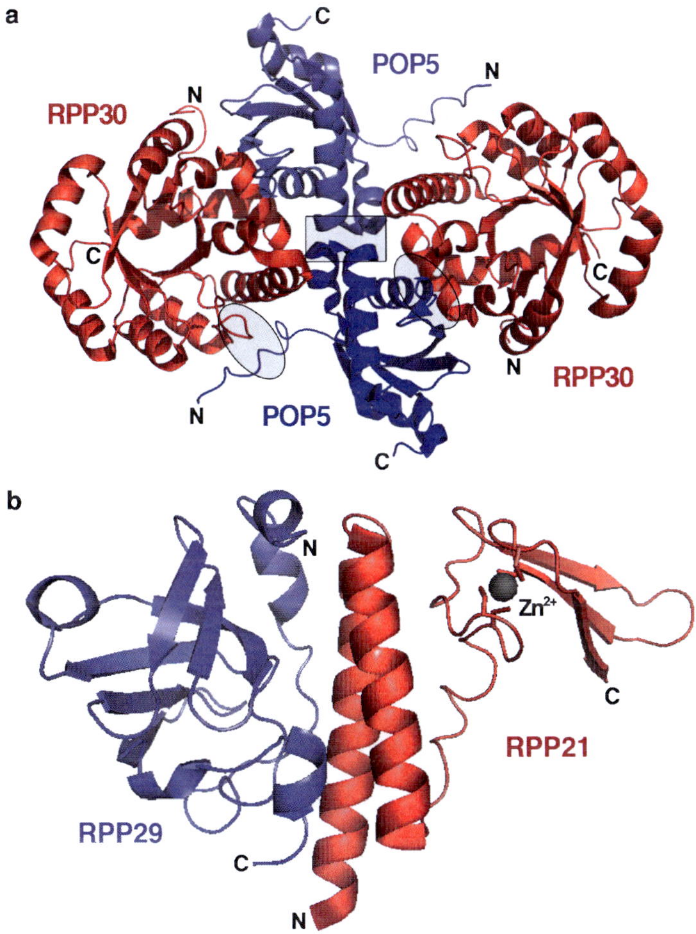

Fig. 9.2 Tertiary structures of archaeal RPP binary complexes. Ribbon diagrams of *Pho* POP5-RPP30 (**a,** PDB ID 2ZCV; Kakuta et al. 2005) and *Pho* RPP21-RPP29 (**b,** PDB ID 2ZAE; Honda et al. 2008). PYMOL was used to generate the images (www.pymol.org). In (**a**), each of the two POP5 molecules is colored either *blue* or *slate*; the POP5 homodimer interface is depicted by a *shaded rectangle*, and the two different POP5-RPP30 interfaces by *shaded ovals*. In (**b**), the *gray sphere* and the *red sticks* in RPP21 indicate the zinc ion and its coordinating Cys side chains

POP5: The tertiary structure of *Pfu* POP5 (alone) and *Pho* POP5 complexed with RPP30 has been solved by X-ray crystallography (Kawano et al. 2006; Wilson et al. 2006). POP5 adopts an α–β sandwich fold with a central four-stranded antiparallel β-sheet surrounded by four α-helices, similar to an RNA recognition motif (RRM). One face of the β-sheet packs against helices α1–α3 to form a hydrophobic core; this face has some solvent-exposed hydrophobic side chains, which NMR chemical-shift perturbations indicated as being part of the RPP30-interacting surface (Wilson et al. 2006). The other face of the β-sheet packs against α4 and has some loosely packed apolar residues that could engage in stacking interactions with RPR/ptRNA. Although it shares no sequence homology to the single bacterial RPP (Stams et al. 1998), the tertiary fold of the two proteins is similar despite different secondary structure connectivities (Wilson et al. 2006). If this structural homology translates to parallels in functional attributes, it would be an interesting example of convergent evolution (see Sect. 4.4).

RPP30: The structure of *Pho* RPP30, the largest among the four RPPs, was solved by X-ray crystallography with and without POP5 (Takagi et al. 2004) and shown to be identical in both cases. It adopts an α/β barrel similar to the well-known triose phosphate isomerase (TIM) structure. It has ten α-helices and seven β-strands, and forms an oblate ellipsoid with the C-terminal helix serving as a lid to the barrel.

POP5-RPP30: Due to the presence of two copies of this binary complex in the asymmetric unit, the crystal structure obtained was that of a heterotetramer (Fig. 9.2a; Kawano et al. 2006). The quaternary structure reveals a homodimer of POP5 in the middle with two symmetrically placed copies of RPP30 on either side; a similar packing arrangement was observed in crystals of free POP5. In addition to homodimerization by a hydrogen-bonding network, each POP5 molecule simultaneously interacts with two molecules of RPP30 through salt bridges and hydrophobic interactions (consistent with chemical-shift perturbation data; Wilson et al. 2006). In RPP30, helices α6–α8 together with the C-terminal strands β6 and β7 form a hydrophobic patch and participate in interactions with POP5. Mutagenesis experiments indicate that formation of a single POP5-RPP30 heterodimer can be uncoupled from heterotetramer formation by abolishing the ability of POP5 to homodimerize (Kawano et al. 2006). Although gel filtration data indicate that the heterotetramer forms in solution, the functional relevance of this oligomerization in vivo is unclear.

RPP21: The structures of RPP21 from *Pho* and *Pfu* were solved by X-ray crystallography and NMR, respectively (Kakuta et al. 2005; Amero et al. 2008). These proteins have a zinc ribbon motif embedded in an L-shaped structure: one arm comprises two α-helices in the N-terminal domain and the other arm three antiparallel β-strands in the C-terminal domain. A linker connects these two arms. Two hairpins provide the scaffold for an ordered four-Cys motif that coordinates a single zinc ion; mutating any of these invariant Cys residues results in loss of structure and activity (Kakuta et al. 2005).

RPP29: The structures of archaeal RPP29 from *Mth, Pho and Archaeoglobus fulgidus (Afu)* were solved by NMR and X-ray crystallography (Boomershine et al. 2003; Sidote and Hoffman 2003; Numata et al. 2004; Sidote et al. 2004). These different high-resolution studies uniformly revealed a twisted, six-stranded antiparallel β-sheet wrapped around a conserved hydrophobic core. The crystal structures additionally reveal the presence of small N- and C-terminal helices, which were not seen in the NMR studies probably due to their being disordered or flexible in solution. Overall, the RPP29 fold resembles Sm/Sm-like proteins that play an integral role in mRNA processing and degradation.

RPP21-RPP29: Chemical-shift perturbations indicate tight *Pfu* RPP21–RPP29 interactions in the absence of the RPR (akin to the POP5–RPP30 complex) and that the primary contacts are mediated by the two N-terminal helices of RPP21 (Amero et al. 2008). Indeed, the recently determined crystal structure of *Pho* RPP21–RPP29 complex (Fig. 9.2b) revealed that RPP21 uses these helices to interact with the N-terminal unstructured segment, strand β2, and C-terminal helix α3 of RPP29 (Honda et al. 2008). The protein–protein interface is dominated by several hydrogen bonds and salt bridges rather than the hydrophobic interactions found in POP5–RPP30. A change in conformation induced upon binary complex formation is more notable for RPP21 than RPP29.

The structures of the binary complexes permit the design of mutagenesis and footprinting experiments to uncover regions in the RPPs that are essential for RPR and/or ptRNA recognition. In fact, the electrostatic potential maps of both POP5–RPP30 and RPP21–RPP29 complexes show an unequal surface charge distribution and highlight the positively charged faces that are likely involved in RNA binding. Although mutation of various conserved Lys and Arg residues on these surfaces resulted in decreased enzymatic activity, the molecular basis for these defects remains to be determined.

9.4.3 In Vitro Reconstitution of Archaeal RNase P

Successful in vitro reconstitution of archaeal RNase P from recombinant RPR and RPPs was reported for *Pho* and *Pfu*, with notable differences (Kouzuma et al. 2003; Tsai et al. 2006). First, while the *Pho* RPR was inactive when assayed alone, the *Pfu* RPR was active and allowed for kinetic studies to delineate the roles of individual subunits and the coordination between the RPR and multiple RPPs (Table 9.1; Tsai et al. 2006). Second, order-of-addition experiments suggest that the activity of *Pho*

Table 9.1 Steady-state kinetic parameters for processing of *Eco* ptRNA^{Tyr} by *Pfu* RNase P[a]

Reconstituted complexes	k_{cat}, min^{-1}	K_m, μM	Optimal [Mg^{2+}], mM
RPR alone	0.38 ± 0.04	30.5 ± 2.30	500
RPR + RPP21 + RPP29	0.61 ± 0.14	6.53 ± 0.94	120
RPR + POP5 + RPP30	11.5 ± 0.20	11.6 ± 1.09	120
RPR + POP5 + RPP30 + RPP21 + RPP29	9.50 ± 0.74	0.18 ± 0.04	30

[a]Mean values and standard errors were calculated from results of three independent experiments in each case (Tsai et al. 2006).

RNase P requires at least POP5, RPP21, and RPP29 in addition to the RPR (Terada et al. 2006), but studies with *Pfu* RNase P revealed two minimal core complexes, each with the RPR plus either POP5–RPP30 or RPP21–RPP29; no activity was observed when only one RPP was added to the RPR (Tsai et al. 2006). We recently found that each RPP pair (from *Mja*, *Mth* or *Pfu*) can be co-expressed and co-purified to homogeneity without any affinity tag using strong cation-exchange chromatography; in vitro reconstitution with RPR and RPPs, purified either individually or as binary complexes, exhibits comparable k_{cat} and K_m (unpublished results).

While the *Pfu* RPR alone is capable of multiple turnover, addition of all four RPPs results in a 25-fold increase in its k_{cat} and a 170-fold decrease in K_m (Table 9.1; Tsai et al. 2006). POP5-RPP30 increases the k_{cat} of RPR by ~30-fold, an effect similar to that with all four RPPs together, while RPP21-RPP29 elicits only a 1.6-fold change in k_{cat}. Addition of either RPP21-RPP29 or POP5-RPP30 to RPR results in a five- or a three-fold decrease in K_m, respectively.

Since k_{cat} imposes a lower limit on the first-order rate constants for steps subsequent to substrate binding, we inferred that *Pfu* POP5-RPP30 must play a vital role in cleavage and/or product release during archaeal RNase P catalysis under multiple-turnover conditions. By studying the rate of self-cleavage of a $ptRNA^{Tyr}$ conjugated in cis to *Mja* RPR (pt^{Tyr}-Mja RPR), in the absence and presence of *Mja* RPPs, we focused on the bond-breaking step (akin to a single-turnover reaction) without influence from substrate binding or product release (Pulukkunat and Gopalan 2008). The k_{obs} for self-cleavage of pt^{Tyr}-Mja RPR is accelerated ~100-fold by *Mja* POP5-RPP30 but not at all by RPP21-RPP29. As presented for *Eco* RPP based on comprehensive thermodynamic and kinetic studies (Sun et al. 2006), rather than influencing the bond-breaking step directly, POP5-RPP30 is likely to shift the equilibrium from ES to ES*, a conformation in which the ptRNA and catalytic metal ions are positioned optimally for cleavage.

Each binary complex can lower the optimal concentrations of substrate, NH_4^+ and Mg^{2+} required by the RPR to catalyze ptRNA cleavage (Table 9.1; Tsai et al. 2006). Clearly, protein-mediated increases in the affinity of the RPR moiety for Mg^{2+} is exploited by archaeal RNase P in a manner reminiscent of bacterial RNase P (Kurz and Fierke 2002; Sun and Harris 2007) and reinforces a recurring theme in protein-aided RNA catalysis. However, the specific archaeal RPP(s) that influence the binding of active-site metal ions have not been identified.

It is worth exploring if lessons regarding hierarchy of assembly in archaeal RNase P are applicable to eukaryal RNase P, especially because some of the protein–protein interactions are conserved between archaeal and eukaryal RNase P (Jiang and Altman 2001; Houser-Scott et al. 2002; Hall and Brown 2004; Kifusa et al. 2005). The observations that human RPR (1) supports cleavage of select ptRNAs and model substrates unaided by RPPs (Kikovska et al. 2007), and (2) exhibits trace activity when reconstituted with human RPP21 + RPP29 (Mann et al. 2003), also provide a framework to begin reconstitution studies with human RNase P.

9.4.4 Mapping RPR–RPP Interactions and Delineating Subunit Coordination in Archaeal RNase P

The simple and rapid reconstitution of different archaeal RNase P holoenzymes has enabled mapping the RPP-binding sites on the RPR. As RNase T1 cleaves 3′ to unpaired Gs, T1-digestion patterns of an RPR with and without RPPs help identify some of the RPR–RPP interaction sites in the holoenzyme. Such an experiment with S domain-deleted RPR of *Pfu* led to the inference that POP5–RPP30 protects regions either at or encompassing the conserved nucleotides in the C domain, a striking parallel to the footprint of bacterial RPP on its cognate RPR (Tsai et al. 2006). Our ongoing footprinting studies with full-length RPRs indicate that POP5-RPP30 and RPP21-RPP29 exclusively interact with the C and S domains, respectively (unpublished results). This demarcation is also consistent with the finding that addition of RPP21-RPP29 to the S domain-deleted *Pfu* and *Mja* RPRs changes neither the rate nor the NH_4^+/Mg^{2+} requirement, unlike its effects on the full-length archaeal RPRs (Tsai et al. 2006; Pulukkunat and Gopalan 2008). Further experiments employing site-specific-tethered footprinting or crosslinking reagents will provide distance constraints among subunits and help build three-dimensional models of archaeal RNase P that might reveal structural and mechanistic insights, as was illustrated for bacterial RNase P (Chen et al. 1998; Massire et al. 1998).

Results from these initial footprinting and kinetic studies on archaeal RNase P collectively suggest a working model to understand the functional coordination among the various subunits. Although an overarching theme in all RPRs is likely to be the exquisite cooperation of the S and C domains for overall structure and optimal function, the manner in which this is accomplished might be different. In bacterial RNase P, the crystal structures of two different RPRs illustrate elegantly how intramolecular interactions stabilize the tertiary structure and facilitate RPR-alone catalysis (Kazantsev et al. 2005; Torres-Larios et al. 2005), while biochemical studies indicate that a single RPP interacts with the C domain and enhances the RPR's catalytic potential and affinity for catalytic metal ions (Crary et al. 1998; Niranjanakumari et al. 1998; Kurz and Fierke 2002; Sun et al. 2006; Sun and Harris 2007). In archaeal/eukaryal RNase P, lack of interdomain structural braces in the RPRs suggests a possible division of labor among their cognate RPPs, a subset substituting for the missing intramolecular struts and another aiding the cleavage step directly (Gopalan 2007).

The two archaeal binary complexes indeed appear to fulfill different roles. Our studies with *Pfu* and *Mja* RNase P suggest that POP5-RPP30 enhances RPR-mediated catalysis similar to the role played by the sole bacterial RPP. In this regard, it is interesting that POP5 shares structural homology with the bacterial RPP, which has been demonstrated to confer uniformity to binding and catalysis by the RPR whether or not a ptRNA contains consensus recognition elements (Sun et al. 2006). The bacterial RPP alters its own energetic contributions to leader binding and enhances the rate of RPR-mediated cleavage by 3- to 1000-fold to ensure that processing of different substrates by the holoenzyme occurs at an almost invariable rate. It remains to be seen if POP5 performs similar roles. Although none of the archaeal RPPs (individually) can substitute for the bacterial RPP in vivo (Gossringer and Hartmann 2007), that there is some functional conservation among RPPs is borne out by the fact that bacterial RPP promotes ptRNA cleavage by the archaeal RPR (Pannucci et al. 1999) as does archaeal POP5-RPP30 for the bacterial RPR (unpublished results). RPP21-RPP29 binds to RPR in the S domain, which in bacterial RPR recognizes the T stem-loop (TSL) region in the ptRNA. The TSL–S domain interaction then leads to a conformational change that aids catalysis by positioning the chemical groups and Mg^{2+} near the cleavage site in the C domain (Pan et al. 1995; Krasilnikov et al. 2004; Brannvall et al. 2007). Based on its ability to decrease NH_4^+/Mg^{2+} requirement, we postulate that binding of RPP21-RPP29 to the S domain is vital for RPR interdomain cooperation and for promoting RPR-mediated recognition of the TSL.

9.5 *Pyrobaculum* RNase P Exemplifies the Extraordinary Divergence in Thermoproteaceae

Our recent study on *Pyrobaculum* RNase P (Lai et al. 2009) conveys evidence contrary to the observations that *Pae* has neither a conserved RPR nor a 5′ ptRNA-processing activity (Li and Altman 2004; Randau et al. 2008). As mentioned earlier, we found in all five sequenced *Pyrobaculum* species, an RPR sequence with a conserved C domain, but with a radically minimized S domain (Fig. 9.1b). Nevertheless, this mostly-C-domain RPR is active in processing a *Pae* $ptRNA^{Phe}$ in vitro, again illustrating that the activity of archaeal RPRs is attributable to the C domain. We also found a 5′ ptRNA-processing activity in both *Pae* and *Pca* cell extracts, partially purified on ion-exchange matrices. The *Pca* activity was further characterized and found to cleave $ptRNA^{Phe}$ at the canonical RNase P cleavage site, generating a 5′ phosphate and a 3′ hydroxyl group. Moreover, nuclease and proteinase treatment prior to RNase P assay abolished the 5′ ptRNA-processing activity. Therefore, a typical RNP-based RNase P exists in *Pyrobaculum* and must be responsible for processing the 20–48% of tRNAs found in four *Pyrobaculum* species by tRNA sequencing to be transcribed with short (1–6-nt) leaders.

Further computational searches among sequenced archaeal genomes revealed *Pyrobaculum*-like RPRs only in *Caldivirga maquilingensis* and *Vulcanisaeta distributa,* both belonging to the family Thermoproteaceae, same as *Pyrobaculum.* Moreover, these Thermoproteaceae (dubbed type T) RPRs have the GGU sequence in L15 that is absent in all seven crenarchaeal RPRs identified in the RNase P database (Brown 1999). These findings hint at a family of divergent RNase P in Thermoproteaceae. Since POP5-RPP30 binds to the C domain and RPP21-RPP29 to S domain, the diminutive S domain of type T RPRs suggest that there might be changes in the RPPs as well, especially in RPP21-RPP29. Indeed, extensive searches in these Thermoproteaceae genomes uncover homologs for only POP5, RPP30 and RPP29, but not RPP21. Hence, characterization of a native type T

RNase P holoenzyme will provide further insights into its subunit makeup and the dynamic co-evolution of RPR and RPPs.

9.6 Concluding Remarks

The study of archaeal RNase P has revealed that this RNP is an interesting mosaic of its bacterial and eukaryal relatives. Archaeal RPRs are more similar to their relatives from bacteria than from eukarya. In contrast, none of the currently known archaeal RPPs is a homolog of the single bacterial RPP, but all are homologous to four of at least nine eukaryal RPPs. Despite the vast difference in the RPR catalytic potential and subunit composition among the three domains of life, the turnover numbers of native holoenzymes are comparable (Andrews et al. 2001; Xiao et al. 2005; Sun et al. 2006). Clearly, the individual contribution of the RPR and RPPs must vary in each holoenzyme and remains to be delineated. The intermediate nature of archaeal RNase P in both RPP-independent activity and protein content makes it an appealing enzyme for such investigations as discussed below.

Despite advances in reconstituting archaeal RNase P holoenzymes, the k_{cat}/K_m for in vitro reconstituted *Pfu* RNase P (~$10^6\,M^{-1}s^{-1}$) is still lower compared to ~$10^7\,M^{-1}s^{-1}$ reported for partially purified native *Mth* and *Mja* RNase P (Andrews et al. 2001; Tsai et al. 2006). Moreover, the temperature for maximal activity of the reconstituted *Pfu* and *Pho* RNase P is ~15°C lower than that for the respective partially purified native holoenzyme (Kouzuma et al. 2003). Although these differences could be due to variations in enzyme sources, assay conditions, and substrates, it is likely that the in vitro assembly lacks RPP(s) present in vivo. Consistently, a buoyant density of 1.4 g/mL for *Mth* and *Mja* RNase P in Cs_2SO_4 gradients suggests an approximate 1:1 mass ratio of RPR:RPPs (Andrews et al. 2001). With the *Mth* RPR being ~100 kDa and the four known RPPs totaling ~65 kDa, unidentified RPP(s) must account for the difference of 35 kDa, although this discrepancy could also be explained if some RPPs function as oligomers. Therefore, an immediate priority is to purify and characterize native archaeal RNase P to establish the number, identity, and stoichiometry of RPPs; this will also permit elucidation of native modifications both post-transcriptional (in RPR) and post-translational (in RPPs). In addition, comparing different classes of archaeal RNase P (type A vs. M, mesophilic vs. thermophilic, and euryarcheal vs. crenarchaeal) will shed light on how cellular milieu and lineage have influenced its subunit makeup.

Dramatic variations (~10^6-fold) in the catalytic potential indicate that, despite the presence of a conserved catalytic core, not all RPRs are equal (activity of RPRs from bacteria > archaea > eukarya). Notably, the decrease in RPR activity parallels the increased RPP content in the corresponding RNP. Different structural elements in bacterial RPRs, demonstrated to be essential for substrate binding/positioning, catalysis, and global stability, were either never acquired or lost during evolution of archaeal/eukaryal RPRs, thus accounting for their lower stability/activity in the absence of cognate RPPs. While we cannot retrace this path in evolution, determining how substrate specificity and regulation of the archaeal RNase P holoenzyme are influenced by different RPPs is feasible and should be pursued. The availability of a robust reconstitution of partial and complete holoenzymes permits the dissection of the role of different archaeal RPPs in RNase P catalysis in terms of binding affinities, cleavage rates, and fidelity of processing for different ptRNAs. Understanding how the activity and long-term stability of archaeal RNase P are affected by organic osmolytes and unique polyamines, which protect macromolecules in extremophiles from chemical/physical duress (Lentzen and Schwarz 2006), will facilitate kinetic and structural studies.

Down-regulating RNase P activity in vivo to allow its substrates to accumulate in archaea will help identify substrates other than ptRNAs. While screening for small-molecule inhibitors will aid this goal, a genetic tool will be even more attractive. In fact, the lack of gene knockdown tools in archaea prompted us to examine the use of endogenous RNase P as a functional genomics tool. Any cellular RNA could be targeted for degradation by endogenous RNase P, if the binding of the target RNA to an external guide sequence (EGS) forms a sequence- and structure-specific complex resembling a ptRNA (Guerrier-Takada and Altman 2000). This idea merits exploration, especially since we recently obtained evidence that in vitro reconstituted *Pfu*, *Mja,* and *Methanococcus maripaludis (Mma)* RNase P cleave such bipartite substrates and that the EGS recycles between rounds of cleavage, thereby permitting efficient cleavage of even highly expressed target RNAs (unpublished results).

Various reports have suggested that archaeal RNase P is part of a higher-order network and is subject to co-regulation with other macromolecular assemblages based on the cellular metabolic status. The placement of the archaeal RPP29 gene in a large conserved operon encoding ribosomal proteins implies coupled synthesis of RNase P and ribosome (Hartmann and Hartmann 2003). A parallel exists in bacteria where the sole RPP is typically co-transcribed with the ribosomal protein L34 (Hansen et al. 1985; Ellis and Brown 2003), and biochemical data suggest interactions between ribosome and RNase P (Vioque et al. 1988; Barrera and Pan 2004). In some archaea, a comparative genomic approach revealed the presence of tandem POP5 and RPP30 genes in a superoperon with genes encoding exosome, ribosome, and proteasome subunits (Koonin et al. 2001). In fact, the POP5 gene was shown to be co-transcribed with the proteasomal gene *pmsA* in *Hvo* (Gil et al. 2007). Together, these observations have led to the hypothesis that in archaea there is some coordination among RNA processing (mediated by RNase P), translation, and RNA and protein degradation. Therefore, mapping the physical and functional interactions between archaeal RNase P and other cellular machineries is needed in order to understand the crosstalk between tRNA biogenesis and RNA/protein synthesis and turnover.

A list of future directions would be incomplete if it did not include determining high-resolution structures of RNase P holoenzymes from the three domains of life, which are essential to ultimately uncover how a uniform active-site architecture results amid the diversity in subunit makeup. As more genomes are sequenced and a picture of the diversity of this ancient enzyme is expanded, we might also better appreciate if Nature has left adequate clues in the form of RNase P variants, signifying unique snapshots in evolution, to construct a hypothetical journey of this remnant from the RNA world.

Acknowledgments The authors are grateful to Mark P. Foster (OSU) for assistance with preparation of Fig. 9.2 and for valuable discussions. The author, Venkat Gopalan, acknowledges support from the NSF (MCB-0238233 and MCB-0843543) and the NIH (GM067807 to MPF and VG).

References

Amero CD, Boomershine WP, Xu Y, Foster M (2008) Solution structure of *Pyrococcus furiosus* RPP21, a component of the archaeal RNase P holoenzyme, and interactions with its RPP29 protein partner. Biochemistry 47:11704–11710

Andrews AJ, Hall TA, Brown JW (2001) Characterization of RNase P holoenzymes from *Methanococcus jannaschii* and *Methanothermobacter thermoautotrophicus*. Biol Chem 382:1171–1177

Aravind L, Iyer LM, Anantharaman V (2003) The two faces of Alba: the evolutionary connection between proteins participating in chromatin structure and RNA metabolism. Genome Biol 4:R64

Barrera A, Pan T (2004) Interaction of the *Bacillus subtilis* RNase P with the 30 S ribosomal subunit. RNA 10:482–492

Boomershine WP, McElroy CA, Tsai HY, Wilson RC, Gopalan V, Foster MP (2003) Structure of Mth11/Mth Rpp 29, an essential protein subunit of archaeal and eukaryotic RNase P. Proc Natl Acad Sci USA 100:15398–15403

Brannvall M, Kikovska E, Wu S, Kirsebom LA (2007) Evidence for induced fit in bacterial RNase P RNA-mediated cleavage. J Mol Biol 372:1149–1164

Brown JW (1999) The ribonuclease P database. Nucleic Acids Res 27:314

Chamberlain JR, Lee Y, Lane WS, Engelke DR (1998) Purification and characterization of the nuclear RNase P holoenzyme complex reveals extensive subunit overlap with RNase MRP. Genes Dev 12:1678–1690

Chen JL, Nolan JM, Harris ME, Pace NR (1998) Comparative photocross-linking analysis of the tertiary structures of *Escherichia coli* and *Bacillus subtilis* RNase P RNAs. EMBO J 17:1515–1525

Christian EL, Smith KM, Perera N, Harris ME (2006) The P4 metal binding site in RNase P RNA affects active site metal affinity through substrate positioning. RNA 12:1463–1467

Crary SM, Niranjanakumari S, Fierke CA (1998) The protein component of *Bacillus subtilis* ribonuclease P increases catalytic efficiency by enhancing interactions with the 5′ leader sequence of pre-tRNA^{Asp}. Biochemistry 37:9409–9416

Darr SC, Pace B, Pace NR (1990) Characterization of ribonuclease P from the archaebacterium *Sulfolobus solfataricus*. J Biol Chem 265:12927–12932

Eder PS, Kekuda R, Stolc V, Altman S (1997) Characterization of two scleroderma autoimmune antigens that copurify with human ribonuclease P. Proc Natl Acad Sci USA 94:1101–1106

Ellis JC, Brown JW (2003) Genes within genes within bacteria. Trends Biochem Sci 28:521–523

Ellis JC, Barnes J, Brown JW (2007) Is Alba an RNase P subunit? RNA Biol 4:169–172

Fukuhara H, Kifusa M, Watanabe M, Terada A, Honda T, Numata T, Kakuta Y, Kimura M (2006) A fifth protein subunit Ph1496p elevates the optimum temperature for the ribonuclease P activity from *Pyrococcus horikoshii* OT3. Biochem Biophys Res Commun 343:956–964

Gil MA, Sherwood KE, Maupin-Furlow JA (2007) Transcriptional linkage of *Haloferax volcanii* proteasomal genes with non-proteasomal gene neighbours including RNase P, MOSC domain and SAM-methyltransferase homologues. Microbiology 153:3009–3022

Gopalan V, Altman S (2006) Ribonuclease P: structure and catalysis. In: Gesteland RF, Cech TR, Atlkins JF (eds) The RNA world. Cold Spring Harbor Laboratory Press, New York (only online at http://rna.cshl.edu)

Gopalan V (2007) Uniformity amid diversity in RNase P. Proc Natl Acad Sci USA 104:2031–2032

Gossringer M, Hartmann RK (2007) Function of heterologous and truncated RNase P proteins in *Bacillus subtilis*. Mol Microbiol 66:801–813

Guerrier-Takada C, Gardiner K, Marsh T, Pace N, Altman S (1983) The RNA moiety of ribonuclease P is the catalytic subunit of the enzyme. Cell 35:849–857

Guerrier-Takada C, Altman S (2000) Inactivation of gene expression using ribonuclease P and external guide sequences. Methods Enzymol 313:442–456

Hada K, Nakashima T, Osawa T, Shimada H, Kakuta Y, Kimura M (2008) Crystal structure and functional analysis of an archaeal chromatin protein Alba from the hyperthermophilic archaeon *Pyrococcus horikoshii* OT3. Biosci Biotechnol Biochem 72:749–758

Hall TA, Brown JW (2001) The ribonuclease P family. Methods Enzymol 341:56–77

Hall TA, Brown JW (2002) Archaeal RNase P has multiple protein subunits homologous to eukaryotic nuclear RNase P proteins. RNA 8:296–306

Hall TA, Brown JW (2004) Interactions between RNase P protein subunits in archaea. Archaea 1:247–254

Hamilton MG, Kivie M, Lawrence G (1971) Isodensity equilibrium centrifugation of ribosomal particles; the calculation of the protein content of ribosomes and other ribonucleo-proteins from buoyant density measurements. Methods Enzymol 20:512–521

Hansen FG, Hansen EB, Atlung T (1985) Physical mapping and nucleotide sequence of the rnpA gene that encodes the protein component of ribonuclease P in *Escherichia coli*. Gene 38:85–93

Harris JK, Haas ES, Williams D, Frank DN, Brown JW (2001) New insight into RNase P RNA structure from comparative analysis of the archaeal RNA. RNA 7:220–232

Harris ME, Christian EL (2003) Recent insights into the structure and function of the ribonucleoprotein enzyme ribonuclease P. Curr Opin Struct Biol 13:325–333

Hartmann E, Hartmann RK (2003) The enigma of ribonuclease P evolution. Trends Genet 19:561–569

Holzmann J, Frank P, Loffler E, Bennett KL, Gerner C, Rossmanith W (2008) RNase P without RNA: identification and functional reconstitution of the human mitochondrial tRNA processing enzyme. Cell 135:462–474

Honda T, Kakuta Y, Kimura K, Saho J, Kimura M (2008) Structure of an archaeal homolog of the human protein complex Rpp 21-Rpp29 that is a key core component for the assembly of active ribonuclease P. J Mol Biol 384:652–662

Houser-Scott F, Xiao S, Millikin CE, Zengel JM, Lindahl L, Engelke DR (2002) Interactions among the protein and RNA subunits of *Saccharomyces cerevisiae* nuclear RNase P. Proc Natl Acad Sci USA 99:2684–2689

Jarrous N, Wolenski JS, Wesolowski D, Lee C, Altman S (1999) Localization in the nucleolus and coiled bodies of protein subunits of the ribonucleoprotein ribonuclease P. J Cell Biol 146:559–572

Jarrous N (2002) Human ribonuclease P: subunits, function, and intranuclear localization. RNA 8:1–7

Jiang T, Altman S (2001) Protein-protein interactions with subunits of human nuclear RNase P. Proc Natl Acad Sci USA 98:920–925

Kakuta Y, Ishimatsu I, Numata T, Kimura K, Yao M, Tanaka I, Kimura M (2005) Crystal structure of a ribonuclease P protein Ph1601p from Pyrococcus horikoshii OT3: an archaeal homologue of human nuclear ribonuclease P protein Rpp 21. Biochemistry 44:12086–12093

Kawamoto SA, Sudhahar CG, Hatfield CL, Sun J, Behrman EJ, Gopalan V (2008) Studies on the mechanism of inhibition of bacterial ribonuclease P by aminoglycoside derivatives. Nucleic Acids Res 36:697–704

Kawano S, Nakashima T, Kakuta Y, Tanaka I, Kimura M (2006) Crystal structure of protein Ph1481p in complex with protein Ph1877p of archaeal RNase P from *Pyrococcus horikoshii* OT3: implication of dimer formation of the holoenzyme. J Mol Biol 357:583–591

Kaye NM, Zahler NH, Christian EL, Harris ME (2002) Conservation of helical structure contributes to functional metal ion interactions in the catalytic domain of ribonuclease P RNA. J Mol Biol 324:429–442

Kazantsev AV, Krivenko AA, Harrington DJ, Holbrook SR, Adams PD, Pace NR (2005) Crystal structure of a bacterial ribonuclease P RNA. Proc Natl Acad Sci USA 102:13392–13397

Kifusa M, Fukuhara H, Hayashi T, Kimura M (2005) Protein-protein interactions in the subunits of ribonuclease P in the hyperthermophilic archaeon Pyrococcus horikoshii OT3. Biosci Biotechnol Biochem 69:1209–1212

Kikovska E, Svard SG, Kirsebom LA (2007) Eukaryotic RNase P RNA mediates cleavage in the absence of protein. Proc Natl Acad Sci USA 104:2062–2067

Kirsebom LA, Svard SG (1994) Base pairing between *Escherichia coli* RNase P RNA and its substrate. EMBO J 13:4870–4876

Kirsebom LA (2007) RNase P RNA mediated cleavage: substrate recognition and catalysis. Biochimie 89:1183–1194

Koonin EV, Wolf YI, Aravind L (2001) Prediction of the archaeal exosome and its connections with the proteasome and the translation and transcription machineries by a comparative-genomic approach. Genome Res 11:240–252

Kouzuma Y, Mizoguchi M, Takagi H, Fukuhara H, Tsukamoto M, Numata T, Kimura M (2003) Reconstitution of archaeal ribonuclease P from RNA and four protein components. Biochem Biophys Res Commun 306:666–673

Krasilnikov AS, Xiao Y, Pan T, Mondragon A (2004) Basis for structural diversity in homologous RNAs. Science 306:104–107

Kurz JC, Fierke CA (2002) The affinity of magnesium binding sites in the *Bacillus subtilis* RNase P x pre-tRNA complex is enhanced by the protein subunit. Biochemistry 41:9545–9558

LaGrandeur TE, Darr SC, Haas ES, Pace NR (1993) Characterization of the RNase P RNA of *Sulfolobus acidocaldarius*. J Bacteriol 175:5043–5048

Lai LB, Chan PP, Cozen AE, Bernick DL, Brown JW, Gopalan V, Lowe T (2009) Discovery of the elusive *Pyrobaculum* RNase P: An unexpected form of an ancient RNA. Manuscript under review

Lawrence N, Wesolowski D, Gold H, Bartkiewicz M, Guerrier-Takada C, McClain WH, Altman S (1987) Characteristics of ribonuclease P from various organisms. Cold Spring Harb Symp Quant Biol 52:233–238

Lentzen G, Schwarz T (2006) Extremolytes: Natural compounds from extremophiles for versatile applications. Appl Microbiol Biotechnol 72:623–634

Li D, Willkomm DK, Hartmann RK (2009) Minor changes largely restore catalytic activity of archaeal RNase P RNA from *Methanothermobacter thermoautotrophicus*. Nucleic Acids Res 37(1):231–242

Li Y, Altman S (2004) In search of RNase P RNA from microbial genomes. RNA 10:1533–1540

Loria A, Pan T (1996) Domain structure of the ribozyme from eubacterial ribonuclease P. RNA 2:551–563

Mann H, Ben-Asouli Y, Schein A, Moussa S, Jarrous N (2003) Eukaryotic RNase P: role of RNA and protein subunits of a primordial catalytic ribonucleoprotein in RNA-based catalysis. Mol Cell 12:925–935

Marquez SM, Evans D, Kazantsev AV, Pace NR (2008) A structural analysis of ribonuclease P. In: Lilley DMJ, Eckstein F (eds) Ribozymes and RNA catalysis. RSC Publishing, Cambridge, pp 153–177

Massire C, Jaeger L, Westhof E (1998) Derivation of the three-dimensional architecture of bacterial ribonuclease P RNAs from comparative sequence analysis. J Mol Biol 279:773–793

Nieuwlandt DT, Haas ES, Daniels CJ (1991) The RNA component of RNase P from the archaebacterium *Haloferax volcanii*. J Biol Chem 266:5689–5695
Niranjanakumari S, Stams T, Crary SM, Christianson DW, Fierke CA (1998) Protein component of the ribozyme ribonuclease P alters substrate recognition by directly contacting precursor tRNA. Proc Natl Acad Sci USA 95:15212–15217
Numata T, Ishimatsu I, Kakuta Y, Tanaka I, Kimura M (2004) Crystal structure of archaeal ribonuclease P protein Ph1771p from Pyrococcus horikoshii OT3: an archaeal homolog of eukaryotic ribonuclease P protein Rpp 29. RNA 10:1423–1432
Oh BK, Pace NR (1994) Interaction of the 3′-end of tRNA with ribonuclease P RNA. Nucleic Acids Res 22:4087–4094
Pan T, Loria A, Zhong K (1995) Probing of tertiary interactions in RNA: 2′-hydroxyl-base contacts between the RNase P RNA and pre-tRNA. Proc Natl Acad Sci USA 92:12510–12514
Pannucci JA, Haas ES, Hall TA, Harris JK, Brown JW (1999) RNase P RNAs from some Archaea are catalytically active. Proc Natl Acad Sci USA 96:7803–7808
Pulukkunat DK, Gopalan V (2008) Studies on *Methanocaldococcus jannaschii* RNase P reveal insights into the roles of RNA and protein cofactors in RNase P catalysis. Nucleic Acids Res 36:4172–4180
Randau L, Schroder I, Soll D (2008) Life without RNase P. Nature 453:120–123
Sidote DJ, Hoffman DW (2003) NMR structure of an archaeal homologue of ribonuclease P protein Rpp 29. Biochemistry 42:13541–13550
Sidote DJ, Heideker J, Hoffman DW (2004) Crystal structure of archaeal ribonuclease P protein aRpp 29 from Archaeoglobus fulgidus. Biochemistry 43:14128–14138
Stams T, Niranjanakumari S, Fierke CA, Christianson DW (1998) Ribonuclease P protein structure: evolutionary origins in the translational apparatus. Science 280:752–755
Sun L, Campbell FE, Zahler NH, Harris ME (2006) Evidence that substrate-specific effects of C5 protein lead to uniformity in binding and catalysis by RNase P. EMBO J 25:3998–4007
Sun L, Harris ME (2007) Evidence that binding of C5 protein to P RNA enhances ribozyme catalysis by influencing active site metal ion affinity. RNA 13:1505–1515
Takagi H, Watanabe M, Kakuta Y, Kamachi R, Numata T, Tanaka I, Kimura M (2004) Crystal structure of the ribonuclease P protein Ph1877p from hyperthermophilic archaeon *Pyrococcus horikoshii* OT3. Biochem Biophys Res Commun 319:787–794
Terada A, Honda T, Fukuhara H, Hada K, Kimura M (2006) Characterization of the archaeal ribonuclease P proteins from *Pyrococcus horikoshii* OT3. J Biochem 140:293–298
Terada A, Yoshida T, Kimura M (2007) Identification of nucleotide residues essential for RNase P activity from the hyperthermophilic archaeon *Pyrococcus horikoshii* OT3. Biosci Biotechnol. Biochem. 71:1940–1945
Torres-Larios A, Swinger KK, Krasilnikov AS, Pan T, Mondragon A (2005) Crystal structure of the RNA component of bacterial ribonuclease P. Nature 437:584–587
Tsai HY, Masquida B, Biswas R, Westhof E, Gopalan V (2003) Molecular modeling of the three-dimensional structure of the bacterial RNase P holoenzyme. J Mol Biol 325:661–675
Tsai HY, Pulukkunat DK, Woznick WK, Gopalan V (2006) Functional reconstitution and characterization of *Pyrococcus furiosus* RNase P. Proc Natl Acad Sci USA 103:16147–16152
Ushida C, Muramatsu T, Mizushima H, Ueda T, Watanabe K, Stetter KO, Crain PF, McCloskey JA, Kuchino Y (1996) Structural feature of the initiator tRNA gene from *Pyrodictium occultum* and the thermal stability of its gene product, tRNA(imet). Biochimie 78:847–855
Vioque A, Arnez J, Altman S (1988) Protein-RNA interactions in the RNase P holoenzyme from *Escherichia coli*. J Mol Biol 202:835–848
Walker SC, Engelke DR (2006) Ribonuclease P: the evolution of an ancient RNA enzyme. Crit Rev Biochem Mol Biol 41:77–102
Wilson RC, Bohlen CJ, Foster MP, Bell CE (2006) Structure of *Pfu* Pop5, an archaeal RNase P protein. Proc Natl Acad Sci USA 103:873–878
Xiao S, Day-Storms JJ, Srisawat C, Fierke CA, Engelke DR (2005) Characterization of conserved sequence elements in eukaryotic RNase P RNA reveals roles in holoenzyme assembly and tRNA processing. RNA 11:885–896

Chapter 10
Eukaryote RNase P and RNase MRP

Scott C. Walker, Michael C. Marvin, and David R. Engelke

10.1 Introduction: Increased Complexity in the Eukaryote

Eukaryotes are characterized by their membrane bound nucleus and other organelles, each of which contributes towards a greater degree of organization and specialization of function than is observed in bacteria and archaea. In the yeast *Saccharomyces cerevisiae*, a model eukaryote, the transcription and maturation of transfer RNA (tRNA) illustrates this complexity. In yeast, tRNA biogenesis begins with the tRNA genes which are found to be spatially organized at the nucleolus, a dense subnuclear region (Thompson et al. 2003). The transcription of precursor tRNAs (pre-tRNAs) is carried out from the tRNA genes by one of three specialized RNA polymerases (RNA pol III). The eukaryotic RNA polymerases conserve the core catalytic subunits of the bacterial RNA polymerases but are vastly expanded in the eukaryote by additional protein subunits (totaling 10–17 subunits each) and other interacting transcription factors that serve to diversify the regulatory control over its function (Cramer et al. 2008). Following transcription the pre-tRNAs require several maturation steps; processing of their 5′ and 3′ ends, addition of a 3′-CCA, removal of introns and the chemical modification of various nucleotides specific to individual tRNAs [reviewed in (Hopper and Phizicky 2003)]. These maturation steps are also highly organized both temporally and spatially, with specific steps taking place in either the nucleus or cytoplasm. Maturation of pre-tRNAs begins at the site of transcription, the nucleolus, with the removal of the 5′ leader sequence by ribonuclease P (RNase P). This endonuclease is required in almost all living cells, and consists of both RNA and protein subunits. RNase P contains a single small protein in bacteria (~10% by mass), at least four in archaea (~45% by mass) and at least nine in eukaryotic nuclei (~70% by mass).[1]

S.C. Walker, M.C. Marvin, and D.R. Engelke (✉)
Department of Biological Chemistry, University of Michigan, 1150 W. Medical Center Drive, Ann Arbor, MI 48109-0606, USA
e-mail: engelke@umich.edu

[1]Estimation of mass assumes a single RNA subunit and that each protein is present only once in the holoenzyme. Subunit stoichiometries are currently not well defined in the eukaryote systems.

F. Liu and S. Altman (eds.), *Ribonuclease P*, Protein Reviews 10,
DOI 10.1007/978-1-4419-1142-1_10, © Springer Science+Business Media, LLC 2010

Despite the differences in subunit composition, these activities are clearly related as each contains a fundamentally similar and catalytically active RNA subunit. The catalytic RNA is presumed to be a remnant from the hypothesized “RNA World” where RNA is thought to have preceded protein as the functional biological molecule (Orgel 1986). The majority of work that has been performed on both the eukaryotic RNase P and MRP ribonucleoproteins has taken place in the yeast (*S. cerevisiae*) and human (*H. sapiens*) systems. Information on the human enzymes is provided (for continuity with Chap. 12), however this chapter will focus on the yeast system.

10.2 Multiple RNase P Enzymes Exist in Eukaryotes

In addition to the increase in the subunit number and protein content of the eukaryote RNase P, eukaryotes have different forms of the enzyme that are specific to particular organelles. The main cellular RNase P (nuclear RNase P) in the eukaryote can also coexist alongside other specific chloroplast and mitochondrially encoded versions of RNase P (see Chap. 11). The eukayotic nuclear RNase P has also evolved a closely related parologous enzyme, RNase MRP. Although RNase MRP is not technically an RNase P enzyme, meaning that it does not perform 5′ pre-tRNA cleavage, it has clearly evolved from RNase P in that it retains significant elements of the RNA subunit structure and has a considerable overlap in protein subunit content (Chamberlain et al. 1998; Salinas et al. 2005).

10.2.1 Yeast Nuclear RNase P

Within the eukaryotic nucleus, the nucleolus is a dense subnuclear region which is the main site of ribosomal biogenesis. In yeast, the major cellular pre-tRNA processing activity is referred to as the “nuclear” RNase P and is predominantly localized to the nucleolus but with some nucleoplasmic foci (Bertrand et al. 1998). It has also been shown that the yeast tRNA genes, as well as the bulk of the pre-tRNAs and RNase P, are positioned at the nucleolus and so the RNase P holoenzyme is localized at the site of transcription of its pre-tRNA substrate (Bertrand et al. 1998; Thompson et al. 2003; Wang et al. 2005; Haeusler et al. 2008). Given the increased complexity in eukaryotes, the localization of processing activities within the appropriate cellular compartments is particularly important.

The composition of yeast (*S. cerevisiae*) nuclear RNase P was established through biochemical purification and the identification of a single RNA, *RPR1* (369 nucleotides), and nine protein subunits: Pop1, Pop3, Pop4, Pop5, Pop6, Pop7, Pop8, Rpr2 and Rpp1 (Chamberlain et al. 1998) (Table 10.1). Most proteins in the complex are moderately sized (15–33 kDa) except for the comparatively large Pop1 (100 kDa). Four of the eukaryotic proteins are homologs of archaeal proteins, Pop4, Pop5, Rpp1 and Rpr2 (Table 10.1, shaded) indicating that eukaryotic and

Table 10.1 Subunit composition of the eukaryote nuclear RNase P and RNase MRP

Yeast[a] RNase P/RNase MRP					Human[b] RNase P/RNase MRP				
	Properties		Subunit			Properties		Subunit	
	M_r	(pI)	P	MRP		M_r^c	(pI)	P	MRP
RNA					RNA				
RPR1	120		•		*H1*	112		•	
NME1	112			•	*7-2*	88			•
Protein					Protein				
POP1	100.5	9.8	•	•	*hPOP1*	115	9.6	•	•
POP3	22.6	9.6	•	•	*RPP38*	38	9.6	•	•
POP4	32.9	9.3	•	•	*RPP29*	29	10.2	•	•
POP5	19.6	7.8	•	•	*hPOP5*	19	7.9	•	•
POP6[d]	18.2	9.3	•	•					
POP7	15.8	9.3	•	•	*RPP20*	20	8.6	•	•
POP8	15.5	4.6	•	•					
RPP1	32.2	9.8	•	•	*RPP30*	30	9.2	•	•
RPR2	16.3	10.0	•		*RPP21*	21	9.6	•	
SNM1	22.5	9.8		•					
RMP1	23.6	9.8		•					
					RPP40	40	5.2	•	
					RPP25[d]	25	9.7	•	•
					RPP14	14	7.6	•	

[a]Data taken from Xiao et al. (2002)
[b]Data taken from Jarrous (2002)
[c]Molecular weights of the human proteins are based upon gel migrations in many cases the mass differs from that calculated from the sequence
[d]The human Rpp25 protein has been proposed to be the homolog of the yeast Pop6 protein (Rosenblad et al. 2006). Proteins with known archaeal homologs are shown with a shaded background

archaeal RNase P share a common ancestor (Hall and Brown 2002; Hartmann and Hartmann 2003). The single bacterial protein and archaeal Pop5 proteins have similar structures (RNR-fold) although these are topologically distinct and are derived from unrelated sequences (Stams et al. 1998; Spitzfaden et al. 2000; Kazantsev et al. 2003; Kawano et al. 2006; Wilson et al. 2006). Convergent evolution appears possible, however it has not been proven that these proteins serve the same function in bacteria and archaea. Currently, it is not clear why the eukaryotic enzyme has acquired so many additional protein subunits, neither have specific functions for these proteins been established (see Sect. 10.5). The additional protein subunits in the eukaryote holoenzyme could potentially play a role in defining its subcellular localization, stabilizing the RNA structure, interacting with substrates or facilitating conformational change.

Depletion of the yeast RNase P RNA and protein subunits in vivo has shown that each of the ten eukaryote subunits is essential for pre-tRNA processing and cell viability (Lee et al. 1991b; Lygerou et al. 1994; Chu et al. 1997; Dichtl and Tollervey 1997; Stolc and Altman 1997; Chamberlain et al. 1998). The essential nature of every individual subunit is particularly interesting given that a purified precursor of yeast RNase P RNP complex was shown to have little or no Rpr2 and Pop3 proteins while still active for pre-tRNA cleavage in vitro (Srisawat et al. 2002). The Pop3 protein is a member of at least one other essential complex (i.e., RNase MRP) and so the direct relevance of its absence from the purified pre-RNase P is more difficult to interpret. However, the Rpr2 protein is both essential and unique to RNase P (i.e., not found to be present in RNase MRP) and this data could indicate that it is required for other functions independent of pre-tRNA cleavage. The potential for nuclear RNase P to participate in the cleavage of additional substrates is discussed later (see Sect. 10.3.2.).

10.2.2 Mitochondrial RNase P

In addition to the main cellular RNase P, eukaryotes also have distinct pre-tRNA processing activities specific to their mitochondria and chloroplasts (Schon 1999; Xiao et al. 2002). In eukaryotes, mitochondria and chloroplasts are thought to have originated from endosymbiotic bacteria or cyanobacteria respectively (Gray 1993). Although many examples of organellar RNase P are similar in structure and composition to the ancestral bacterial enzymes, others have evolved considerable diversity. For example, the yeast (*S. cerevisiae*) mitochondrial RNase P (mtRNase P) consists of a mitochondrially encoded RNA, *RPM1* (423 nucleotides), with a structure that appears to be significantly diverged from the bacterial RNAs (Seif et al. 2003). The Rpm1 RNA forms a holoenzyme with a single nuclear encoded protein, Rpm2 (105 kDa), which is produced in the cytoplasm and then transported into the mitochondria (Dang and Martin 1993). An entirely distinct enzyme appears to have developed for the human mtRNase P. The activity has recently been convincingly demonstrated to be a protein-only complex, whose activity can be reconstituted in vitro from three subunits : MRPP1 (47 kDa), MRPP2 (26 kDa) and MRPP3 (60 kDa) (Holzmann et al. 2008; Walker and Engelke 2008). Since organelle RNase P is addressed elsewhere in this volume (Chap. 11), we will not further discuss these enzymes or the possible evolutionary paths here.

10.2.3 Yeast RNase MRP

When discussing RNase P in the eukaryote, it is necessary to include its apparent paralog, RNase MRP, which is only found in eukaryotes. The RNase P and MRP holoenzymes each have an essential RNA subunit and although these are distinct the RNase MRP RNA, *NME1* (339 nucleotides), has sequences and structural features that clearly relate it to all forms of RNase P RNA (see Fig. 10.1 and Sect. 10.4). RNase MRP has three known substrates in yeast; mitochondrial RNA (mtRNA), precursor ribosomal RNA (pre-rRNA) and at least one specific messenger RNA (mRNA) (see Sect. 10.3.3). The nucleolar activity of RNase MRP involves the processing of the

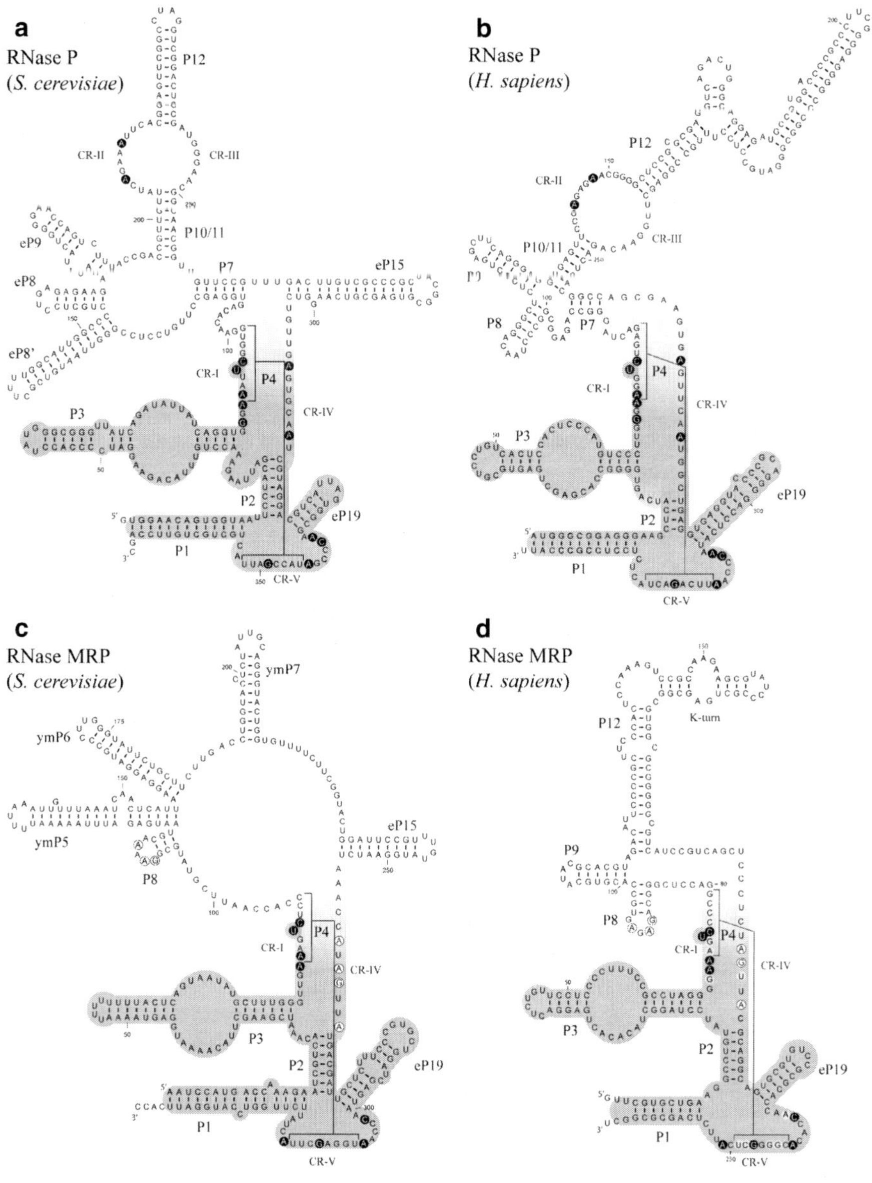

Fig. 10.1 Comparison of eukaryotic nuclear RNase P RNAs and RNase MRP RNAs. Structures are shown for the *Saccharomyces cerevisiae* and *Homo sapiens* RNase P RNAs (**a**) and (**b**) and RNase MRP RNAs (**c**) and (**d**), respectively. RNase MRP structures are based on previous structural data and are updated according to (Piccinelli et al. 2005). Nucleotides conserved in all known RNase P RNAs or RNase MRP RNAs are shown inside *black circles*. The *shaded domains* in both RNase P and RNase MRP RNAs show similarity to the bacterial "catalytic domain" (C-domain). In the MRP RNAs, the conserved region (CR-IV) has the general consensus (5′-ANAGNNA-3′) and the P8 loop has the general consensus (5′-GARAR-3′) [R=purine], and highly conserved nucleotides are circled (*white background*)

5.8S rRNA from within the internally transcribed spacer of the 35S precursor RNA (pre-rRNA). This substrate specificity is reminiscent of the processing of some tRNAs from within the internal spacers of bacterial pre-rRNA transcripts by bacterial RNase P and this may provide some insight into the likely origins of RNase MRP in eukaryotes (Morrissey and Tollervey 1995). At present, it is not clear if all of these activities are carried out by the same enzyme or by subsets that retain the MRP RNA but with differing protein compositions. The composition of the main cellular RNase MRP from yeast has been established by biochemical purification (Salinas et al. 2005). In addition to an RNA subunit it was shown to have ten proteins, eight of which are also components of the yeast nuclear RNase P; Pop1, Pop3-8 and Rpp1 (Table 10.1) (Lygerou et al. 1994; Dichtl and Tollervey 1997; Stolc and Altman 1997; Chamberlain et al. 1998; Salinas et al. 2005). RNase MRP does not contain the Rpr2 protein found in RNase P but has two unique proteins (Snm1 and Rmp1), both of which are essential in yeast (Schmitt and Clayton 1994; Salinas et al. 2005). Since both Rpr2 and Snm1 have a predicted zinc binding site, it has been suggested that these proteins may be paralogs (Schmitt and Clayton 1994; Hartmann and Hartmann 2003). The holoenzyme is predominantly localized to the nucleolus where a role in the processing of pre-rRNA has been established (Schmitt and Clayton 1993; Lygerou et al. 1996a). Additional substrates and localization patterns for RNase MRP are discussed in Sect. 10.3.3. The nucleolar localization is similar to that of RNase P and in both cases, this is the site of transcription of its major RNA substrate. Despite having such a significant overlap in localization and subunit compositions, RNase P and MRP have distinct substrates and affinity isolations do not suggest that the holoenzymes are stably associated with each other in either yeast or human cells (Srisawat and Engelke 2001; Li and Altman 2002). With such a significant overlap in both RNA structure and protein subunits, the two holoenzymes are likely to have related catalytic mechanisms. The differences in substrate specificities between RNase P and RNase MRP could be directed by the unique protein subunits, differences between the RNA structures, or a combination of both.

10.2.4 Comparison with the Human Enzymes

The RNase P and RNase MRP enzymes have been investigated in human cell lines. The composition of the human RNase P has been determined and shown to contain a single RNA subunit *H1* (339 nucleotides) (Bartkiewicz et al. 1989) and at least ten proteins Rpp14, 20, 21, 25, 29, 30, 38, 40, hPop5 and hPop1 (Table 10.1) (Eder et al. 1997; Jarrous and Altman 2001). Seven of these proteins are homologs of the yeast proteins and several were initially identified due to their sequence homology (Lygerou et al. 1996b; Jarrous et al. 1999a; van Eenennaam et al. 1999, 2001a). The relationships between Rpp14, Rpp25 and Rpp40 and the yeast proteins are less clear, although there are some weak sequence homologies that indicate potential associations for these proteins (Hartmann and Hartmann 2003; Rosenblad et al. 2006). A weak sequence homology between the human Rpp14 and hPop5, has led to the suggestion that Rpp14 may be a paralog of hPop5 (Koonin et al. 2001; Hartmann and Hartmann 2003). Further sequence analysis has also linked the yeast Pop8 proteins to the mammalian Rpp14 and hPop5 proteins, and it has been suggested that the fungal Pop8 may be the ortholog to Rpp14 (Rosenblad et al. 2006). The human Rpp20, Rpp25 and the yeast Pop7 proteins each have predicted sequence similarity to the Alba superfamily of proteins, related to an RNA binding family of proteins (Aravind et al. 2003). The yeast Pop7 and human Rpp20 were placed in a distinct branch of Alba proteins, distinguishing them from Rpp25 and consistent with their predicted homology. The Rpp25 protein has been proposed to be related to the yeast Pop6 protein (Rosenblad et al. 2006). The positively charged regions on basic proteins are common to most ribonucleoproteins and these regions frequently bind the negatively charged RNA subunits, however both the yeast and human enzymes possess a single acidic protein, Pop8 (pI 4.6) and Rpp40 (pI 5.2), respectively. Although these proteins do not appear to be linked by sequence, the unique acidic nature of these proteins may indicate that they serve similar functions in the holoenzyme.

The yeast and human RNase P enzymes have related RNA subunits and at least seven homologous proteins. As discussed above, the remaining proteins are significantly diverged so as to hinder confident homology predictions between yeast and humans. The significance of an additional protein subunit in the human enzyme (ten vs. nine in yeast) is not known and it would not be entirely surprising if the differences in subunit composition also reflected subtle differences in function between the two enzymes. It is not currently clear if the localization and organization of tRNA transcription and maturation in human cells directly mirrors that observed in yeast [reviewed in (Hopper and Phizicky 2003; Hopper 2006)]. Components of the human RNase P may also be linked to transcription by RNA polymerases I and III in human cells (Reiner et al. 2006, 2008), discussed in Chap. 12. In yeast RNase P RNA is predominantly nuclear and although this does not exclude the possibility that smaller amounts may exist elsewhere, the main pool of RPR1 RNA is nucleolar (Bertrand et al. 1998). In human cells the H1 RNA appears to only transiently enter the nucleolus before diffusing into the nucleoplasm although the RNA has also been detected in the perinuclear compartment and the cytoplasm (Jacobson et al. 1994; Wolin and Matera 1999). Additionally, various human protein components, Rpp14, Rpp29, Rpp38, hPop1 and hPop5 have been localized to the nucleolus, however Rpp29 and Rpp38 are also detected in Cajal bodies (Lygerou et al. 1996b; Jarrous et al. 1999b; Savino et al. 1999; van Eenennaam et al. 2001b). The presence of these subunits in both RNase P and RNase MRP makes the protein localization data more difficult to interpret.

The human RNase MRP RNA possesses a unique RNA subunit (*7-2*) (Hashimoto and Steitz 1983), which either binds to, or coprecipitates with at least seven of the ten proteins (Rpp20, 25, 29, 30, 38, hPop5 and hPop1) identified in the human RNase P suggesting that they are also components of the human RNase MRP (Table 10.1) (Stolc and Altman 1997; Pluk et al. 1999; van Eenennaam et al. 1999, 2001a; Welting et al. 2004). There is also evidence that supports different pools of the enzyme existing with differing protein compositions (Welting et al. 2006).

The human RNase MRP RNA sediments as both 12S and 60–80S in glycerol gradients, the heavier MRP particles most likely represent association with preribosomal particles. The Rpp20, Rpp21, Rpp25 and hPop4 proteins could not be detected in 60–80S fractions, suggesting that these proteins may not be present in ribosome-associated RNase MRP. The composition of RNase MRP in human cells has not been determined by biochemical purification. It is particularly notable that there are currently no predicted homologs of the MRP specific proteins (Snm1 and Rmp1) which are present in yeast (Rosenblad et al. 2006).

The human RNase MRP RNA gene (*RMRP*) has been linked to a number of developmental disorders including cartilage-hair hypoplasia (CHH) and anauxetic dysplasia (Ridanpaa et al. 2002; Thiel et al. 2005). There are many causative mutations and these are found within both the promoter region and the mature RNA [reviewed in (Martin and Li 2007)]. If unique proteins are discovered for the human MRP it is equally likely that mutations within these subunits could contribute to a disease state. Known mutations that fall within the mature RNA have been investigated in both yeast and human cells and have been shown to have different effects on known MRP targets (5.8S rRNA and CLB2 mRNA) (Thiel et al. 2005). Since different mutations can result in the aberrant processing of at least two substrates, the relationship between the mutations present in patients and the exact disease manifestation would appear to be complex. It is also possible that RNase MRP could have further, as yet uncharacterized, substrates that are involved in the manifestation of disease states (see Sect. 10.3.3).

10.3 Substrate Specificity and Mechanism

The relative simplicity of the bacterial RNase P has facilitated its study in vitro. The bacterial RNA subunit alone (the ribozyme) is quite active at high salt and reconstitution with its single protein (forming the holoenzyme) is straightforward. The ability to access and manipulate both the RNA and protein subunits has allowed the dissection of the roles of both RNA and protein within a true multiple-turnover ribonucleoprotein enzyme [reviewed in (Smith et al. 2007)]. Similar mechanistic studies of the eukaryote enzymes have been hindered by the inability to fully reconstitute authentic eukaryotic holoenzymes. An obvious challenge is deciphering the subunit interactions between the numerous eukaryote proteins and the RNA (Sects. 10.4 and 10.5). Some initial progress has been made towards deciphering the mechanism of the yeast RNase P using purified enzyme (Sect. 10.3.1). Studies using the intact, purified, enzyme will ultimately provide a baseline upon which to assess a true reconstitution of the eukaryote RNase P in vitro.

In addition to the well-studied pre-tRNA substrates, the bacterial RNase P is known to process a number of non-tRNA substrates; 4.5S RNA, tmRNA, viral RNAs, mRNAs, ribo-switches, ColE1 replication origin control RNAs, and C4 antisense RNA from phages P1 and P7 (Peck-Miller and Altman 1991; Giege et al.

1993; Alifano et al. 1994; Komine et al. 1994; Hartmann et al. 1995; Gimple and Schon 2001; Li and Altman 2003; Li et al. 2003; Altman et al. 2005). It would seem likely that the eukaryotic RNase P could also have additional substrates. The evolution of nuclear RNase P into RNase MRP expanded the repertoire to multiple non-tRNA substrates (Sect. 10.3.3). Beyond that, there is a growing body of evidence that the nuclear RNase P itself has additional substrates in addition to pre-tRNAs (Sect. 10.3.2).

10.3.1 RNase P Mechanism

Studying reaction kinetics can provide some insights into enzyme mechanism, and this has been especially true in the bacterial RNase P system where each subunit can be individually manipulated and the resulting effects studied [reviewed in (Smith et al. 2007)]. Kinetic studies of the bacterial holoenzymes show a burst of product formation under presteady state conditions (Reich et al. 1988; Tallsjo and Kirsebom 1993; Beebe and Fierke 1994). A similar burst in product formation was shown using purified yeast nuclear RNase P suggesting that they have similar kinetic mechanisms, despite their differing protein compositions (Hsieh et al. 2008). A burst in product formation indicates that the initial turnover of substrate by RNase P is faster than the subsequent turnovers. Therefore, the acquisition and cleavage of the initial substrate molecule occurs quickly and the observation of subsequent cleavages is slowed by a bottleneck in the reaction. The bottleneck could be simply due to the slow release of the product or more complex scenarios, such as conformational change to facilitate chemistry or the regeneration of the active site. An exploratory analysis of the yeast nuclear RNase P has shown that the rate of steady state substrate turnover is largely temperature dependent, the energy dependence is most consistent with a conformational change at catalysis, although other models could also apply (Hsieh et al. 2008).

10.3.2 RNase P Substrate Specificity

There are subtle differences between bacterial and eukaryote pre-tRNA transcripts that could potentially influence substrate recognition in the respective holoenzymes. In bacteria, pre-tRNAs often have a transcribed 3′-CCA and this can interact with the CCA binding loop (within P15) that is present in some, but not all, of the bacterial RNAs (Guerrier-Takada et al. 1989; Kirsebom and Svard 1994). Unlike most bacterial tRNAs, the 3′CCA is not encoded in eukaryote tRNAs and is added to the tRNA posttranscriptionally [reviewed in (Hopper and Phizicky 2003)]. In yeast the tRNA primary transcripts end with oligo U (U_{4-6}), the terminator sequence for RNA polymerase III and this sequence often forms a short bulged stem structure with purine-rich 5′ leader sequences. The presence of the 3′ trailer sequence slightly enhances substrate cleavage rates and a bulge at the beginning of the 5′/3′ stem structure, leaving the nucleotide before the RNase P cleavage site (−1 position) unpaired, appears to be a required feature for efficient cleavage (Lee et al. 1997). Despite these subtle differences in the primary transcripts, tRNA structures are extremely well conserved and RNase P from bacteria, archaea and eukaryotes can each recognize and cleave each other's pre-tRNAs.

The bacterial RNase P protein interacts with the 5′ leader of the pre-tRNA substrate and enhances the overall rate of cleavage in the holoenzyme (Crary et al. 1998; Kurz et al. 1998; Niranjanakumari et al. 1998; Hsieh et al. 2004). The bacterial protein is also required for recognition of non-tRNA substrates and helps broaden the specificity of the bacterial enzyme (Gopalan et al. 1997). Additional proteins in the eukaryote could perform similar functions and, given their increased number, perhaps play a greater role in catalysis or the selection of additional substrates. A number of observations indicate that the bacterial and eukaryote enzyme respond differently to non-tRNA substrates. The human enzyme does not cleave the same minimal substrates as the bacterial enzyme, requiring an additional loop between the acceptor stem and T stem-loop domains (Yuan and Altman 1995). The yeast enzyme binds to, and is strongly inhibited by, single-stranded homoribopolymers [poly(G) and poly(U)>poly(A) >>poly(C)] and mixed-sequence RNAs, but the bacterial holoenzyme was not affected by these RNAs (Ziehler et al. 2000, unpublished observations). The yeast and bacterial enzymes also show quite different cleavage patterns when they are allowed to cleave in vitro transcribed ribosomal RNA precursors (Chamberlain et al. 1996). Although the physiological relevance of these non-tRNA substrates is unclear, they demonstrate that the eukaryote and bacterial enzymes have different determinants for interacting with non-tRNA substrates.

Studies have indicated various potential non-tRNA substrates for the yeast nuclear RNase P, these include the pre-rRNA, small nucleolar RNAs (snoRNAs), noncoding RNAs (ncRNAs) and a number of other more tentative observations (Chamberlain et al. 1996; Samanta et al. 2006; Yang and Altman 2007; Coughlin et al. 2008). A temperature sensitive mutant of the RPR1 RNA accumulates a 3′ extended form of the 5.8S rRNA (approx +35 nt) at the nonpermissive temperature (Chamberlain et al. 1996). It was not established if this was a direct or indirect effect of RNase P function, however the colocalization of RNase P and ribosomal biogenesis at the nucleolus makes a direct involvement possible. A broader search for RNase P substrates has utilized a multipronged approach to determine RNAs that accumulate upon disruption of RNase P function and also those RNAs that copurify with a fully active holoenzyme (Coughlin et al. 2008). A large number of potential substrates for the yeast nuclear RNase P were revealed in this study and one subset was explored in greater detail. The maturation of the intron-encoded box C/D small nucleolar RNAs (snoRNAs) can proceed by either splicing-dependent or splicing-independent pathways (Villa et al. 1998, 2000). RNase P mutants were found to accumulate larger forms of these intron encoded box C/D snoRNAs and the enzyme appears to function within the splicing-independent pathway. These RNAs were shown to be substrates for yeast nuclear RNase P in vitro although a consistent cleavage pattern across the group did not emerge and it was not possible to confirm the sites of cleavage in vivo. A number of other potential mRNA substrates were also identified in this study, including ORFs that encode ribosomal protein subunits, translation initiation factors, box C/D snoRNP proteins and common subunits of RNA polymerases I, II and III. One caution in interpreting these data is that they did not distinguish sense vs. antisense strandedness and at least some of these RNAs are known to have antisense transcripts. A similar genome wide study has also identified various novel noncoding RNAs that were perturbed upon the depletion of the Rpp1 protein subunit (Samanta et al. 2006). The data sets from these two independent studies did not significantly overlap; however since Rpp1 is a protein subunit common to both RNase P and RNase MRP this is more difficult to interpret. One of the ncRNAs identified (HRA1) is antisense to the DRS2 gene that encodes a protein involved in 18S rRNA maturation. The level of HRA1 RNA in yeast cells was shown to increase upon depletion of the Rpp1 protein (common to RNase P and MRP). Subsequently, the HRA1 RNA was found to be a substrate for the yeast RNase P in vitro with at least two canonical RNase P recognition sites, however, verification of these cleavages in vivo was not possible due to low levels of endogenous HRA1 RNA (Yang and Altman 2007).

10.3.3 RNase MRP Substrate Specificity

Currently, the yeast RNase MRP is known to have at least three entirely different kinds of RNA substrates. RNase MRP was originally identified as a mitochondrial RNA processing activity, where it was shown to cleave mitochondrial RNAs that serve as RNA primers in the replication of mitochondrial DNA (Chang and Clayton 1987; Lee and Clayton 1998). The mitochondrial processing activity, was linked to the RNase MRP RNA subunit and it should be noted that, at present, the mitochondrial RNase MRP has not been purified to homogeneity. The protein content of the mitochondrial enzyme could therefore, differ from the cellular RNase MRP. As previously discussed, visualization of the MRP RNA has shown that the RNA is predominantly found in the nucleolus and a role in the processing of preribosomal RNA (pre-rRNA) was established. In yeast, three of the four eukaryotic ribosomal RNAs (18S, 5.8S and 23S rRNA) are transcribed as a single large precursor RNA (35S pre-rRNA) and subsequently assembled and processed into the respective large (60S) and small (40S) ribosomal subunits [reviewed in (Fatica and Tollervey 2002)]. RNase MRP performs a specific endonucleolytic cleavage at site A3 between the 18S and 5.8S rRNAs of the 35S pre-rRNA (Schmitt and Clayton 1993; Chu et al. 1994; Lygerou et al. 1996a). Subsequent exonucleolytic cleavage results in the mature 5′ end of the short form of the $5.8S_{(S)}$ rRNA (Henry et al. 1994). However the situation is complex; in the absence of RNase MRP cleavage at site A3 another pathway is utilized and results in the long $5.8S_{(L)}$ rRNA (6nt longer at the 5′ end). In normal yeast cells both the short and long forms of the 5.8S rRNA exist with the short form in excess. Both forms of 5.8S rRNA are viable and it is not known if they have any functional differences. A role in the cell cycle was shown in yeast carrying a mutation of the Snm1 protein that displayed a defect in exit from mitosis (Cai et al. 2002). It was demonstrated that RNase MRP cleaves the cyclin B2 (CLB2) mRNA within its 5′ untranslated region (UTR) leading to the degradation of the mRNA by the 5′ to 3′exonuclease Xrn1 (Schmitt et al. 2002; Gill et al. 2004). The depletion of cyclin B2 protein is a required step for the progression from anaphase into G1 (Charles et al. 1998). The involvement of RNase MRP in cell cycle progression has been further explored in yeast. RNase MRP is found to relocalize from the nucleolus after the initiation of mitosis (Gill et al. 2006). During this stage, RNase MRP is found throughout the entire nucleus and also in a discreet cytoplasmic spot, which is localized to the daughter cell that disappears following telophase. The cytoplasmic bodies have been named TAM bodies (temporal asymmetric MRP bodies) and appear to be distinct versions of the cytoplasmic P-bodies which are thought to be sites of the mRNA degradation machinery [P-bodies reviewed in (Parker and Sheth 2007)]. It was also shown that the Xrn1 exonuclease colocalizes with the TAM-bodies (Gill et al. 2006). There have been no reports of specific genome wide searches for further RNase MRP substrates. However, as previously discussed, the depletion of a common protein subunit (Rpp1) affected a number of novel noncoding RNAs and these could be potential substrates for either RNase P or RNase MRP (Samanta et al. 2006).

10.4 The Eukaryote RNA Subunits

10.4.1 The RNA Remains the Catalytic Subunit of Nuclear RNase P

The RNA subunit of the bacterial holoenzyme is unquestionably the catalytic subunit, as demonstrated by robust cleavage of pre-tRNAs in vitro (180–360 min^{-1}, single turn over conditions) (Beebe and Fierke 1994). Similarly, a number of the archaeal RNA subunits are also capable of ribozyme activity in vitro (Pannucci et al. 1999). The efficiency of the archaeal ribozymes is significantly reduced when compared to the bacterial ribozymes, suggesting that the archaeal RNA subunit is more reliant on proteins in vivo to maintain its structure and promote catalysis. In the eukaryote, the conservation of the RNA subunit and its key structural elements (i.e., the catalytic domain) has made its continued role as the catalytic subunit seem highly likely and there has been some evidence from mutated RNA subunits to support this (Pagan-Ramos et al. 1996a, b). Comparison of the bacterial and eukaryote enzymes using pre-tRNAs with phosphothiorate substitutions at their cleavage sites shows similar inhibitory effects which is consistent with a similar catalytic mechanism (Pfeiffer et al. 2000; Thomas et al. 2000). However, the direct demonstration of the catalytic properties of a eukaryote RNA has posed a significant technical challenge that has only recently been met. By screening several eukaryote

RNAs and selecting solution conditions in vitro to minimize basal RNA hydrolysis, ribozyme activity was demonstrated in some eukaryote RNAs, including the human nuclear RNase P RNA (Kikovska et al. 2007). The relative activity of the human RNase P RNA was very low (0.000026 min^{-1}, single turn over conditions) when compared to rates for the bacterial RNA under identical conditions (8 min^{-1}, single turn over conditions) and only a few 100-fold above the spontaneous hydrolysis of RNA under the same buffer conditions. Nevertheless, these eukaryote RNAs were demonstrated to perform pre-tRNA cleavage in vitro with authentic reaction products and appropriate controls to rule out the possibility of contaminating bacterial RNase P RNAs (Kikovska et al. 2007). The RNA component of *Schizosaccharomyces pombe* nuclear RNase P has not been shown to be capable of pre-tRNA cleavage in vitro however, the RNA subunit has been demonstrated to capable of binding to pre-tRNAs under moderate conditions (Marquez et al. 2006). This data implies that the eukaryote RNA subunit can adopt a similar structure to the bacterial RNA that, in the case of the *S. pombe* RNA, is capable of substrate binding.

10.4.2 The RNA Subunit of RNase P

Examples of nuclear RNase P RNAs from both yeast (*RPR1* RNA, *S. cerevisiae*) and mammals (*H1* RNA, *H. sapiens*) are shown in Fig. 10.1. The majority of the sequence data for the RNase P RNA subunits has been archived online at the RNase P database (Brown 1999; http://www.mbio.ncsu.edu/RNaseP/home.html). Sequence comparisons from many eukaryotes have revealed a conserved core RNA structure which retains many of the helices found in the bacterial and archaeal RNAs; P1, P2, P3, P4, P7, P8, P9, P10/12 and P12 (Altman et al. 1993; Chen and Pace 1997; Pitulle et al. 1998; Frank et al. 2000; Marquez et al. 2005; Piccinelli et al. 2005). This minimal consensus structure also contains each of the conserved regions of sequence (CR-I to CR-V) found in both the bacterial and archaeal RNAs. A subset of helices are designated "eukaryal" (eP8, eP9, eP15 and eP19), as these occupy similar positions to those in bacteria, but it is not clear if these are performing analogous functions in the eukaryote. Phylogenetic data shows that an eP19 stem is present in most eukaryotes while an eP15 stem is present in most fungal lineages but not generally found in other eukaryotes (Marquez et al. 2005).

Despite being largely similar, the eukaryote RNAs appear to lack many of the peripheral elements found in the bacterial type A (P6, P13, P14 and P18) and type B (P5.1, P10.1, P15.1 and P18) RNAs (Haas et al. 1996). The construction of a minimal bacterial RNase P RNA (Micro P) has demonstrated that such peripheral structures are not absolutely required, however their presence allows for ribozyme activity at lower salt concentrations than the Micro P RNA (Siegel et al. 1996). Comparisons of both the bacterial type A and type B RNA crystal structures also place these additional helices at the periphery of the structure, where they appear to aid in the stabilization and correct orientation of the core structure (Krasilnikov et al. 2003, 2004; Kazantsev et al. 2005; Torres-Larios et al. 2005). It is likely that the absence of similar structures in the eukaryotic RNA is compensated by increased protein content where proteins would help to facilitate the correct folding and stabilization of the RNA. This conclusion is somewhat supported by a tertiary structural model of the *S. pombe* RNase P RNA (Marquez et al. 2006). Although the tertiary model lacks the P3 and P12 helices, it does show that the P5.1–P15.1 interaction in the bacterial RNA is adjacent to the likely position of the eukaryote P3 stem. The eukaryote P3 stem is a site of protein binding in the eukaryote (Pop1 and Pop6/7, see Sect. 10.5.2) and these proteins would thus be ideally positioned to assume the equivalent structural roles of the bacterial P5.1–P15.1 interaction.

In addition to the absence of structures that can stabilize the RNA core, the eukaryote RNAs also lack conserved features of the bacterial RNA that have been implicated in binding the pre-tRNA substrate. The CCA binding loop within the P15 helix and the highly conserved bulged nucleotides around the four-helix junction are known to directly interact or crosslink to the tRNA substrate (Pan et al. 1995; Loria and Pan 1997, 1999; Chen et al. 1998; Christian et al. 2002). The absence of these pre-tRNA binding features in the eukaryote RNAs may suggest a more involved role for the protein subunits in binding substrates in the eukaryote enzymes.

The eukaryote P3 stem has a large internal loop structure with limited sequence conservation, and is required for the correct assembly and function of the holoenzyme in vivo, in both yeast and human cells (Lindahl et al. 2000; Li and Altman 2002). The deletion of the P3 region from the human RNase P RNA has been shown to result in the mislocalization of the RNA when micro-injected into human cells (Jacobson et al. 1997). In addition, the purification of a human RNase P with a deleted P3 stem does not result in the recovery of an active enzyme (Li and Altman 2002). In the yeast system the nucleotides within the P3a/P3b loop have been randomized and selected for function in vivo (Ziehler et al. 2001). A number of temperature sensitive mutations were revealed and characterization of the resulting defects in vivo showed an accumulation of the precursor-RPR1 RNA, which is indicative of an assembly defect. Consistent with a role in holoenzyme assembly is the observation that the P3 stem is also the binding site of at least three of the eukaryote proteins (see Sect. 10.5.2). Conversely, in the bacterial RNA the removal or truncation of the P3 stem in bacteria has only modest effects on the ribozyme. The bulged P3 stem appears to be a distinct feature that is specific to the eukaryote RNase P RNA.

More specific studies of the eukaryote RNase P RNAs have been carried out in the yeast system, where the genetic manipulation of the RNA subunit and the study of the subsequent effects of mutations in vivo are easier than in mammalian systems. In yeast the, RPR1 RNA and the pre-tRNA processing profile can be examined by northern blot. In majority of cases, mutations either affect the stability of the RNase P RNA (interpreted as a defect in holoenzyme assembly/stability) or result in a deficiency in pre-tRNA cleavage (interpreted as a defect in catalytic proficiency). Each of the conserved regions CR-I, CR-II, CR-IV and CR-V have been investigated by the randomization of conserved nucleotides and in vivo screening for conditional phenotypes (Pagan-Ramos et al. 1996a, b; Ziehler et al. 1998).

Conditional mutants within the CR-I, CR-IV and CR-V regions were isolated and showed reduction in the catalytic efficiency (k_{cat}/K_m) of the holoenzyme in a pre-tRNA assay in vitro. In the majority of cases these mutations were shown to have a greater influence on catalysis (k_{cat}) than substrate recognition (K_m). These results are consistent with what has been learned from the crystal structures of the bacterial RNA subunits, which place the CR-I, CR-IV and CR-V sequences adjacent to the proposed catalytic site. The CR-II and CR-III sequences form part of the J11/12-J12/11 module, flanked by the P10/11 and P12 stems forming part of the "specificity" domain. This region has also been analyzed by random mutation of the CR-II sequence and the mutations that were isolated and characterized showed a dependence on magnesium for growth and an increased magnesium requirement for catalytic activity in vitro (Pagan-Ramos et al. 1996a; Ziehler et al. 1998).

There are also other more loosely conserved sequence regions that may be specific to the yeast RNase P RNAs and have been analyzed in vivo. In yeast species the sequence between P4 and P7 (jP4/7) shows some conservation adjacent to the P4 stem (5′-GAAC-3′). The mutation of these nucleotides affects growth, holoenzyme assembly and localization. The mutant enzymes also show severe defects in pre-tRNA processing in vitro and a tenfold reduction in the catalytic efficiency (k_{cat}/K_m) in vitro (Xiao et al. 2005a). These conserved nucleotides are in close proximity to the P4 stem and close to the proposed catalytic region in the bacterial crystal structures, they may serve a specific function in the eukaryote that is distinct to the bacterial enzymes. In yeast, a high degree of sequence conservation is shown in the loops of stems eP8 (5′-NUGA-3′) and eP9 (5′-GNAA-3′). Mutation of the loop sequences affects growth, pre-tRNA processing (in vivo) and holoenzyme assembly. The eP9 mutations were more severe and also affected enzyme localization in vivo. The mutant enzymes also showed aberrant kinetics in pre-tRNA assays in vitro. The eP9 mutant showed a reduction in catalytic efficiency (k_{cat}/K_m), whereas the eP8 mutant showed an unexpected increase in catalytic efficiency. The increase in catalytic efficiency appears to be due to a loss of specific substrate selectivity as the enzyme is more efficiently inhibited by total yeast RNA than the wild type enzyme (Xiao et al. 2005a). The eP8 and eP9 stems play key roles in the maturation and function of the holoenzyme and the conservation of sequence within the stem-loop structures suggests that the loops may be involved in RNA–RNA or RNA–protein interactions. This is consistent with the observation that the loop structures are protected from nuclease attack in footprinting studies of the *S. cerevisiae* holoenzyme (Tranguch et al. 1994).

10.4.3 The RNA Subunit of RNase MRP

Examples of RNase MRP RNAs from both yeast (*NME1* RNA, *S. cerevisiae*) and mammals (*7-2* RNA, *H. sapiens*) are shown in Fig. 10.1. The MRP RNAs possess many structural features that are common to the bacterial, archaeal and eukaryote RNase P RNAs. Generally, conserved structures within related RNAs are an indicator that these structures are functionally relevant. The structural homology between RNase MRP and P RNAs includes stems P1, P2, P3, P4, and eP19 (also eP15 in some yeasts) and forms a structure that is consistent with the catalytic domain of the RNase P RNAs (Fig. 10.1, shaded) (Forster and Altman 1990; Schmitt et al. 1993; Paluh and Clayton 1995). The MRP RNAs also contain regions of conserved sequence similar to those found in RNase P RNA; CR-I, CR-V and a more weakly conserved CR-IV (Piccinelli et al. 2005). Conservation of the catalytic domain strongly suggests that the RNA is also the catalytic core in the RNase MRP holoenzyme. Given the wealth of data regarding the RNase P RNAs, the catalytic-like domain of the MRP RNAs has been relatively easy to define. Prediction of the secondary structure in the remainder of the MRP RNA has been more complicated. Initial phylogentic analyzes did not show homology to the RNase P specificity domain and this region also differed between the two groups of MRP RNA sequences that were studied (Schmitt et al. 1993; Li et al. 2002). Each of the predicted structures was supported by biochemical probing of the RNA secondary structure from representative yeast species (*S. cerevisiae*) and vertebrates (*Mus musculus* and *H. sapiens*) (Topper and Clayton 1990; Li et al. 2002; Walker and Avis 2004, 2005). Subsequent studies over a broader range of eukaryote species have demonstrated that the arrangement of the MRP-specific domain is not fixed and appears to vary between species (Piccinelli et al. 2005; Woodhams et al. 2007). The most commonly identified arrangement is the three stems; P8, P9 and P12 as shown in the *H.sapiens* MRP RNA (Fig. 10.1). The P8 stem had been predicted in some earlier models but had initially been overlooked as this was not well supported by structure probing data or covariation within the short stem. However, the addition of more eukaryote MRP RNA sequences showed that the stem and the stem-loop are both highly conserved within most MRP RNAs (Piccinelli et al. 2005). The conserved sequence forms a pentaloop structure (5′-GARAR-3′) and occasionally shows a single 3′ deletion to a tetraloop (5′-GARA-3′). Although the P8 stem is well conserved, it is not clear how the other stems are related or if they have any functional relevance (Fig. 10.1, compare *S. cerevisiae* [ymP5, ymP6 and ymP7] with *H. sapiens* [P9 and P12]). The differences between the MRP RNAs within the MRP-specific domain may indicate an adaptation towards a wider range of substrates, perhaps with specialization within phylogenetic branches (Li et al. 2002; Walker and Avis 2005). In this respect, it is interesting that many verterbrate species have a conserved structural motif within their P12 stems, the kink-turn (K-turn) (Klein et al. 2001; Rosenblad et al. 2006). The K-turn RNA structures are known to bind the L7Ae archaeal protein which is a predicted structural homolog of the Pop3/Rpp38 proteins (Rozhdestvensky et al. 2003; Hamma and Ferre-D'Amare 2004; Dlakic 2005).

Given the conservation in most MRP RNAs, the P8 stem is clearly an important structural feature of the MRP RNA. The loop sequence of the P8 stem from *S. cerevisiae* (5′-GAAAA-3′) has been previously characterized within the context of the coliphage HK022 protein-lambda-phage boxB RNA complex (Faber et al. 2001). In this related structure, the sequence forms a variant of the GNRA tetraloop, where the fourth base is flipped out and does not participate in contacts with

RNA or protein. The P8 stem is not well supported by structure probing of in vitro transcribed MRP RNAs (human and yeast) and it is possible that a protein interaction may be required in vivo to stabilize the short stem. Another observation is that the conserved sequence within the P8 loop can participate in a P7-like stem in the yeast MRP RNA in vitro, however, extensive mutational analysis only provided support for the stem in vitro and not in vivo (Walker and Avis 2004). It remains possible that the P7-like stem could represent a valid intermediate structure during assembly of the yeast RNase MRP since it would increase the homology between the MRP and P RNAs within yeasts.

Further mutational analysis of the yeast MRP RNA has defined a minimal sequence for the MRP RNA (mini 2) that is functional in vivo (Li et al. 2004). It was demonstrated that the ymP5, ymP6, ymP7 and eP19 stems can be substantially truncated without significantly affecting the enzyme. The additional truncation of the eP15 stem resulted in temperature sensitivity and a region of sequence within ymP5 (loosely conserved in yeasts) was also sensitive to mutation. The P8 stem had not been confirmed and was not included in this study.

10.4.4 Coevolution of RNase P and RNase MRP RNAs

In keeping with the extensive overlap in protein subunits between the RNase P and RNase MRP holoenzymes, regions of the RNA subunits appear to have coevolved. Some stem structures and sequence elements are more frequently retained between RNase P and MRP RNAs within the same species than they are across species. The P3 stem from the yeast RNase P RNA can functionally replace that of the MRP RNA within the same species of yeast, although it could not functionally replace the P3 stem of the same enzyme (i.e., RNase P) from another closely related yeast species (Lindahl et al. 2000). There is also some loose conservation of sequence with the P3a/P3b loop region of the P3 stem that appears to be greater between P and MRP RNAs within a species and does not persist well when examined across a wide range of species (Perederina et al. 2007). The P3 stem with an internal loop is a eukaryote-specific RNA structure and a key protein binding region for at least three proteins including the largest eukaryote protein subunit, Pop1, and also the Pop6/Pop7 heterodimer (Ziehler et al. 2001; Perederina et al. 2007).

Another structure that appears to be coevolved is the eP15 stem which is present in the RNA subunits of both P and MRP in yeasts but not found in either RNase P or MRP RNAs in verterbrate sequences (Fig. 10.2) (Li et al. 2002; Marquez et al. 2005). Unlike the P3 stem, there is no sequence conservation within the eP15 stem and mutational analysis does not support a significant role in enzyme function (Li et al. 2004). Nevertheless, the eP15 stem structure is conserved between yeast RNase P and MRP RNAs and this may represent a need for these RNAs to retain a degree of structural similarity in order to aid the assembly of these RNA subunits with the eight protein subunits that are common to each holoenzyme.

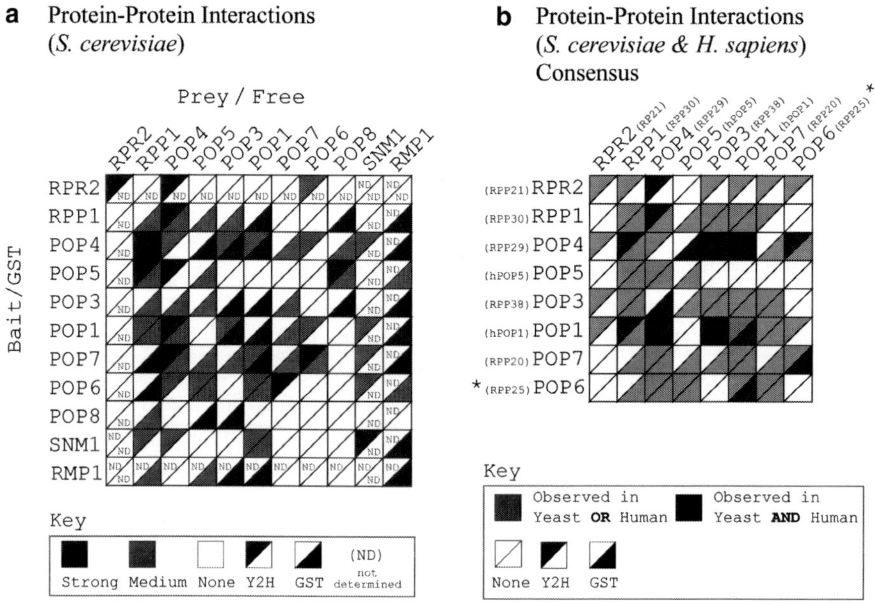

Fig. 10.2 (**a**) A summary of interactions demonstrated in the yeast system (*Saccharomyces cerevisiae*). The Y2H interactions are shown as described in the original text (Houser-Scott et al. 2002). The GST pull-downs results shown are those obtained under the more stringent salt conditions (300 mM NaCl) as described in the original text (Aspinall et al. 2007). A similar summary of Y2H and GST pull-down studies in the human system can be found elsewhere (Walker and Engelke 2006). (**b**) A summary of consistent interactions from the combined yeast and human studies (Jiang and Altman 2001; Houser-Scott et al. 2002; Welting et al. 2004; Aspinall et al. 2007). An interaction can be detected up to eight times representing reciprocal interactions detected in both the Y2H and GST methods and also in both the yeast and human systems

10.5 The Protein Subunits: Holoenzyme Architecture

Although the RNA is the catalytic subunit of the nuclear RNase P, it is clear that the protein subunits are required to assist the RNA and facilitate enzymatic function. In yeast, the individual subunits can be manipulated genetically and this has allowed investigations into the holoenzyme assembly in vivo (Sect. 10.5.1). The protein subunits must assemble with each other and also the RNA subunit to build a functional enzyme and the individual subunit interactions have been studied in vitro (discussed in Sects. 10.5.2 and 10.5.3). The ultimate goals of this research are to develop a thorough understanding of the holoenzyme architecture and how this relates to substrate selection and catalytic function. It would seem reasonable to assume that the conserved archaeal protein homologs (Pop4, Pop5, Rpp1 and Rpr2, or possibly Snm1 in RNase MRP) could form the core of the eukaryote holoenzyme. Although some progress has been made towards reconstitution (Mann et al. 2003), one obstacle to in vitro reconstitution has been in producing and purifying the individual protein subunits as a number of these are difficult to express, insoluble and only transiently stable (Houser-Scott et al. 2002; Aspinall et al. 2007; Perederina et al. 2007). There has been significant progress towards reconstituting the archaeal holoenzymes (see Chap. 9) and given the conservation of subunits it is likely that these endeavors will further inform studies of the eukaryote enzymes. Given the subunit complexity of the eukaryote systems, compared to the bacterial and archaeal systems, our knowledge regarding the architecture of the eukaryote RNase P and RNase MRP holoenzymes is still at an early stage.

10.5.1 Holoenzyme Assembly (in Vivo)

The maturation and assembly of nuclear RNase P in vivo has been best characterized in yeast. In *S. cerevisiae*, the RPR1 RNA is initially transcribed by RNA polymerase III as a large primary transcript (pre-RPR1 RNA) with an additional 84 nucleotides at the 5′ end, and up to 30 nucleotides at the 3′ end (Lee et al. 1991b). Similar precursor RNAs have not been found in the *H. sapiens* system and this maturation pathway may be specific to *Saccharomycetes*. The precursor transcript in *S. cerevisiae* undergoes a single 5′ cleavage and multiple 3′ cleavages to yield the mature 369-nt RPR1 RNA (Lee et al. 1991a). The precursor form of the enzyme has been specifically purified by utilizing an RNA affinity tag "aptamer" inserted into the 5′ leader sequence (Srisawat et al. 2002). The affinity purified precursor RNase P was found to be similar to the mature holoenzyme, but lacked the Pop3 and Rpr2 subunits suggesting that these proteins have roles that are late in assembly. Thus, the precursor RNA is assembled into a substantial ribonucleoprotein prior to the maturation of the RPR1 RNA. The purified precursor complex showed similar steady state kinetics in the processing of pre-tRNAs in vitro suggesting that the Pop3 and Rpr2 proteins are not absolutely required for pre-tRNA cleavage, despite being essential for growth in yeast. The pre-RPR1 RNA (but not the mature RPR1 RNA) seems to be associated with a complex of Sm-like (Lsm) proteins, specifically the Lsm 2–8 complex (Salgado-Garrido et al. 1999; Pannone et al. 2001; Kufel et al. 2002, reviewed in Beggs 2005). The Lsm 2-8 complex presumably interacts with the pre-RPR1 pol III transcript via its U-rich 3′ tail, although it is not absolutely certain that this interaction takes place in vivo, as well as in cell lysates. Archaeal homologs of the Sm proteins have also been shown to interact with their corresponding RNase P RNAs (Toro et al. 2001).

The structure of intact RNase P and RNase MRP holoenzymes, purified from yeast, has been investigated using structure sensitive RNA footprinting assays (Tranguch et al. 1994; Esakova et al. 2008). These studies show similar patterns of nucleotides that are solvent exposed or protected by protein in the holoenzymes and indicate similar architectures in the "catalytic" domain of both RNase P and RNase MRP holoenzymes, consistent with their extensive overlap of protein subunits. In both the RNase P and RNase MRP holoenzymes, the stem-loop structures of the P3 and eP15 stems are solvent exposed and the P1, P2, P4 and eP19 stem structures are protected in the presence of protein. In the "specificity" domain of yeast RNase P RNA the stems and loops of the P8 and P9 structures are protected in the holoenzyme, consistent with an active role for these structures that has been indicated by mutagenesis studies (Tranguch et al. 1994; Xiao et al. 2005b). Similarly, the ymP5 and P8 stems of the yeast RNase MRP RNA are also protected by protein, consistent with active roles for these structures that are suggested by observed sequence conservation and the results of mutagenesis studies (Shadel et al. 2000; Li et al. 2004; Piccinelli et al. 2005; Esakova et al. 2008).

10.5.2 RNA–Protein Interactions

The yeast three-hybrid system has been used to demonstrate individual interactions between the yeast RNase P RNA and the Pop1 and Pop4 proteins respectively (Houser-Scott et al. 2002). The RPR1–Pop1 interaction was shown to be specific to the P3 stem and additionally, the binding was abolished by specific mutations within the P3a/P3b loop which were previously shown to affect holoenzyme assembly (Ziehler et al. 2001). Given the demonstrated functional equivalence of the P3 stem between RNase P and MRP within *S. cerevisiae*, it is also likely that the Pop1–P3 interaction is present in the RNase MRP system (Lindahl et al. 2000). A similar approach, utilizing the yeast three-hybrid system has also been used to demonstrate interactions between the human RNase P RNA with the human proteins Rpp38 (Pop3), Rpp29 (Pop4), Rpp30 (Rpp1) and Rpp21 (Rpr2) (Jiang et al. 2001). This three-hybrid study did not establish an interaction between the hPop1 protein and the human RNase P RNA, however, a subsequent study utilizing GST pull-downs supported the interaction in both the RNase P and MRP RNAs and also implicated the P3 domain as the likely binding site (Jiang et al. 2001; Welting et al. 2004). Present data from RNase P and RNase MRP in both the yeast and human systems supports the P3 stem as an essential structure for holoenzyme assembly that is specifically bound by the Pop1 protein. The P3 stem has also been implicated as a specific site of binding for the yeast Pop6-Pop7 heterodimer complex that binds with nanomolar affinity (Perederina et al. 2007). The individual Pop6 protein was not observed to bind the RNA subunits, however, the use of a Pop6-Pop7 heterodimer was necessary due to the poor solubility and low stability of the Pop7 monomer. In the human system, the Pop7 homolog, Rpp20, was also shown to bind the P3 region, as did the Rpp25 protein (Welting et al. 2004). The human Rpp20 and Rpp25 proteins also interacted with each other strongly, this may provide additional support for the suggestion that Rpp25 may be the homolog of the yeast Pop6 protein (Welting et al. 2004; Rosenblad et al. 2006).

Other studies have used in vitro binding techniques, such as GST pull-downs or UV crosslinking to explore the binding of individual proteins to both the human and yeast RNase P and RNase MRP RNAs (Pluk et al. 1999; Jiang et al. 2001; Van Eenennaam et al. 2002; Welting et al. 2004; Aspinall et al. 2007; Perederina et al. 2007).

The human Rpp29 and yeast Pop4 proteins have each been implicated as core proteins in the yeast and human holoenzymes and are shown to bind their respective RNase P and RNase MRP RNAs (Jiang et al. 2001; Houser-Scott et al. 2002; Welting et al. 2004; Aspinall et al. 2007). The Rpp38 protein appears to bind the human RNase MRP RNA within the P12 region of the RNA, which bears a K-turn motif (Welting et al. 2004). This observation is consistent with the prediction that Rpp38 is a homolog of the archaeal L7Ae protein which is known to bind to the K-turn motif of archaeal snoRNAs (Rozhdestvensky et al. 2003; Hamma and Ferre-D'Amare 2004). It is noted that K-turn motifs have been found in the equivalent P12 stems in examples of both RNase P and RNase MRP RNAs (Rosenblad et al. 2006). However, despite the presence of the Pop3 protein in yeasts the K-turn motifs are absent in the yeast RNAs and this is potentially a lineage specific RNA–protein interaction whose significance is currently unknown (Walker and Avis 2005; Rosenblad et al. 2006). A GST pull-down study that focused on the yeast RNase MRP established binding between the RNA and the Pop3, Pop4, Pop5, Rmp1 and Snm1 proteins, additionally controls suggested a preference for single stranded RNA over double stranded RNA (Aspinall et al. 2007). This study also addressed the binding of the yeast protein subunits to the pre-rRNA substrate of the RNase MRP holoenzyme and showed interactions with Pop4, Pop6, Rmp1 and Snm1 subunits. An interaction between Pop3 and the RNase P substrate, pre-tRNA, has also been reported (Brusca et al. 2001). A large number of RNA–protein interactions have been reported in both the RNase P and RNase MRP systems from yeast and humans, using various approaches. Presently, we have the highest confidence in interactions being formed between Pop1 and the P3 domain, Pop6/Pop7 and the P3 domain, Pop4 to the RNA and Pop3 to the P12 K-turn (if present).

10.5.3 Protein–Protein Interactions

In addition to RNA–protein interactions, protein–protein interactions have been investigated in the yeast and human systems. The binary interactions between the protein subunits have been probed in both cases using the yeast two-hybrid system (Y2H) (Jiang and Altman 2001; Houser-Scott et al. 2002) and also using an in vitro affinity pull-down assay (Welting et al. 2004; Aspinall et al. 2007). The pull-down studies did not include tagged versions of the Rpp30 (human), Rpp40 (human) and also the Rpr2 (yeast) subunits, whereas the Y2H studies lacked the hPop5 (human), Rpp25 (human) and the Rmp1 (yeast) subunits. A summary of the protein–protein interaction data from both studies of the yeast proteins is given in Fig. 10.2a, a similar summary of both studies performed with the human proteins has been published elsewhere (Walker and Engelke 2006). In an effort to clarify consistent interactions, the entire data set from both yeast and humans is summarized in Fig. 10.2b. There are 11 well supported interactions (detected at least four times in the combined studies) and although this does not preclude other interactions, we limit the following discussion to these examples.

A subset of the eukaryotic data can first be compared with that from earlier Y2H studies of the four archaeal core proteins (Hall and Brown 2004; Kifusa et al. 2005). The archaeal Y2H studies do not entirely agree with each other, however, both studies show an interaction between the archaeal Pop4 (Rpp29) and Rpr2 (Rpp30) homologs in addition to an interaction between the archaeal Pop5 (hPop5) and Rpp1 (Rpp30) homologs. These interactions have now been well established via crystallography and NMR studies (Kawano et al. 2006; Wilson et al. 2006; Honda et al. 2008). These two protein–protein interactions are also detected in the eukaryotic studies (Fig. 10.2); however, they are not among those we have defined as well supported and this may reflect the limitations in the Y2H and GST pull-down techniques. Given the subunit homology and conservation of key amino acids at the protein–protein surface it seems likely that these subunit interactions persist in the eukaryote holoenzymes.

The eukaryote data set strongly supports interactions between the Rpp1 (Rpp30) subunit and the Pop4 (Rpp29), Pop1 (hPop1), and also the Pop3 (Rpp38) subunits. Additional interactions of the Pop4 (Rpp29) subunit that are well supported are with the Pop1 (hPop1), Pop3 (Rpp38), Pop6 (Rpp25) and also the Pop5 (hPop5) subunits. There are fewer well supported interactions for the Pop5 (hPop5) and the Rpr2 (Rpp21) proteins and this may be partially due to under representation of these subunits as they are absent from certain studies. The interactions between the eukaryote Pop4 (Rpp29) and Rpp1 (Rpp30) subunits as well as between the eukaryote Pop5 (hPop5) and Pop4 (Rpp29) are also supported by one of the two archaeal Y2H studies (Hall and Brown 2004). The four core archaeal homologs are likely to form key interactions in the eukaryote holoenzymes, however, several well supported interactions are detected in the remaining "eukaryote-specific" subunits. The best supported interaction is with the largest protein subunit, Pop1 (hPop1), shown to interact with the Pop4 (Rpp29) subunit in all four of the studies. Other well supported interactions for the Pop1 (hPop1) subunit are with the Pop3 (Rpp38), Pop7 (Rpp20) and also the Pop6 (Rpp25) subunits. Both the Pop1 (hPop1) and Pop4 (Rpp29) subunits have multiple binding partners and previously demonstrated RNA binding properties. The Pop1/Pop4 (hPop1/Rpp29) complex would, therefore, appear to be a good candidate for an early building block of the eukaryotic holoenzymes. The remaining well supported interaction is between Pop7 (Rpp20) and Pop6 (Rpp25) and this protein–protein complex has been previously verified and characterized as an RNA binding subcomplex (Perederina et al. 2007).

It should be noted that only binary interactions are described, and the subunit stoichiometry of the holoenzymes is not known. Preliminary analysis of RNase MRP from *S. cerevisiae* suggests that the subunit composition of the eukaryotic enzymes might be quite complex (Salinas et al. 2005). Working models of these protein–protein interaction and protein–RNA interactions have been drawn from the data (Jiang and Altman 2001; Houser-Scott et al. 2002; Welting et al. 2004). However, it is likely that more detailed knowledge of the individual subcomplexes will be required in order to refine our current models of the RNase P and RNase MRP holoenzymes.

10.6 Perspective

Ribonuclease P contains an essential and catalytic RNA subunit that is widely presumed to be a remnant of the prebiotic "RNA World". This RNA-based catalytic activity has been retained throughout the evolution of cellular systems and in all cases the RNA subunit requires protein cofactors to function in vivo. The retention of RNA as the catalytic subunit may be linked to the strict conservation of tRNA structures in all kingdoms of life, since there would be pressure to conserve a preexisting and highly efficient RNA-based pre-tRNA processing activity. The multisubunit eukaryote RNase P enzymes do not appear to be any more efficient in the processing pre-tRNAs than their two-subunit bacterial counterparts. Thus, the vast increase in protein content of the eukaryote enzymes and the essential nature of each subunit has been interpreted as an adaptation of the holoenzymes towards new functions in the complex eukaryotic nuclear environment. To date, the eukaryote RNase P has been implicated in the turnover of a number of non-tRNA substrates including intron-encoded Box C/D snoRNAs and other noncoding RNAs and it is likely that other substrate classes or entirely new functions may be uncovered in future studies. A further complicating factor is the existence of the related RNase MRP which retains a structurally similar RNA subunit and has a significant overlap in protein subunit composition. The substrate specificity of RNase MRP is significantly altered from that of RNase P and this must by determined by changes in the RNA subunit, the unique protein subunits or a combination of both. Given the high degree of similarity between RNase P and RNase MRP studies of both enzymes will ultimately be useful in determining the relative contributions of both the RNA and protein subunits in localization, substrate selection and catalysis.

References

Alifano P, Rivellini F, Piscitelli C, Arraiano CM, Bruni CB, Carlomagno MS (1994) Ribonuclease E provides substrates for ribonuclease P-dependent processing of a polycistronic mRNA. Genes Dev 8:3021–3031

Altman S, Wesolowski D, Puranam RS (1993) Nucleotide sequences of the RNA subunit of RNase P from several mammals. Genomics 18:418–422

Altman S, Wesolowski D, Guerrier-Takada C, Li Y (2005) RNase P cleaves transient structures in some riboswitches. Proc Natl Acad Sci U S A 102:11284–11289

Aravind L, Iyer LM, Anantharaman V (2003) The two faces of Alba: the evolutionary connection between proteins participating in chromatin structure and RNA metabolism. Genome Biol 4:R64

Aspinall TV, Gordon JM, Bennett HJ, Karahalios P, Bukowski JP, Walker SC, Engelke DR, Avis JM (2007) Interactions between subunits of *Saccharomyces cerevisiae* RNase MRP support a conserved eukaryotic RNase P/MRP architecture. Nucleic Acids Res 35:6439–6450

Bartkiewicz M, Gold H, Altman S (1989) Identification and characterization of an RNA molecule that copurifies with RNase P activity from HeLa cells. Genes Dev 3:488–499

Beebe JA, Fierke CA (1994) A kinetic mechanism for cleavage of precursor tRNA(Asp) catalyzed by the RNA component of *Bacillus subtilis* ribonuclease P. Biochemistry 33:10294–10304

Beggs JD (2005) Lsm proteins and RNA processing. Biochem Soc Trans 33:433–438

Bertrand E, Houser-Scott F, Kendall A, Singer RH, Engelke DR (1998) Nucleolar localization of early tRNA processing. Genes Dev 12:2463–2468

Brown JW (1999) The ribonuclease P database. Nucleic Acids Res 27:314

Brusca EM, True HL, Celander DW (2001) Novel RNA-binding properties of Pop3p support a role for eukaryotic RNase P protein subunits in substrate recognition. J Biol Chem 276:42543–42548

Cai T, Aulds J, Gill T, Cerio M, Schmitt ME (2002) The *Saccharomyces cerevisiae* RNase mitochondrial RNA processing is critical for cell cycle progression at the end of mitosis. Genetics 161:1029–1042

Chamberlain JR, Pagan R, Kindelberger DW, Engelke DR (1996) An RNase P RNA subunit mutation affects ribosomal RNA processing. Nucleic Acids Res 24:3158–3166

Chamberlain JR, Lee Y, Lane WS, Engelke DR (1998) Purification and characterization of the nuclear RNase P holoenzyme complex reveals extensive subunit overlap with RNase MRP. Genes Dev 12:1678–1690

Chang DD, Clayton DA (1987) A novel endoribonuclease cleaves at a priming site of mouse mitochondrial DNA replication. EMBO J 6:409–417

Charles JF, Jaspersen SL, Tinker-Kulberg RL, Hwang L, Szidon A, Morgan DO (1998) The Polo-related kinase Cdc5 activates and is destroyed by the mitotic cyclin destruction machinery in *S. cerevisiae*. Curr Biol 8:497–507

Chen JL, Pace NR (1997) Identification of the universally conserved core of ribonuclease P RNA. RNA 3:557–560

Chen JL, Nolan JM, Harris ME, Pace NR (1998) Comparative photocross-linking analysis of the tertiary structures of *Escherichia coli* and *Bacillus subtilis* RNase P RNAs. EMBO J 17:1515–1525

Christian EL, Zahler NH, Kaye NM, Harris ME (2002) Analysis of substrate recognition by the ribonucleoprotein endonuclease RNase P. Methods 28:307–322

Chu S, Archer RH, Zengel JM, Lindahl L (1994) The RNA of RNase MRP is required for normal processing of ribosomal RNA. Proc Natl Acad Sci U S A 91:659–663

Chu S, Zengel JM, Lindahl L (1997) A novel protein shared by RNase MRP and RNase P. RNA 3:382–391

Coughlin DJ, Pleiss JA, Walker SC, Whitworth GB, Engelke DR (2008) Genome-wide search for yeast RNase P substrates reveals role in maturation of intron-encoded box C/D small nucleolar RNAs. Proc Natl Acad Sci U S A 105:12218–12223

Cramer P, Armache KJ, Baumli S, Benkert S, Brueckner F, Buchen C, Damsma GE, Dengl S, Geiger SR, Jasiak AJ et al (2008) Structure of eukaryotic RNA polymerases. Annu Rev Biophys 37:337–352

Crary SM, Niranjanakumari S, Fierke CA (1998) The protein component of *Bacillus subtilis* ribonuclease P increases catalytic efficiency by enhancing interactions with the 5′ leader sequence of pre-tRNAAsp. Biochemistry 37:9409–9416

Dang YL, Martin NC (1993) Yeast mitochondrial RNase P. Sequence of the RPM2 gene and demonstration that its product is a protein subunit of the enzyme. J Biol Chem 268:19791–19796

Dichtl B, Tollervey D (1997) Pop3p is essential for the activity of the RNase MRP and RNase P ribonucleoproteins in vivo. EMBO J 16:417–429

Dlakic M (2005) 3D models of yeast RNase P/MRP proteins Rpp 1p and Pop3p. RNA 11:123–127

Eder PS, Kekuda R, Stolc V, Altman S (1997) Characterization of two scleroderma autoimmune antigens that copurify with human ribonuclease P. Proc Natl Acad Sci U S A 94:1101–1106

Esakova O, Perederina A, Quan C, Schmitt ME, Krasilnikov AS (2008) Footprinting analysis demonstrates extensive similarity between eukaryotic RNase P and RNase MRP holoenzymes. RNA 14:1558–1567

Faber C, Scharpf M, Becker T, Sticht H, Rosch P (2001) The structure of the coliphage HK022 Nun protein-lambda-phage boxB RNA complex. Implications for the mechanism of transcription termination. J Biol Chem 276:32064–32070

Fatica A, Tollervey D (2002) Making ribosomes. Curr Opin Cell Biol 14:313–318

Forster AC, Altman S (1990) Similar cage-shaped structures for the RNA components of all ribonuclease P and ribonuclease MRP enzymes. Cell 62:407–409

Frank DN, Adamidi C, Ehringer MA, Pitulle C, Pace NR (2000) Phylogenetic-comparative analysis of the eukaryal ribonuclease P RNA. RNA 6:1895–1904

Giege R, Florentz C, Dreher TW (1993) The Tymv transfer RNA-like structure. Biochimie 75:569–582

Gill T, Cai T, Aulds J, Wierzbicki S, Schmitt ME (2004) RNase MRP cleaves the CLB2 mRNA to promote cell cycle progression: novel method of mRNA degradation. Mol Cell Biol 24:945–953

Gill T, Aulds J, Schmitt ME (2006) A specialized processing body that is temporally and asymmetrically regulated during the cell cycle in *Saccharomyces cerevisiae*. J Cell Biol 173:35–45

Gimple O, Schon A (2001) In vitro and in vivo processing of cyanelle tmRNA by RNase P. Biol Chem 382:1421–1429

Gopalan V, Baxevanis AD, Landsman D, Altman S (1997) Analysis of the functional role of conserved residues in the protein subunit of ribonuclease P from *Escherichia coli*. J Mol Biol 267:818–829

Gray MW (1993) Origin and evolution of organelle genomes. Curr Opin Genet Dev 3:884–890

Guerrier-Takada C, Lumelsky N, Altman S (1989) Specific interactions in RNA enzyme-substrate complexes. Science 246:1578–1584

Haas ES, Banta AB, Harris JK, Pace NR, Brown JW (1996) Structure and evolution of ribonuclease P RNA in Gram-positive bacteria. Nucleic Acids Res 24:4775–4782

Haeusler RA, Pratt-Hyatt M, Good PD, Gipson TA, Engelke DR (2008) Clustering of yeast tRNA genes is mediated by specific association of condensin with tRNA gene transcription complexes. Genes Dev 22:2204–2214

Hall TA, Brown JW (2002) Archaeal RNase P has multiple protein subunits homologous to eukaryotic nuclear RNase P proteins. RNA 8:296–306

Hall TA, Brown JW (2004) Interactions between RNase P protein subunits in archaea. Archaea 1:247–254

Hamma T, Ferre-D'Amare AR (2004) Structure of protein L7Ae bound to a K-turn derived from an archaeal box H/ACA sRNA at 1.8 A resolution. Structure (Camb) 12:893–903

Hartmann E, Hartmann RK (2003) The enigma of ribonuclease P evolution. Trends Genet 19:561–569

Hartmann RK, Heinrich J, Schlegl J, Schuster H (1995) Precursor of C4 antisense RNA of bacteriophages P1 and P7 is a substrate for RNase P of *Escherichia coli*. Proc Natl Acad Sci U S A 92:5822–5826

Hashimoto C, Steitz JA (1983) Sequential association of nucleolar 7–2 RNA with two different autoantigens. J Biol Chem 258:1379–1382

Henry Y, Wood H, Morrissey JP, Petfalski E, Kearsey S, Tollervey D (1994) The 5′ end of yeast 5.8S rRNA is generated by exonucleases from an upstream cleavage site. EMBO J 13:2452–2463

Holzmann J, Frank P, Loffler E, Bennett KL, Gerner C, Rossmanith W (2008) RNase P without RNA: identification and functional reconstitution of the human mitochondrial tRNA processing enzyme. Cell 135:462–474

Honda T, Kakuta Y, Kimura K, Saho J, Kimura M (2008) Structure of an archaeal homolog of the human protein complex Rpp 21-Rpp29 that is a key core component for the assembly of active ribonuclease P. J Mol Biol 384:652–662

Hopper AK (2006) Cellular dynamics of small RNAs. Crit Rev Biochem Mol Biol 41:3–19

Hopper AK, Phizicky EM (2003) tRNA transfers to the limelight. Genes Dev 17:162–180

Houser-Scott F, Xiao S, Millikin CE, Zengel JM, Lindahl L, Engelke DR (2002) Interactions among the protein and RNA subunits of *Saccharomyces cerevisiae* nuclear RNase P. Proc Natl Acad Sci U S A 99:2684–2689

Hsieh J, Andrews AJ, Fierke CA (2004) Roles of protein subunits in RNA–protein complexes: lessons from ribonuclease P. Biopolymers 73:79–89

Hsieh J, Walker SC, Fierke CA, Engelke DR (2008) Pre-tRNA turnover catalyzed by the yeast nuclear RNase P holoenzyme is limited by product release. RNA
Jacobson MR, Cao LG, Wang YL, Pederson T (1994) The RNA subunit of ribonuclease-P rapidly localizes in the nucleolus after microinjection into the nucleus of living cells. J Cell Biochem 115–115
Jacobson MR, Cao LG, Taneja K, Singer RH, Wang YL, Pederson T (1997) Nuclear domains of the RNA subunit of RNase P. J Cell Sci 110(Pt 7):829–837
Jarrous N (2002) Human ribonuclease P: subunits, function, and intranuclear localization. RNA 8:1–7
Jarrous N, Altman S (2001) Human ribonuclease P. Methods Enzymol 342:93–100
Jarrous N, Eder PS, Wesolowski D, Altman S (1999a) Rpp 14 and Rpp29, two protein subunits of human ribonuclease P. RNA 5:153–157
Jarrous N, Wolenski JS, Wesolowski D, Lee C, Altman S (1999b) Localization in the nucleolus and coiled bodies of protein subunits of the ribonucleoprotein ribonuclease P. J Cell Biol 146:559–572
Jiang T, Altman S (2001) Protein–protein interactions with subunits of human nuclear RNase P. Proc Natl Acad Sci U S A 98:920–925
Jiang T, Guerrier-Takada C, Altman S (2001) Protein–RNA interactions in the subunits of human nuclear RNase P. RNA 7:937–941
Kawano S, Nakashima T, Kakuta Y, Tanaka I, Kimura M (2006) Crystal structure of protein Ph1481p in complex with protein Ph1877p of archaeal RNase P from *Pyrococcus horikoshii* OT3: implication of dimer formation of the holoenzyme. J Mol Biol 357:583–591
Kazantsev AV, Krivenko AA, Harrington DJ, Carter RJ, Holbrook SR, Adams PD, Pace NR (2003) High-resolution structure of RNase P protein from *Thermotoga maritima*. Proc Natl Acad Sci U S A 100:7497–7502
Kazantsev AV, Krivenko AA, Harrington DJ, Holbrook SR, Adams PD, Pace NR (2005) Crystal structure of a bacterial ribonuclease P RNA. Proc Natl Acad Sci U S A 102:13392–13397
Kifusa M, Fukuhara H, Hayashi T, Kimura M (2005) Protein–protein interactions in the subunits of ribonuclease P in the hyperthermophilic archaeon *Pyrococcus horikoshii* OT3. Biosci Biotechnol Biochem 69:1209–1212
Kikovska E, Svard SG, Kirsebom LA (2007) Eukaryotic RNase P RNA mediates cleavage in the absence of protein. Proc Natl Acad Sci U S A 104:2062–2067
Kirsebom LA, Svard SG (1994) Base pairing between *Escherichia coli* RNase P RNA and its substrate. EMBO J 13:4870–4876
Klein DJ, Schmeing TM, Moore PB, Steitz TA (2001) The kink-turn: a new RNA secondary structure motif. EMBO J 20:4214–4221
Komine Y, Kitabatake M, Yokogawa T, Nishikawa K, Inokuchi H (1994) A transfer-RNA-like structure is present in 10sa RNA, a small stable RNA from *Escherichia coli*. Proc Natl Acad Sci U S A 91:9223–9227
Koonin EV, Wolf YI, Aravind L (2001) Prediction of the archaeal exosome and its connections with the proteasome and the translation and transcription machineries by a comparative-genomic approach. Genome Res 11:240–252
Krasilnikov AS, Yang X, Pan T, Mondragon A (2003) Crystal structure of the specificity domain of ribonuclease P. Nature 421:760–764
Krasilnikov AS, Xiao Y, Pan T, Mondragon A (2004) Basis for structural diversity in homologous RNAs. Science 306:104–107
Kufel J, Allmang C, Verdone L, Beggs JD, Tollervey D (2002) Lsm proteins are required for normal processing of pre-tRNAs and their efficient association with La-homologous protein Lhp1p. Mol Cell Biol 22:5248–5256
Kurz JC, Niranjanakumari S, Fierke CA (1998) Protein component of *Bacillus subtilis* RNase P specifically enhances the affinity for precursor-tRNAAsp. Biochemistry 37:2393–2400
Lee DY, Clayton DA (1998) Initiation of mitochondrial DNA replication by transcription and R-loop processing. J Biol Chem 273:30614–30621
Lee JY, Evans CF, Engelke DR (1991a) Expression of RNase P RNA in *Saccharomyces cerevisiae* is controlled by an unusual RNA polymerase III promoter. Proc Natl Acad Sci U S A 88:6986–6990

Lee JY, Rohlman CE, Molony LA, Engelke DR (1991b) Characterization of RPR1, an essential gene encoding the RNA component of *Saccharomyces cerevisiae* nuclear RNase P. Mol Cell Biol 11:721–730

Lee Y, Kindelberger DW, Lee JY, McClennen S, Chamberlain J, Engelke DR (1997) Nuclear pre-tRNA terminal structure and RNase P recognition. RNA 3:175–185

Li Y, Altman S (2002) Partial reconstitution of human RNase P in HeLa cells between its RNA subunit with an affinity tag and the intact protein components. Nucleic Acids Res 30:3706–3711

Li Y, Altman S (2003) A specific endoribonuclease, RNase P, affects gene expression of polycistronic operon mRNAs. Proc Natl Acad Sci U S A 100:13213–13218

Li X, Frank DN, Pace N, Zengel JM, Lindahl L (2002) Phylogenetic analysis of the structure of RNase MRP RNA in yeasts. RNA 8:740–751

Li Y, Cole K, Altman S (2003) The effect of a single, temperature-sensitive mutation on global gene expression in *Escherichia coli*. RNA 9:518–532

Li X, Zaman S, Langdon Y, Zengel JM, Lindahl L (2004) Identification of a functional core in the RNA component of RNase MRP of budding yeasts. Nucleic Acids Res 32:3703–3711

Lindahl L, Fretz S, Epps N, Zengel JM (2000) Functional equivalence of hairpins in the RNA subunits of RNase MRP and RNase P in *Saccharomyces cerevisiae*. RNA 6:653–658

Loria A, Pan T (1997) Recognition of the T stem-loop of a pre-tRNA substrate by the ribozyme from *Bacillus subtilis* ribonuclease P. Biochemistry 36:6317–6325

Loria A, Pan T (1999) The cleavage step of ribonuclease P catalysis is determined by ribozyme-substrate interactions both distal and proximal to the cleavage site. Biochemistry 38:8612–8620

Lygerou Z, Mitchell P, Petfalski E, Seraphin B, Tollervey D (1994) The POP1 gene encodes a protein component common to the RNase MRP and RNase P ribonucleoproteins. Genes Dev 8:1423–1433

Lygerou Z, Allmang C, Tollervey D, Seraphin B (1996a) Accurate processing of a eukaryotic precursor ribosomal RNA by ribonuclease MRP in vitro. Science 272:268–270

Lygerou Z, Pluk H, van Venrooij WJ, Seraphin B (1996b) hPop1: an autoantigenic protein subunit shared by the human RNase P and RNase MRP ribonucleoproteins. EMBO J 15:5936–5948

Mann H, Ben-Asouli Y, Schein A, Moussa S, Jarrous N (2003) Eukaryotic RNase P: role of RNA and protein subunits of a primordial catalytic ribonucleoprotein in RNA-based catalysis. Mol Cell 12:925–935

Marquez SM, Harris JK, Kelley ST, Brown JW, Dawson SC, Roberts EC, Pace NR (2005) Structural implications of novel diversity in eucaryal RNase P RNA. RNA 11:739–751

Marquez SM, Chen JL, Evans D, Pace NR (2006) Structure and function of eukaryotic Ribonuclease P RNA. Mol Cell 24:445–456

Martin AN, Li Y (2007) RNase MRP RNA and human genetic diseases. Cell Res 17:219–226

Morrissey JP, Tollervey D (1995) Birth of the snoRNPs: the evolution of RNase MRP and the eukaryotic pre-rRNA-processing system. Trends Biochem Sci 20:78–82

Niranjanakumari S, Stams T, Crary SM, Christianson DW, Fierke CA (1998) Protein component of the ribozyme ribonuclease P alters substrate recognition by directly contacting precursor tRNA. Proc Natl Acad Sci U S A 95:15212–15217

Orgel LE (1986) RNA catalysis and the origins of life. J Theor Biol 123:127–149

Pagan-Ramos E, Lee Y, Engelke DR (1996a) A conserved RNA motif involved in divalent cation utilization by nuclear RNase P. RNA 2:1100–1109

Pagan-Ramos E, Lee Y, Engelke DR (1996b) Mutational analysis of *Saccharomyces cerevisiae* nuclear RNase P: randomization of universally conserved positions in the RNA subunit. RNA 2:441–451

Paluh JL, Clayton DA (1995) *Schizosaccharomyces pombe* RNase MRP RNA is homologous to metazoan RNase MRP RNAs and may provide clues to interrelationships between RNase MRP and RNase P. Yeast 11:1249–1264

Pan T, Loria A, Zhong K (1995) Probing of tertiary interactions in RNA: 2′-hydroxyl-base contacts between the RNase P RNA and pre-tRNA. Proc Natl Acad Sci U S A 92:12510–12514

Pannone BK, Kim SD, Noe DA, Wolin SL (2001) Multiple functional interactions between components of the Lsm2–Lsm8 complex, U6 snRNA, and the yeast La protein. Genetics 158:187–196

Pannucci JA, Haas ES, Hall TA, Harris JK, Brown JW (1999) RNase P RNAs from some Archaea are catalytically active. Proc Natl Acad Sci U S A 96:7803–7808
Parker R, Sheth U (2007) P bodies and the control of mRNA translation and degradation. Mol Cell 25:635–646
Peck-Miller KA, Altman S (1991) Kinetics of the processing of the precursor to 4.5 S RNA, a naturally occurring substrate for RNase P from *Escherichia coli*. J Mol Biol 221:1–5
Perederina A, Esakova O, Koc H, Schmitt ME, Krasilnikov AS (2007) Specific binding of a Pop6/Pop7 heterodimer to the P3 stem of the yeast RNase MRP and RNase P RNAs. RNA 13:1648–1655
Pfeiffer T, Tekos A, Warnecke JM, Drainas D, Engelke DR, Seraphin B, Hartmann RK (2000) Effects of phosphorothioate modifications on precursor tRNA processing by eukaryotic RNase P enzymes. J Mol Biol 298:559–565
Piccinelli P, Rosenblad MA, Samuelsson T (2005) Identification and analysis of ribonuclease P and MRP RNA in a broad range of eukaryotes. Nucleic Acids Res 33:4485–4495
Pitulle C, Garcia-Paris M, Zamudio KR, Pace NR (1998) Comparative structure analysis of vertebrate ribonuclease P RNA. Nucleic Acids Res 26:3333–3339
Pluk H, van Eenennaam H, Rutjes SA, Pruijn GJ, van Venrooij WJ (1999) RNA–protein interactions in the human RNase MRP ribonucleoprotein complex. RNA 5:512–524
Reich C, Olsen GJ, Pace B, Pace NR (1988) Role of the protein moiety of ribonuclease P, a ribonucleoprotein enzyme. Science 239:178–181
Reiner R, Ben-Asouli Y, Krilovetzky I, Jarrous N (2006) A role for the catalytic ribonucleoprotein RNase P in RNA polymerase III transcription. Genes Dev 20:1621–1635
Reiner R, Krasnov-Yoeli N, Dehtiar Y, Jarrous N (2008) Function and assembly of a chromatin-associated RNase P that is required for efficient transcription by RNA polymerase I. PLoS ONE 3:e4072
Ridanpaa M, Sistonen P, Rockas S, Rimoin DL, Makitie O, Kaitila I (2002) Worldwide mutation spectrum in cartilage-hair hypoplasia: ancient founder origin of the major70A–>G mutation of the untranslated RMRP. Eur J Hum Genet 10:439–447
Rosenblad MA, Lopez MD, Piccinelli P, Samuelsson T (2006) Inventory and analysis of the protein subunits of the ribonucleases P and MRP provides further evidence of homology between the yeast and human enzymes. Nucleic Acids Res 34:5145–5156
Rozhdestvensky TS, Tang TH, Tchirkova IV, Brosius J, Bachellerie JP, Huttenhofer A (2003) Binding of L7Ae protein to the K-turn of archaeal snoRNAs: a shared RNA binding motif for C/D and H/ACA box snoRNAs in Archaea. Nucleic Acids Res 31:869–877
Salgado-Garrido J, Bragado-Nilsson E, Kandels-Lewis S, Seraphin B (1999) Sm and Sm-like proteins assemble in two related complexes of deep evolutionary origin. EMBO J 18:3451–3462
Salinas K, Wierzbicki S, Zhou L, Schmitt ME (2005) Characterization and purification of *Saccharomyces cerevisiae* RNase MRP reveals a new unique protein component. J Biol Chem 280:11352–11360
Samanta MP, Tongprasit W, Sethi H, Chin CS, Stolc V (2006) Global identification of noncoding RNAs in *Saccharomyces cerevisiae* by modulating an essential RNA processing pathway. Proc Natl Acad Sci U S A 103:4192–4197
Savino TM, Bastos R, Jansen E, Hernandez-Verdun D (1999) The nucleolar antigen Nop52, the human homolog of the yeast ribosomal RNA processing RRP1, is recruited at late stages of nucleologenesis. J Cell Sci 112(Pt 12):1889–1900
Schmitt ME, Clayton DA (1993) Nuclear RNase MRP is required for correct processing of pre-5.8S rRNA in *Saccharomyces cerevisiae*. Mol Cell Biol 13:7935–7941
Schmitt ME, Clayton DA (1994) Characterization of a unique protein component of yeast RNase MRP: an RNA-binding protein with a zinc-cluster domain. Genes Dev 8:2617–2628
Schmitt ME, Bennett JL, Dairaghi DJ, Clayton DA (1993) Secondary structure of RNase MRP RNA as predicted by phylogenetic comparison. FASEB J 7:208–213
Schmitt ME, Gill T, Aulds J, Cai T, Wierzbicki S (2002) RNase MRP specifically cleaves the mRNA for a B-type cyclin to allow cell cycle progression. Int J Cancer 343–343
Schon A (1999) Ribonuclease P: the diversity of a ubiquitous RNA processing enzyme. FEMS Microbiol Rev 23:391–406

Seif ER, Forget L, Martin NC, Lang BF (2003) Mitochondrial RNase P RNAs in ascomycete fungi: lineage-specific variations in RNA secondary structure. RNA 9:1073–1083
Shadel GS, Buckenmeyer GA, Clayton DA, Schmitt ME (2000) Mutational analysis of the RNA component of *Saccharomyces cerevisiae* RNase MRP reveals distinct nuclear phenotypes. Gene 245:175–184
Siegel RW, Banta AB, Haas ES, Brown JW, Pace NR (1996) Mycoplasma fermentans simplifies our view of the catalytic core of ribonuclease P RNA. RNA 2:452–462
Smith JK, Hsieh J, Fierke CA (2007) Importance of RNA–protein interactions in bacterial ribonuclease P structure and catalysis. Biopolymers 87:329–338
Spitzfaden C, Nicholson N, Jones JJ, Guth S, Lehr R, Prescott CD, Hegg LA, Eggleston DS (2000) The structure of ribonuclease P protein from *Staphylococcus aureus* reveals a unique binding site for single-stranded RNA. J Mol Biol 295:105–115
Srisawat C, Engelke DR (2001) Streptavidin aptamers: affinity tags for the study of RNAs and ribonucleoproteins. RNA 7:632–641
Srisawat C, Houser-Scott F, Bertrand E, Xiao S, Singer RH, Engelke DR (2002) An active precursor in assembly of yeast nuclear ribonuclease P. RNA 8:1348–1360
Stams T, Niranjanakumari S, Fierke CA, Christianson DW (1998) Ribonuclease P protein structure: evolutionary origins in the translational apparatus. Science 280:752–755
Stolc V, Altman S (1997) Rpp 1, an essential protein subunit of nuclear RNase P required for processing of precursor tRNA and 35S precursor rRNA in *Saccharomyces cerevisiae*. Genes Dev 11:2926–2937
Tallsjo A, Kirsebom LA (1993) Product release is a rate-limiting step during cleavage by the catalytic RNA subunit of *Escherichia coli* RNase P (Vol 21, Pg 51, 1993). Nucleic Acids Res 21:1686
Thiel CT, Horn D, Zabel B, Ekici AB, Salinas K, Gebhart E, Ruschendorf F, Sticht H, Spranger J, Muller D et al (2005) Severely incapacitating mutations in patients with extreme short stature identify RNA-processing endoribonuclease RMRP as an essential cell growth regulator. Am J Hum Genet 77:795–806
Thomas BC, Chamberlain J, Engelke DR, Gegenheimer P (2000) Evidence for an RNA-based catalytic mechanism in eukaryotic nuclear ribonuclease P. RNA – A Publication of the RNA Society 6:554–562
Thompson M, Haeusler RA, Good PD, Engelke DR (2003) Nucleolar clustering of dispersed tRNA genes. Science 302:1399–1401
Topper JN, Clayton DA (1990) Secondary structure of the RNA component of a nuclear/mitochondrial ribonucleoprotein. J Biol Chem 265:13254–13262
Toro I, Thore S, Mayer C, Basquin J, Seraphin B, Suck D (2001) RNA binding in an Sm core domain: X-ray structure and functional analysis of an archaeal Sm protein complex. EMBO J 20:2293–2303
Torres-Larios A, Swinger KK, Krasilnikov AS, Pan T, Mondragon A (2005) Crystal structure of the RNA component of bacterial ribonuclease P. Nature 437:584–587
Tranguch AJ, Kindelberger DW, Rohlman CE, Lee JY, Engelke DR (1994) Structure-sensitive RNA footprinting of yeast nuclear ribonuclease P. Biochemistry 33:1778–1787
van Eenennaam H, Pruijn GJ, van Venrooij WJ (1999) hPop4: a new protein subunit of the human RNase MRP and RNase P ribonucleoprotein complexes. Nucleic Acids Res 27:2465–2472
van Eenennaam H, Lugtenberg D, Vogelzangs JHP, van Venrooij WJ, Pruijn GJM (2001a) hPop5, a protein subunit of the human RNase MRP and RNase P endoribonucleases. J Biol Chem 276:31635–31641
van Eenennaam H, van der Heijden A, Janssen RJ, van Venrooij WJ, Pruijn GJ (2001b) Basic domains target protein subunits of the RNase MRP complex to the nucleolus independently of complex association. Mol Biol Cell 12:3680–3689
Van Eenennaam H, Vogelzangs JH, Lugtenberg D, Van Den Hoogen FH, Van Venrooij WJ, Pruijn GJ (2002) Identity of the RNase MRP- and RNase P-associated Th/To autoantigen. Arthritis Rheum 46:3266–3272
Villa T, Ceradini F, Presutti C, Bozzoni I (1998) Processing of the intron-encoded U18 small nucleolar RNA in the yeast *Saccharomyces cerevisiae* relies on both exo- and endonucleolytic activities. Mol Cell Biol 18:3376–3383

Villa T, Ceradini F, Bozzoni I (2000) Identification of a novel element required for processing of intron-encoded box C/D small nucleolar RNAs in *Saccharomyces cerevisiae*. Mol Cell Biol 20:1311–1320

Walker SC, Avis JM (2004) A conserved element in the yeast RNase MRP RNA subunit can participate in a long-range base-pairing interaction. J Mol Biol 341:375–388

Walker SC, Avis JM (2005) Secondary structure probing of the human RNase MRP RNA reveals the potential for MRP RNA subsets. Biochem Biophys Res Commun 335:314–321

Walker SC, Engelke DR (2006) Ribonuclease P: the evolution of an ancient RNA enzyme. Crit Rev Biochem Mol Biol 41:77–102

Walker SC, Engelke DR (2008) A protein-only RNase P in human mitochondria. Cell 135:412–414

Wang L, Haeusler RA, Good PD, Thompson M, Nagar S, Engelke DR (2005) Silencing near tRNA genes requires nucleolar localization. J Biol Chem 280:8637–8639

Welting TJ, van Venrooij WJ, Pruijn GJ (2004) Mutual interactions between subunits of the human RNase MRP ribonucleoprotein complex. Nucleic Acids Res 32:2138–2146

Welting TJ, Kikkert BJ, van Venrooij WJ, Pruijn GJ (2006) Differential association of protein subunits with the human RNase MRP and RNase P complexes. RNA 12:1373–1382

Wilson RC, Bohlen CJ, Foster MP, Bell CE (2006) Structure of Pfu Pop5, an archaeal RNase P protein. Proc Natl Acad Sci U S A 103:873–878

Wolin SL, Matera AG (1999) The trials and travels of tRNA. Genes Dev 13:1–10

Woodhams MD, Stadler PF, Penny D, Collins LJ (2007) RNase MRP and the RNA processing cascade in the eukaryotic ancestor. BMC Evol Biol 7(Suppl 1):S13

Xiao SH, Scott F, Fierke CA, Engelke DR (2002) Eukaryotic ribonuclease P: A plurality of ribonucleoprotein enzymes. Annu Rev Biochem 71:165–189

Xiao S, Day-Storms JJ, Srisawat C, Fierke CA, Engelke DR (2005a) Characterization of conserved sequence elements in eukaryotic RNase P RNA reveals roles in holoenzyme assembly and tRNA processing. RNA 11:885–896

Xiao SH, Day-Storms JJ, Srisawat C, Fierke CA, Engelke DR (2005b) Characterization of conserved sequence elements in eukaryotic RNase P RNA reveals roles in holoenzyme assembly and tRNA processing. RNA – A Publication of the RNA Society 11:885–896

Yang L, Altman S (2007) A noncoding RNA in *Saccharomyces cerevisiae* is an RNase P substrate. RNA 13:682–690

Yuan Y, Altman S (1995) Substrate recognition by human RNase P: identification of small, model substrates for the enzyme. EMBO J 14:159–168

Ziehler WA, Yang J, Kurochkin AV, Sandusky PO, Zuiderweg ER, Engelke DR (1998) Structural analysis of the P10/11–P12 RNA domain of yeast RNase P RNA and its interaction with magnesium. Biochemistry 37:3549–3557

Ziehler WA, Day JJ, Fierke CA, Engelke DR (2000) Effects of 5′ leader and 3′ trailer structures on pre-tRNA processing by nuclear RNase P. Biochemistry 39:9909–9916

Ziehler WA, Morris J, Scott FH, Millikin C, Engelke DR (2001) An essential protein-binding domain of nuclear RNase P RNA. RNA 7:565–575

Chapter 11
RNase P from Organelles

Agustín Vioque

11.1 Introduction

Thirty years of research on RNase P has resulted in a very detailed knowledge of this fascinating enzyme, and other chapters in this book nicely illustrate this point. The RNA and protein components of RNase P from the three domains of life are known. The enzyme from Bacteria, Archaea and Eukarya (nucleus) contains an essential RNA that is clearly homologous across the three domains of life (Frank and Pace 1998). The protein subunits from archaeal RNase P are homologous to a subset of the protein subunits from nuclear RNase P, but the protein subunit of bacterial RNase P is unrelated to archaeal or eukaryotic proteins. The crystal structures of the protein (P protein), and RNA (P RNA) subunits from bacterial RNase P and of the several RNase P protein subunits from Archaea are known, and the catalytic mechanism has been studied in great detail. In contrast, the knowledge on RNase P from organelles (mitochondria and chloroplast) is more limited. Only in a few instances have the genes coding for RNase P components in organelles been identified. Currently available information indicate a larger diversity in enzyme structure for the organelle RNase P. What is more, in some cases organellar RNase P might be a pure protein enzyme, in other words, a "non-ribozymal" enzyme, as shown by the recent in vitro reconstitution of the protein-only human mitochondrial RNase P (Holzmann et al. 2008).

Mitochondria and chloroplasts were acquired by endosymbiosis from bacteria. Most of the endosymbiont genes were subsequently transferred to the nucleus, with only a small fraction of the original endosymbiont genes remaining in the organelle genome. The macromolecular components of these organelles, encoded either in their own genome or in the nucleus, show in many cases, the similarity to their bacterial ancestor. In other cases, a gene derived from the host or acquired

A. Vioque (✉)
Instituto de Bioquímica Vegetal y Fotosíntesis, Universidad de Sevilla and CSIC, Américo Vespucio 49, 41092 Sevilla, Spain
e-mail: vioque@us.es

F. Liu and S. Altman (eds.), *Ribonuclease P*, Protein Reviews 10,

DOI 10.1007/978-1-4419-1142-1_11, © Springer Science+Business Media, LLC 2010

by horizontal gene transfer has replaced the bacterial encoded protein or RNA. The organelle genome is characterized by its rapid evolution and plasticity, which could explain the diversity in structure and composition of RNase P.

In some cases, the structure of the organelle RNase P is reminiscent of the bacterial ancestor. For instance, the P RNA in *Reclinomonas americana* mitochondria is most similar to α-proteobacterial P RNA (Lang et al. 1997), and the P RNA in some algal chloroplasts is most similar to P RNA from the cyanobacterial ancestor (Baum et al. 1996). In other cases, P RNA is highly degenerated, and its gene (*rnpB*) is difficult to recognize in sequenced organelle genomes, as in Ascomycete fungi (Seif et al. 2003). Finally, no *rnpB* is identified in the genomes of most sequenced organelles.

In this review, I will summarize the accumulated knowledge on organelle RNase P and what are the important challenges that remain with respect to its characterization.

11.2 Mitochondria

Table 11.1 presents data on a selection of the RNA subunits (mtP RNA) of mitochondrial RNase P (mtRNase P) that have been described. More information is available in the Mitochondrial RNase P Database (http://megasun.bch.umontreal.ca/People/lang/rnpB/). Table 11.2 summarizes present knowledge on protein subunits (mtP protein) of mitochondrial RNase P. Yeast and human mtRNase P are the only ones in which the protein subunits have been identified and have been studied in more detail.

11.2.1 Yeast Mitochondria

Very soon after the discovery of RNase P, RNase P activity was identified in yeast mitochondria (Hollingsworth and Martin 1986). Early genetic and biochemical analyses demonstrated that yeast mtRNase P contains both mitochondrial and nuclear-encoded components (Hollingsworth and Martin 1986; Miller and Martin 1983; Underbrink-Lyon et al. 1983). The enzyme is sensitive to micrococcal nuclease, pointing to the presence of an essential RNA subunit (mtP RNA). This RNA is encoded in the mitochondrial genome (*RPM1* gene) and is 490 nucleotides long (Fig. 11.1) (Morales et al. 1989). In partially purified RNase P fractions, the full length RNA is not recovered, suggesting that a fragmented RNA is still active (Morales et al. 1989). Analysis of the RNA fragments remaining in the purified fractions indicate that a fragment of about 90 nucleotides from the 5′ end of the original RNA, and another fragment of about 70 nucleotides from the 3′ end are the only RNA fragments present in the active fraction, indicating that the rest of the molecule is dispensable for activity. These RNA fragments correspond to the conserved

Table 11.1 Some RNA subunits of mitochondrial RNase P

Organisms	Size (nt)	Missing helices[a]	References
Jakobids			
Reclinomonas americana	318	P6, P16, P17[b]	Lang et al. 1997; Seif et al. 2006
Green algae			
Nephroselmis olivacea	332	P6, P13, P16, P17[b]	Turmel et al. 1999b
Ostreococcus tauri	268	P6, P12, P16, P17[b]	Robbens et al. 2007
Ascomyceta			
Saccharomyces cerevisiae	490	All except P1, P4, and P18	Seif et al. 2003
Saccharomyces fibuligera	140	All except P1 and P4	Sbisa et al. 1996
Schizosaccharomyces pombe	273	P5-P14, P16, P17	Wise and Martin 1991b
Aspergillus nidulans	232	P3, P13-P18	Lee et al. 1996b; Seif et al. 2003
Zygomyceta			
Mucor mucedo	388	P6, P13-P17	Seif et al. 2006
Mortierella verticillata	980	P13-P17	Seif et al. 2006
Rhizopus oryzae	830	P6, P13, P14 P16, P17	Seif et al. 2006
Smittium culisetae	188	All except P1 and P4	Seif et al. 2006

[a]Compared with *E. coli* P RNA (Fig. 11.1)
[b]Contains P19

Table 11.2 Protein subunits of mitochondrial RNase P

Organism	Molecular weight (kDa)	Comments	Identification	References
Saccharomyces cerevisiae	100	(Rpm2). No homologs outside yeasts	Genetic and biochemical analysis	Dang and Martin 1993
Aspergillus nidulans	55, 41,40, 26, 24, 18, and 16	Not characterized	Partial purification	Lee et al. 1996a
Potato	47, 34, 22	Not characterized	Partial purification	Marchfelder and Brennicke 1994
Homo sapiens	42 (MRPP1)	tRNA m^1G methyltransferase (MRPP1)	Purification, proteomics, and in vitro reconstitution	Holzmann et al. 2008
	26 (MRPP2)	Short-chain dehydrogenase/ reductase (MRPP2)		
	60 (MRPP3)	Pentatricopeptide repeats and metal binding (MRPP3)		

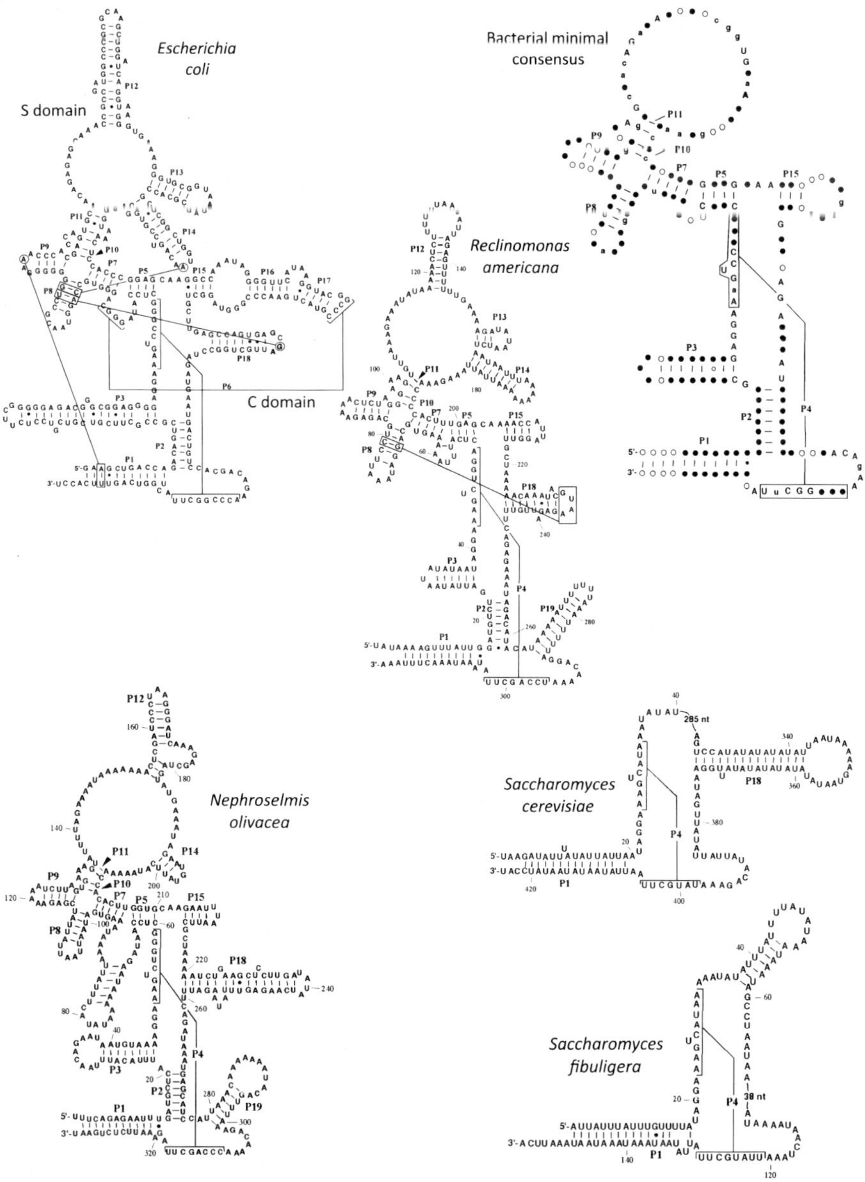

Fig. 11.1 Secondary structure models of *E. coli* P RNA, the bacterial minimun consensus for P RNA (Siegel et al. 1996) and selected mtP RNAs. Long distance tertiary interactions between L14 and L18 with P9 and between L9 and P1 are indicated in the *E. coli* structure. See text for details and references

helices P1, P4, and P18, which are the only elements of the bacterial consensus present in the yeast mtP RNA (Fig. 11.1). The buoyant density of mtRNase P active fractions in $CsSO_4$ gradients is 1.28 g/cm^3, indicating a higher proportion of protein than in the bacterial RNase P (Morales et al. 1989). The only protein subunit so far identified as being part of yeast mitochondrial RNase P is nuclear encoded. A 105 kDa protein (Rpm2) was identified in highly purified fractions and the corresponding gene (*RPM2* gene) cloned (Morales et al. 1992). Disruption of this gene generates a respiratory deficient phenotype and 5′-extended pre-tRNAs accumulate. Antibodies against Rpm2 immunoprecipitate mtP RNA and RNase P activity (Dang and Martin 1993). Therefore, there is a strong biochemical and genetic evidence that Rpm2 is an RNase P subunit. In vitro reconstitution of yeast mtRNase P activity from recombinant purified subunits has not been reported so far. *RPM2* is found only in closely related Saccharomycetales and has no sequence homology with bacterial, archaeal or nuclear RNase P proteins. Rpm2 seems to be a multifunctional protein with other functions in addition to RNase P. Rpm2 acts as a multicopy suppressor of a temperature-sensitive mutant allele of the mitochondrial protein import channel component Isp42p, and is required for normal growth, even in fermentable medium (Kassenbrock et al. 1995). This second essential function is related to mitochondrial biogenesis (Stribinskis et al. 2001a). Rpm2 is also required for processing of the precursor of the mtP RNA (Stribinskis et al. 1996). tRNA processing and mtP RNA processing can be separated on different domains of the protein (Stribinskis et al. 2001b). A fraction of the total cellular Rpm2 localizes to the nucleus, where it acts as a transcriptional activator of genes coding for some mitochondrial proteins (Stribinskis et al. 2005), notably proteins related to the import complex. This regulatory role could explain several of the non RNase P functions of this protein. Finally, Rpm2 has also been found localized to cytoplasmic processing bodies (Stribinskis and Ramos 2007).

11.2.2 Other Fungi

rnpB has been identified in the mitochondrial genome of many other budding yeasts and other Ascomycete (Petersen et al. 2002; Ragnini et al. 1991; Sbisa et al. 1996; Seif et al. 2003; Wise and Martin 1991a, b). It has also been identified in Zygomycete fungi (Seif et al. 2005). However, in many Ascomycete mitochondrial genomes (thus far sequenced), no RNase P RNA gene is found; the same is true in all studied Basidiomycete and Chytridiomycete fungi (Seif et al. 2003). Compared to the bacterial consensus, Ascomycete mtP RNAs have reduced secondary structures and are very AU rich (Seif et al. 2003), making it difficult to define the secondary structure of the domains that are not homologous to the bacterial consensus. There are large variations in size among the sequenced mtP RNAs. *S. cerevisiae* mtP RNA is 490 nucleotides long, while *Saccharomycopsis fibuligera* (140 nucleotides) and *Candida glabrata* (227 nucleotides) mtP RNAs are much smaller (Wise and Martin, 1991b). The availability of several fungal mtP RNA sequences has allowed the definition of its secondary structure, using as a guide, the bacterial consensus (Seif et al. 2003). The only conserved motifs in all fungal mtP RNAs are helices P1 and P4. Other motifs are present or absent in a way correlated with the phylogenetic relationships. P2, P3, P15, and P18 are present in *Schizosaccharomyces* mtP RNAs while P2, P3, and P15 are absent in budding yeasts mtP RNAs, although most of them contain P18. Therefore, there is a clear reduction in structural complexity in fungal mtP RNAs that could be compensated by increased relevance of the protein cofactor(s) and could explain the absence of activity of these RNAs in vitro. The characterization of mtP proteins from fungi other than *S. cerevisiae* would be an important contribution to the understanding of mtRNase P.

Zygomycete fungi also encode *rnpB* in the mitochondria (Seif et al. 2005). They show a high diversity in length and structure. *Smittium culisetae* mtP RNA is 188 nt long, containing only helices P1 and P4. *Mortierella verticillata* and *Rhizopus oryzae* mtP RNAs are the largest known (980 and 830 nt, respectively). *R. oryzae* P RNA is closely similar to bacterial P RNA lacking only P13, P14 and P19. *M. verticilata* P RNA lacks in addition P15. Their long size is due to the presence of two insertions, in P12 and between P5 and P15 in *R. oryzae*, and in P12 and between P5 and P18 in *M. verticillata*. The presence of variable size insertions in P12 that increase the size of P RNA has also been described in cyanobacteria (Vioque 1997).

Some biochemical studies have been performed in *Aspergillus nidulans,* one of the Ascomyceta where an *rnpB* gene has been identified in the mitochondria (Lee et al. 1996a). This RNA lacks P13, P14, P15, and P18 but contains P2 and P3. Purification of mtRNase P from *A. nidulans* (Lee et al. 1996a) has been facilitated by the difference in substrate specificity between nuclear and mitochondrial enzyme in the processing of histidine tRNA. Histidine tRNAs have the unique property of containing an additional G residue at −1 position. In eukaryotic nuclei, this G is not encoded but is added post-RNase P cleavage by a specific tRNA guanylyl transferase. In bacteria, chloroplasts, and plant and fungal mitochondria, the G is genomically encoded and RNase P has modified cleavage specificity for pre-$tRNA^{His}$ so that this substrate is cleaved at −1, generating directly the mature tRNA with the extra G. *A. nidulans* mtRNase P seems to contain seven polypeptides of 55, 41,40, 26, 24, 18, and 16 kDa. These polypeptides are present in highly purified fractions in approximately equimolar amounts. There has been no further characterization published of these proteins, and therefore their identity and function as mtRNase P subunits awaits confirmation.

11.2.3 Plant Mitochondria

Very little progress has been made in the characterization of plant mitochondria RNase P. An *rnpB* gene has been identified only in the mitochondrial genome of some early branching green algae as *Nephroselmis olivacea* (Fig. 11.1) (Turmel et al. 1999a) and *Ostreococcus tauri* (Robbens et al. 2007), and these RNAs are not functional (see below). However, the absence of an *rnpB* gene in the mitochondrial genome does not exclude that there is a nuclear encoded mtP RNA that is imported in the mitochondria. Other RNAs, like tRNAs, are known to be imported in plant mitochondria. Early biochemical studies identified plant mtRNase P activities in wheat (Hanic-Joyce and Gray 1990), *Oenothera berteriana* (Marchfelder et al. 1990), potato (Marchfelder and Brennicke 1994), and wild carrot cell suspensions (Franklin et al. 1995). The presence of an essential RNA subunit was deduced from nuclease sensitivity experiments in potato and carrot cells (Franklin et al. 1995; Marchfelder and Brennicke 1994).

11.2.4 Other Organisms with Bacterial-Like Mitochondrial P RNA

When the mitochondrial genome of the jakobid flagellate *Reclinomonas americana* was sequenced (Lang et al. 1997), a bacterial-like *rnpB* gene was identified that contains all structural elements of the minimal bacterial consensus and also a P19 helix, typical for α-proteobacterial P-RNAs (Fig. 11.1). *R. americana* contains the largest collection of genes so far identified in any mtDNA and most closely resembles the ancestral proto-mitochondrial genome. The presence of a bacterial-like P RNA is one more of these primitive characteristics. Similar mtP RNAs where identified in other jakobids (Seif et al. 2006), containing all the conserved structural elements of the minimal bacterial consensus. The P19 is also present in the mtP RNA of the green algae *Nephroselmis olivacea* (Fig. 11.1) (Turmel et al. 1999a) and *Ostreococcus tauri* (Robbens et al. 2007).

Despite their structural similarity to bacterial P RNAs, the mitochondrial P RNAs from jakobids and *Nephroselmis* are inactive in vitro. Also, no activity is detected in reconstitution assays with *E. coli* P protein (Seif et al. 2006). The same negative results were obtained with *O. tauri* mtP RNA (V. Gopalan, personal communication).

The absence of activity of these RNAs can be ascribed to a number of factors, such as low thermodynamic stability due to the high AT content, the absence of certain GNRA motifs that are involved in long range interactions (Brown et al. 1996; Massire et al. 1997), the absence of the GGU motif in L15/L16 involved in substrate binding, or deviation from consensus at some catalytically important positions. Mutagenesis studies and the construction of chimeras mixing domains of mtP RNAs and *E. coli* P RNA support this conclusion (Seif et al. 2006). A chimeric RNA that contains the S domain from *E. coli* P RNA and the C domain from mtP RNA is active. The reciprocal chimeric RNA with the C domain from *E. coli* P RNA and the S domain from mtP RNA domain is also active, but the reaction rate is tenfold lower. The bacterial P protein can increase this activity. It seems that mtP-RNAs have lost structural elements that make them more dependent on the (unknown) protein subunit(s) for activity. This process of mtP RNA reduction could lead to the loss of the RNA subunit and its replacement by an all-protein enzyme in the course of evolution, as exemplified by the human mtRNase P (see below).

11.2.5 Trypanosomatids

In *Trypanosoma* and *Leishmania*, mitochondria do not encode any tRNA; therefore, they all have to be imported from the cytosol (Schneider 1994; Simpson et al. 1989). If they were imported as mature tRNAs, no RNase P activity would be needed in the mitochondria. There are early reports that 5′-extended precursor tRNAs are imported in *Trypanosoma brucei* mitochondria (Hancock et al. 1992; Yermovsky-Kammerer and Hajduk 1999), but more recent results suggest that mature tRNA is what is actually imported in vivo in *Trypanosoma* (Aphasizhev et al. 1998; Tan et al. 2002) and *Leishmania* (Kapushoc et al. 2000; Rubio et al. 2000). However, pre-tRNAs can be imported in vitro in *Leishmania* mitochondria, and they are processed by an RNase P-like activity (Rubio et al. 2000).

RNase P activity has been partially purified from *Trypanosoma brucei* mitochondria (Salavati et al. 2001). The enzyme was purified more than 10,000-fold and actively processed a *S. cerevisiae* pre-tRNA, generating a 5′-phosphate, indicative of true RNase P activity. This purified fraction was insensitive to micrococcal nuclease, suggesting that it lacks an essential RNA subunit.

11.2.6 Human Mitochondria

Until recently, the nature of human mtRNase P has been a contentious issue. The first studies published (Doersen et al. 1985) showed that human mtRNase P was nuclease sensitive, suggesting the presence of an essential RNA component. In 2001, it was proposed that the human mtRNase P contains an essential RNA subunit that is identical to the nuclear P RNA (H1 RNA) (Puranam and Attardi 2001). This conclusion was reached by rigorous purification of mitochondrial RNase P and analysis of the RNA content of the purified fraction. It was shown that the highly purified mtRNase P activity was enriched in H1 RNA by a factor of two orders of magnitude when compared with other contaminating nuclear RNAs such as U2 and U3 RNA. These results, together with the micrococcal nuclease sensitivity of the activity, supported the claim that H1 RNA is a component of mtRNase P. A single enzyme catalyzing the same reaction in the nucleus and in the mitochondria is not unprecedented. Several tRNA processing enzymes have been shown to be active in nucleus, cytoplasm and mitochondria, and the targeting is dictated by the use of alternative AUG initiation codons (Martin and Hopper 1994). However, in the case of mitochondrial RNase P there is no indication that any of the protein subunits of nuclear RNase P is imported into the mitochondria, and it would be very surprising that new protein subunit(s), unrelated to the nuclear RNase P protein subunits are used in the mitochondria with the nuclear H1 RNA. The proposal that nuclear P RNA is a subunit of mtRNase P was challenged (Rossmanith and Potuschak 2001) because contamination by nuclear RNase P could not be ruled out. Furthermore, the putative H1 RNA-containing mtRNase P activity could process *E.*

coli pre-tRNA^{Tyr} faithfully, like nuclear RNase P could do, while another group had described that mitochondrial mtRNase P activity could not process this *E. coli* pre-tRNA^{Tyr} at the correct site (Rossmanith et al. 1995).

Another report indicated that human mtRNase P lacks an RNA subunit (Rossmanith and Karwan 1998). This proposal was based on the resistance of partially purified mtRNase P to ribonuclease treatment, the protein-like density in CS_2SO_4 gradients, and the absence of RNAs longer than 25 nucleotides in micrococcal nuclease treated active samples. These results were received with skepticism, in an interesting example of shifting paradigms. The initial reports in the 1970s (Stark et al. 1978) that bacterial RNase P contained an essential RNA subunit were received with disbelief (Altman 1989) (see Chap. 1). Twenty years later, the suggestion that in some organelles RNase P might be a protein enzyme was received with much greater skepticism, and a higher standard of biochemical characterization was required before the conclusion, that there is an RNase P enzyme without RNA, was accepted. This goal has now been achieved in the case of human mtRNase P (Holzmann et al. 2008) with the demonstration, beyond reasonable doubt, of its protein-only nature. By performing partial purification under different conditions, and analyzing with proteomic techniques the protein composition of each partially purified preparation, a "partial purification overlap proteome" was obtained with a limited number of candidate proteins to be analyzed. The substrate used in these purifications was a mitochondrial pre-tRNA^{Tyr} that is not processed by the nuclear enzyme that contaminates mitochondrial preparations. Affinity tagging of candidate proteins and purification under mild conditions led, finally, to the identification of the three proteins (named MRPP1, MRPP2, and MRPP3) that constitute mtRNase P. These three proteins are unrelated to protein subunits of previously characterized RNase P enzymes from Bacteria, Archaea or Eukarya, and are also unrelated to yeast Rpm2. MRPP1 and MRPP2 form a stable subcomplex, while MRPP3 interaction with the MRPP1/MRPP2 subcomplex is weak and is disrupted by salt concentrations as low as 150 mM. In retrospect, this finding explains the difficulty in purifying mtRNase P, as many of the chromatography protocols employed high salt elution, which presumably led to dissociation of MRPP3 and loss of RNase P activity. The three recombinant proteins, purified from *E. coli*, can reconstitute a functional RNase P that cleaves mitochondrial precursor tRNAs at the expected site, leaving a 5′-phosphate. This result excludes the need for a human mtP RNA. The possibility that bacterial RNA contaminates the recombinant protein samples and is responsible of the activity was excluded by careful controls. All three subunits are required for activity. Any combination of two subunits is inactive. The involvement of these three proteins in pre-tRNA processing was further proven in vivo by RNAi knock-down experiments. Inhibition of the expression of any of the three proteins resulted in the accumulation of 5′-extended pre-tRNAs. Therefore, this mtRNase P defines a new type of RNase P. Unexpectedly, two of the subunits seem to have another unrelated function. MRPP1 is a tRNA m^1G methyltransferase that methylates G9, a common modification of eukaryal tRNAs. This enzyme is found in Archaea and Eukarya but not in Bacteria. MRPP2 is a well-conserved member of the short-chain dehydrogenase/reductase family.

Its biological role is not clear, but mutations in its gene have been linked to 2-methyl-3-hydroxybutyryl-CoA dehydrogenase deficiency. MRPP2 contains a Rossmann fold (NAD^+ binding domain). MRPP3 has no function ascribed in the databases, but it contains two degenerate in-tandem pentatricopeptide repeats (PPR), and conserved sequence blocks at the carboxyl terminus with three aspartate and one histidine, strictly conserved, that could be involved in metal ion coordination. The mtRNase P has not been characterized further but it has been proposed (Holzmann et al. 2008), based on the properties of the MRPP proteins, that MRPP1 is the substrate binding module, contributing specificity towards tRNA. MRPP2 could contribute to tRNA binding through the Rossmann fold. If it is confirmed experimentally that MRPP3 does bind metal ions, it can be speculated that MRPP3 acts as a metallonuclease through a mechanism similar to RNA-based enzymes.

The demonstration of the protein-only nature of human mtRNase P raises the question of the distribution of this kind of RNase P among mitochondria. Is it a peculiarity of animal mitochondria? or, is it widespread among organisms that do not have a mitochondrial *rnpB*? At this point, this question cannot be solved, and it will require a better sampling of the properties of mtRNase P from a broad set of organisms. Flowering plant genomes encode three homologues of MRPP3, and two homologues are found in trypanosomes. However, m^1G_9 methylation is not found in yeast and plant mitochondria, and the *Arabidopsis* tRNA m^1G methyltransferase is not predicted to be targeted to the mitochondria.

Another interesting question is the evolutionary drive behind the replacement of the RNA based enzyme by a protein-only enzyme in animal mitochondria. This could be related to the fast evolutionary pace of mitochondria and of their tRNA structures, or to the tendency to compactation of metazoan mitochondrial genomes. Also, it is possible that the evolutionary constraints in mtRNase P are more relaxed than in other RNase P enzymes (Walker and Engelke 2008), that must cleave additional substrates in addition to pre-tRNAs. Therefore, it might be easier to replace the RNA-based enzyme by a protein-only enzyme if its only substrates are mitochondrial pre-tRNAs.

11.3 Plastids

11.3.1 Distribution of Plastid rnpB

It is useful first to summarize some aspects of plastid evolution to have in context the distribution of the presence or absence of plastid encoded *rnpB*. Chloroplast arose by the endosymbiosis of a cyanobacterium by a eukaryote. All present day chloroplasts derive from a single endosymbiotic event (Gould et al. 2008; Reyes-Prieto et al. 2007). The engulfed cyanobacterium lost most of its genes due to a massive transfer of genes to the nucleus, and rapidly diversified into three lineages of so called primary endosymbionts: Glaucocystophyceae, Viridiplantae (green algae and plants), and Rhodophyta (red algae). Secondary endosymbionts lineages originated when a eukaryote prototroph engulfed a primary endosymbiont. Secondary endosymbiosis has happened several times and has resulted in a complex pattern of evolution, generating a great diversity of protists and algae, constituting some of the more successful eukaryotic lineages. In some cases, secondary acquired plastids were lost (Oomycetes) (Tyler et al. 2006), or are maintained but have lost photosynthetic capability (Apicomplexans).

So far, *rnpB* has been identified in the plastid genome of the three lineages of primary endosymbionts (Table 11.3). The sample of fully sequenced chloroplasts is strongly biased towards green algae and plants, with almost one hundred plastid genomes of different species available. Of them, only one, *Nephroselmis olivacea*, contains an *rnpB* gene. In contrast, only six red algae plastid genomes are available, and all of them contain an *rnpB* gene. Glaucocystophyceae is a small group with only a few species known. Glaucocystophyceae plastid have some primitive characters that place them close to the cyanobacterial endosymbiont, such as the presence of the peptidoglycan layer in the plastid envelope and the nature of antenna pigments. The only Glaucocystophyceae whose plastid genome has been sequenced is *Cyanophora paradoxa*, and it encodes *rnpB*. All the plastids that encode *rnpB* are rather large and gene rich and are placed close to the base of the plastid evolutionary radiation. None of the secondary endosymbionts (more than 10) whose chloroplast sequence is known encode *rnpB* in the chloroplast. These include secondary endosymbionts derived from green and red algae, representative of all the main lineages.

Table 11.3 RNA subunits of plastid RNase P

Organism	Identification (references)	Activity of RNA (references)
Glaucocystophyceae		
Cyanophora paradoxa	Sequence similarity (Shevelev et al. 1995)	Yes[a] (Li et al. 2007)
	Partial purification (Baum et al. 1996)	
Rhodophyta (red algae)		
Cyanidioschyzon merolae	Sequence similarity (Ohta et al. 2003)	No (A.V., unpublished)
Cyanidium caldarium	Sequence similarity (Seif et al. 2003)	No (A.V., unpublished)
Gracilaria tenuistipitata	Sequence similarity (Hagopian et al. 2004)	No (A.V., unpublished)
Porphyra purpurea	Sequence similarity (Reith and Munholland 1995)	No (De la Cruz and Vioque 2003)
Porphyra yezoensis	Sequence similarity (Accession number AP006715)	Not analyzed
Viridiplantae (green algae, plants)		
Nephroselmis olivacea	Sequence similarity (Turmel et al. 1999b)	No (De la Cruz and Vioque 2003)

[a]Trace activity detected with very high RNA concentration (Li et al. 2007). *C. paradoxa* cpP RNA is the only cpP RNA that has been analyzed under these conditions

Therefore, it seems that the presence of *rnpB* is a general feature of red algae chloroplasts, but is only present in some early branching green algae. *rnpB* was lost independently in secondary endosymbionts derived from red and green algae. What is more, several independent losses of *rnpB* are needed to account for its distribution in green alga because algae that branch earlier than *N. olivacea*, like *Mesostigma viride* do not encode *rnpB* in their chloroplasts (Fig. 11.2).

11.3.2 Structure and Function of Plastid P RNA

Plastid RNase P RNAs (cpP RNA) contain all the secondary structure elements of type B bacterial P RNA (Fig. 11.3). They are very AU rich, and therefore, are expected to have low structural stability. None of them has activity in vitro in the absence of protein under standard assay conditions (Baum et al. 1996; De la Cruz and Vioque 2003) (A.V., unpublished), although recently a low level of activity has been detected with the *C. paradoxa* cpP RNA using very high RNA concentrations (Li et al. 2007). It is possible that some others cpP RNAs have detectable activity when assayed under these conditions that allow detection of low level activity. cpP RNAs, like most cyanobacteria P RNAs (Vioque 1997), lack the conserved GGU sequence in the L15 loop involved in substrate interaction with the substrate 3′-(R)CCA sequence, which correlates with the fact that the CCA is not encoded in plastid tRNAs. The P RNAs from *C. paradoxa*, *N. olivacea*, and *P. purpurea*, taken as representatives of the three lineages of primary chloroplasts, have been studied in vitro (De la Cruz and Vioque 2003). As mentioned above, they lack detectable activity under standard assay conditions. They probably cannot fold in a functional compact conformation in the absence of protein, as shown by their increased sensitivity to Pb^{2+} hydrolysis (De la Cruz and Vioque 2003). In the case of *C. paradoxa*, this conclusion has been further supported by analysis on native acrylamide gels (Li et al. 2007). As a result of their inability to fold in a functional structure, cpP-RNAs are unable to bind efficiently to pre-tRNAs (De la Cruz and Vioque 2003). It is interesting that the main Pb^{2+} cleavage site in the S-domain of P RNA (site Ia) is absent in the three cpP-RNAs assayed (De la Cruz and Vioque 2003). Site Ia is next to nucleotides in the P RNA directly involved in pre-tRNA binding through interaction with the T loop (Kirsebom 2007), in the junction between P10 and P11. The absence of Pb^{2+} hydrolysis at site Ia is in agreement with the inability of the cpP RNAs to fold properly and their inability to bind pre-tRNAs.

Plastid cpP-RNAs fit the bacterial type-A consensus, containing all the secondary structure elements present in the bacterial P RNA. However, they lack some of the conserved GNRA tetraloops involved in tertiary contacts in bacterial P RNA (Fig. 11.3). In the bacterial consensus P RNA structure the loops in helices P9, P14, and P18 (L9, L14, and L18) are GNRA tetraloops involved in tertiary interactions (Brown et al. 1996; Massire et al. 1997), but they are missing in all or in part in cpP RNAs. In *C. merolae* and *C. caldarium*, none of the three loops is a GNRA

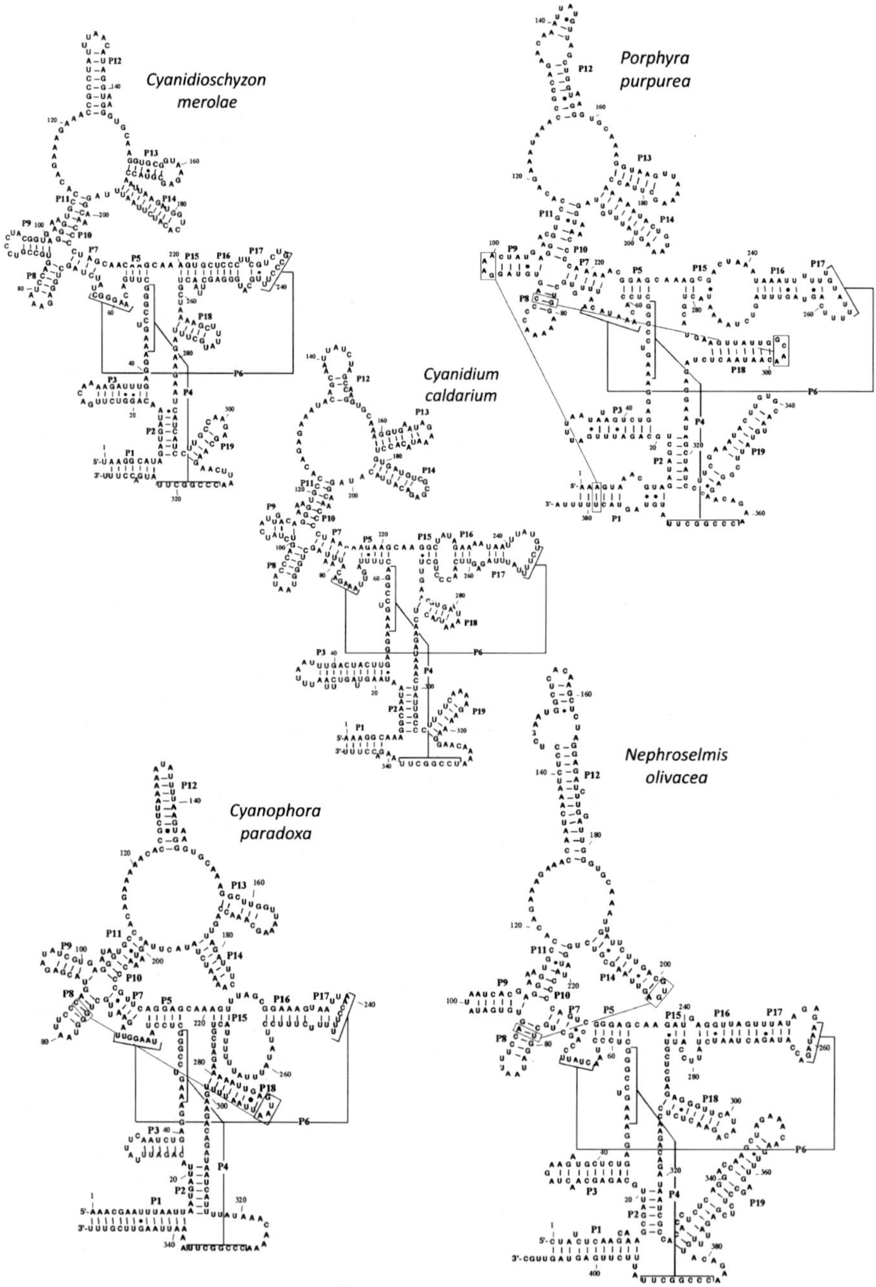

Fig. 11.2 Secondary structure models of cpP RNAs from red alga (*Cyanidioschyzon merolae, Cyanidium caldarium, Porphyra purpurea*), Glaucocystophyceae (*Cyanophora paradoxa*), and green algae (*Nephroselmis olivacea*)

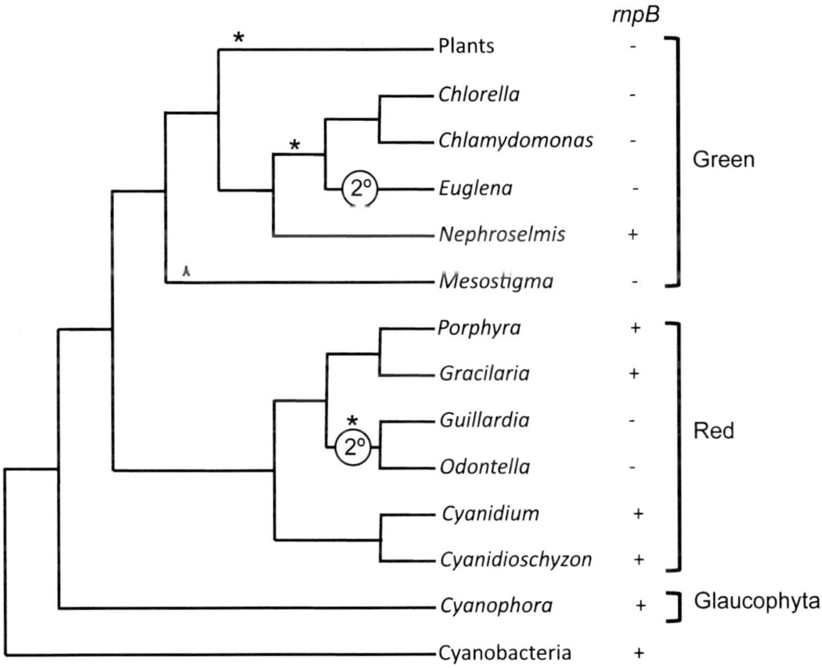

Fig. 11.3 Phylogenetic relationships among chloroplasts. The cladogram is drawn based on information from (Rodriguez-Ezpeleta et al. 2005). The presence or absence of *rnpB* in the plastid genome is indicated by a + or – sign, respectively. Asterisks indicate branches were *rnpB* has been lost. Branches corresponding to secondary endosymbiont are also indicated

tetraloop. In *C. paradoxa,* only L18 is a GNRA tetraloop. In *N. olivacea,* only L14 is a GNRA tetraloop. *P. purpurea* has the three tetraloops, but the sequences in L14 and its interacting base pair in P8 do not match the consensus observed by covariation. The cpP RNA from *C. paradoxa* has been studied in more detail in vitro. The *C. paradoxa* P RNA can reconstitute a functional holoenzyme with the bacterial protein subunit (Li et al. 2007; Pascual and Vioque 1999). Domain swamps with *E. coli* P RNA and site directed mutagenesis of selected nucleotides, and whole domains to make them more similar to the *E. coli* structure has not resulted in significant improvement of activity (Li et al. 2007). Therefore, it seems that the catalytic defect of the cpP-RNAs is due to an overall structural lability, that extends through all domains, making the RNA more dependent on the protein subunits. It is surprising that simple mutations, like the conversion of the L15 loop to a GNRA tetraloop, in order to try to restore a tertiary interaction (Pascual and Vioque 1999), or the replacement of A22 by a conserved G at this position (Li et al. 2007) results in complete inactivation of the RNA in the reconstitution assay with bacterial protein.

Biochemical studies have been performed with the native *C. paradoxa* cpRNase P (Baum et al. 1996; Cordier and Schön 1999). These studies showed that the RNA encoded by the plastid *rnpB* gene is present in purified fractions of the enzyme, and that the cpRNase P is sensitive to micrococcal nuclease, supporting that this RNA is indeed part of the enzyme.

An important unsolved issue is the nature of the protein subunit of RNase P in the plastids that have a bacterial like P RNA. cpP RNA is very similar to bacterial P RNA and cpP RNA from *C. paradoxa* can reconstitute a functional holoenzyme with bacterial P protein (Li et al. 2007; Pascual and Vioque 1999), indicating that this protein can promote correct folding of this RNA, although with low efficiency. Therefore, one would expect that in these cases the cpP protein would be bacterial-like. But no protein subunit of cpRNase P has been identified so far. The nuclear genome of *C. paradoxa* is not yet available, but no bacterial-like RNase P protein is detected in the EST database (http://amoebidia.bcm.umontreal.ca/pepdb/searches/login.php) for *C. paradoxa, N. olivacea,* and several other related organisms. The complete genome of the red algae *Cyanidioschyzon merolae* does not contain a homologue to bacterial P protein. Therefore, it has to be considered that, even though cpP RNAs are very similar to bacterial P RNAs and they can reconstitute activity with bacterial P protein, the protein composition of cpRNase P has changed relative to the original bacterial-like P protein. In this respect, it is relevant that *C. paradoxa* cpRNase P seems to be more protein rich than the bacterial enzyme (Cordier and Schön 1999).

11.3.3 Plant Chloroplast RNase P

Spinach chloroplast RNase P seems to lack an RNA subunit. This was proposed in 1988, based on the resistance to micrococcal nuclease, a buoyant density in CsCl gradients of 1.28 g/mL, and the absence of RNA in purified fractions of the enzyme (Gegenheimer 1995; Wang et al. 1988). This claim was further substantiated by the study of the effect of phosphorothioate substitutions on the phosphate at the cleaved bond (Thomas et al. 2000) in the pre-tRNA, that indicated that a mechanism different form the RNA based mechanism of bacterial or nuclear RNase P is used in cpRNase P. In the bacterial RNase P reaction, magnesium is required to coordinate the non-bridging phosphodiester oxygen(s) of the scissile bond. As a consequence, bacterial RNase P cannot cleave pre-tRNA in which the *pro-*R_p nonbridging oxygen of the scissile bond is replaced by sulfur. Partially purified RNase P from spinach chloroplasts can accurately and efficiently process phosphorothioate-substituted pre-tRNAs; cleavage occurs exclusively at the thiocontaining scissile bond. The chloroplast RNase P reaction mechanism, therefore, does not involve a catalytic magnesium bound to the *pro-*R_p phosphate oxygen, and hence, is distinct from the mechanism of the bacterial ribozyme RNase P. Identification of the components of the plant cpRNase P has not been achieved. Extensive efforts to purify the enzyme have not led to the identification of candidate polypeptides in the active fractions (Peter Gegenheimer, personal communication).

The replacement of a ribonucleoprotein by a protein-only enzyme has a precedent in chloroplast evolution. Signal recognition particle (SRP) is a ribonucleoprotein required for cotranslational export of protein through cytoplasmic membrane in bacteria or endoplasmic membrane in eukaryotes (Pool 2005). SRP contains an RNA subunit (4.5S RNA in bacteria, 7SK RNA in eukaryotes) and one protein in bacteria, or several proteins in eukaryotes. SRP RNA catalyzes the GTP dependent interaction between SRP and SRP receptor in the membrane. Chloroplasts also contain SRP, involved in transport of proteins to the thylakoid membrane or thylakoid lumen (Schunemann 2004). However, plant chloroplast SRP does not contain RNA. Instead, a unique protein (cpSRP43) not present in other SRPs seems to replace the function of SRP RNA. This hypothesis has been confirmed recently by structural studies (Stengel et al. 2008). SRP RNA is present in some algal chloroplasts, with a distribution that overlaps, but is not identical to the P RNA distribution (Jaru-Ampornpan et al. 2007; Stengel et al. 2008).

Therefore, the possibility that the RNA subunit of plastid RNase P in higher plants has been lost, must be seriously considered; in fact, the enzyme might contain only protein, as has been shown for human mtRNase P. Interestingly, plant genomes encode a homologue to MRPP3, the possible catalytic subunit of human mtRNase P, and this plant MRPP3 is predicted to localize to mitochondria and chloroplasts. The proteomic methodology used to identify the components of human mtRNase P could be used to identify the subunit make-up of plant plastid RNase P.

Acknowledgments Dr. Venkat Gopalan is acknowledged for critical reading and comments. Research in the author's laboratory is funded by grants from the Spanish Ministry of Science and Education (BFU2007-60651) and the Andalusian Research Plan (BIO215 and P06-CVI-01692).

References

Altman S (1989) Ribonuclease P: an enzyme with a catalytic RNA subunit. Adv Enzymol Relat Areas Mol Biol 62:1–36

Aphasizhev R, Karmarkar U, Simpson L (1998) Are tRNAs imported into the mitochondria of kinetoplastid protozoa as 5′-extended precursors? Mol Biochem Parasitol 93:73–80

Baum M, Cordier A, Schön A (1996) RNase P from a photosynthetic organelle contains an RNA homologous to the cyanobacterial counterpart. J Mol Biol 257:43–52

Brown JW, Nolan JM, Haas ES, Rubio MA, Major F, Pace NR (1996) Comparative analysis of ribonuclease P RNA using gene sequences from natural microbial populations reveals tertiary structural elements. Proc Natl Acad Sci U S A 93:3001–3006

Cordier A, Schön A (1999) Cyanelle RNase P: RNA structure analysis and holoenzyme properties of an organellar ribonucleoprotein enzyme. J Mol Biol 289:9–20

Dang YL, Martin NC (1993) Yeast mitochondrial RNase P. Sequence of the RPM2 gene and demonstration that its product is a protein subunit of the enzyme. J Biol Chem 268:19791–19796

De la Cruz J, Vioque A (2003) A structural and functional study of plastid RNAs homologous to catalytic bacterial RNase P RNA. Gene 321:47–56

Doersen CJ, Guerrier-Takada C, Altman S, Attardi G (1985) Characterization of an RNase P activity from HeLa cell mitochondria. Comparison with the cytosol RNase P activity. J Biol Chem 260:5942–5949

Frank DN, Pace NR (1998) Ribonuclease P: unity and diversity in a tRNA processing ribozyme. Annu Rev Biochem 67:153–180

Franklin SE, Zwick MG, Johnson JD (1995) Characterization and partial purification of two pre-tRNA 5′-processing activities from *Daucus carrota* (carrot) suspension cells. Plant J 7:553–563
Gegenheimer P (1995) Structure, mechanism and evolution of chloroplast transfer RNA processing systems. Mol Biol Rep 22:147–150
Gould SB, Waller RF, McFadden GI (2008) Plastid evolution. Annu Rev Plant Biol 59:491–517
Hagopian JC, Reis M, Kitajima JP, Bhattacharya D, de Oliveira MC (2004) Comparative analysis of the complete plastid genome sequence of the red alga *Gracilaria tenuistipitata var. liui* provides insights into the evolution of rhodoplasts and their relationship to other plastids. J Mol Evol 59:464–477
Hancock K, LeBlanc AJ, Donze D, Hajduk SL (1992) Identification of nuclear encoded precursor tRNAs within the mitochondrion of *Trypanosoma brucei*. J Biol Chem 267:23963–23971
Hanic-Joyce PJ, Gray MW (1990) Processing of transfer RNA precursors in a wheat mitochondrial extract. J Biol Chem 265:13782–13791
Hollingsworth MJ, Martin NC (1986) RNase P activity in the mitochondria of *Saccharomyces cerevisiae* depends on both mitochondrion and nucleus-encoded components. Mol Cell Biol 6:1058–1064
Holzmann J, Frank P, Löffler E, Bennet KL, Gerner C, Rossmanith W (2008) RNase P without RNA: Identification and functional reconstitution of the human mitochondrial tRNA processing enzyme. Cell 135:462–474
Jaru-Ampornpan P, Chandrasekar S, Shan SO (2007) Efficient interaction between two GTPases allows the chloroplast SRP pathway to bypass the requirement for an SRP RNA. Mol Biol Cell 18:2636–2645
Kapushoc ST, Alfonzo JD, Rubio MA, Simpson L (2000) End processing precedes mitochondrial importation and editing of tRNAs in *Leishmania tarentolae*. J Biol Chem 275:37907–37914
Kassenbrock CK, Gao GJ, Groom KR, Sulo P, Douglas MG, Martin NC (1995) RPM2, independently of its mitochondrial RNase P function, suppresses an ISP42 mutant defective in mitochondrial import and is essential for normal growth. Mol Cell Biol 15:4763–4770
Kirsebom LA (2007) RNase P RNA mediated cleavage: substrate recognition and catalysis. Biochimie 89:1183–1194
Lang BF, Burger G, O'Kelly CJ, Cedergren R, Golding GB, Lemieux C, Sankoff D, Turmel M, Gray MW (1997) An ancestral mitochondrial DNA resembling a eubacterial genome in miniature. Nature 387:493–497
Lee YC, Lee BJ, Hwang DS, Kang HS (1996a) Purification and characterization of mitochondrial ribonuclease P from *Aspergillus nidulans*. Eur J Biochem 235:289–296
Lee YC, Lee BJ, Kang HS (1996b) The RNA component of mitochondrial ribonuclease P from *Aspergillus nidulans*. Eur J Biochem 235:297–303
Li D, Willkomm DK, Schön A, Hartmann RK (2007) RNase P of the *Cyanophora paradoxa* cyanelle: a plastid ribozyme. Biochimie 89:1528–1538
Marchfelder A, Brennicke A (1994) Characterization and partial purification of tRNA processing activities from potato mitochondria. Plant Physiol 105:1247–1254
Marchfelder A, Schuster W, Brennicke A (1990) In vitro processing of mitochondrial and plastid derived tRNA precursors in a plant mitochondrial extract. Nucleic Acids Res 18:1401–1406
Martin NC, Hopper AK (1994) How single genes provide tRNA processing enzymes to mitochondria, nuclei and the cytosol. Biochimie 76:1161–1167
Massire C, Jaeger L, Westhof E (1997) Phylogenetic evidence for a new tertiary interaction in bacterial RNase P RNAs. RNA 3:553–556
Miller DL, Martin NC (1983) Characterization of the yeast mitochondrial locus necessary for tRNA biosynthesis: DNA sequence analysis and identification of a new transcript. Cell 34:911–917
Morales MJ, Wise CA, Hollingsworth MJ, Martin NC (1989) Characterization of yeast mitochondrial RNase P: an intact RNA subunit is not essential for activity in vitro. Nucleic Acids Res 17:6865–6881
Morales MJ, Dang YL, Lou YC, Sulo P, Martin NC (1992) A 105-kDa protein is required for yeast mitochondrial RNase P activity. Proc Natl Acad Sci U S A 89:9875–9879

Ohta N, Matsuzaki M, Misumi O, Miyagishima SY, Nozaki H, Tanaka K, Shin IT, Kohara Y, Kuroiwa T (2003) Complete sequence and analysis of the plastid genome of the unicellular red alga *Cyanidioschyzon merolae*. DNA Res 10:67–77

Pascual A, Vioque A (1999) Functional reconstitution of RNase P activity from a plastid RNA subunit and a cyanobacterial protein subunit. FEBS Lett 442:7–10

Petersen RF, Langkjaer RB, Hvidtfeldt J, Gartner J, Palmen W, Ussery DW, Piskur J (2002) Inheritance and organisation of the mitochondrial genome differ between two *Saccharomyces yeasts*. J Mol Biol 318:627–636

Pool MR (2005) Signal recognition particles in chloroplasts, bacteria, yeast and mammals. Mol Membr Biol 22:3–15

Puranam RS, Attardi G (2001) The RNase P associated with HeLa cell mitochondria contains an essential RNA component identical in sequence to that of the nuclear RNase P. Mol Cell Biol 21:548–561

Ragnini A, Grisanti P, Rinaldi T, Frontali L, Palleschi C (1991) Mitochondrial genome of *Saccharomyces douglasii*: genes coding for components of the protein synthetic apparatus. Curr Genet 19:169–174

Reith M, Munholland J (1995) Complete nucleotide sequence of the *Porphyra purpurea* chloroplast genome. Plant Mol Biol Rep 13:333–335

Reyes-Prieto A, Weber AP, Bhattacharya D (2007) The origin and establishment of the plastid in algae and plants. Annu Rev Genet 41:147–168

Robbens S, Derelle E, Ferraz C, Wuyts J, Moreau H, Van de Peer Y (2007) The complete chloroplast and mitochondrial DNA sequence of *Ostreococcus tauri*: organelle genomes of the smallest eukaryote are examples of compaction. Mol Biol Evol 24:956–968

Rodriguez-Ezpeleta N, Brinkmann H, Burey SC, Roure B, Burger G, Löffelhardt W, Bohnert HJ, Philippe H, Lang BF (2005) Monophyly of primary photosynthetic eukaryotes: green plants, red algae, and glaucophytes. Curr Biol 15:1325–1330

Rossmanith W, Karwan RM (1998) Characterization of human mitochondrial RNase P: novel aspects in tRNA processing. Biochem Biophys Res Commun 247:234–241

Rossmanith W, Potuschak T (2001) Difference between mitochondrial RNase P and nuclear RNase P. Mol Cell Biol 21:8236–8237

Rossmanith W, Tullo A, Potuschak T, Karwan R, Sbisa E (1995) Human mitochondrial tRNA processing. J Biol Chem 270:12885–12891

Rubio MA, Liu X, Yuzawa H, Alfonzo JD, Simpson L (2000) Selective importation of RNA into isolated mitochondria from *Leishmania tarentolae*. RNA 6:988–1003

Salavati R, Panigrahi AK, Stuart KD (2001) Mitochondrial ribonuclease P activity of *Trypanosoma brucei*. Mol Biochem Parasitol 115:109–117

Sbisa E, Pesole G, Tullo A, Saccone C (1996) The evolution of the RNase P- and RNase MRP-associated RNAs: phylogenetic analysis and nucleotide substitution rate. J Mol Evol 43:46–57

Schneider A (1994) Import of RNA into mitochondria. Trends Cell Biol 4:282–286

Schunemann D (2004) Structure and function of the chloroplast signal recognition particle. Curr Genet 44:295–304

Seif ER, Forget L, Martin NC, Lang BF (2003) Mitochondrial RNase P RNAs in ascomycete fungi: lineage-specific variations in RNA secondary structure. RNA 9:1073–1083

Seif E, Leigh J, Liu Y, Roewer I, Forget L, Lang BF (2005) Comparative mitochondrial genomics in zygomycetes: bacteria-like RNase P RNAs, mobile elements and a close source of the group I intron invasion in angiosperms. Nucleic Acids Res 33:734–744

Seif E, Cadieux A, Lang BF (2006) Hybrid *E. coli* – mitochondrial ribonuclease P RNAs are catalytically active. RNA 12:1661–1670

Shevelev EL, Bryant DA, Löffelhardt W, Bohnert HJ (1995) Ribonuclease-P RNA gene of the plastid chromosome from *Cyanophora paradoxa*. DNA Res 2:231–234

Siegel RW, Banta AB, Haas ES, Brown JW, Pace NR (1996) *Mycoplasma fermentans* simplifies our view of the catalytic core of ribonuclease P RNA. RNA 2:452–462

Simpson AM, Suyama Y, Dewes H, Campbell DA, Simpson L (1989) Kinetoplastid mitochondria contain functional tRNAs which are encoded in nuclear DNA and also contain small minicircle and maxicircle transcripts of unknown function. Nucleic Acids Res 17:5427–5445

Stark BC, Kole R, Bowman EJ, Altman S (1978) Ribonuclease P: an enzyme with an essential RNA component. Proc Natl Acad Sci U S A 75:3717–3721

Stengel KF, Holdermann I, Cain P, Robinson C, Wild K, Sinning I (2008) Structural basis for specific substrate recognition by the chloroplast signal recognition particle protein cpSRP43. Science 321:253–256

Stribinskis V, Ramos KS (2007) Rpm2p, a protein subunit of mitochondrial RNase P, physically and genetically interacts with cytoplasmic processing bodies. Nucleic Acids Res 35:1301–1311

Stribinskis V, Gao GJ, Sulo P, Dang YL, Martin NC (1996) Yeast mitochondrial RNase P RNA synthesis is altered in an RNase P protein subunit mutant: insights into the biogenesis of a mitochondrial RNA-processing enzyme. Mol Cell Biol 16:3429–3436

Stribinskis V, Gao GJ, Ellis SR, Martin NC (2001a) Rpm2, the protein subunit of mitochondrial RNase P in *Saccharomyces cerevisiae*, also has a role in the translation of mitochondrially encoded subunits of cytochrome c oxidase. Genetics 158:573–585

Stribinskis V, Gao GJ, Sulo P, Ellis SR, Martin NC (2001b) Rpm2p: separate domains promote tRNA and Rpm1r maturation in *Saccharomyces cerevisiae* mitochondria. Nucleic Acids Res 29:3631–3637

Stribinskis V, Heyman HC, Ellis SR, Steffen MC, Martin NC (2005) Rpm2p, a component of yeast mitochondrial RNase P, acts as a transcriptional activator in the nucleus. Mol Cell Biol 25:6546–6558

Tan TH, Pach R, Crausaz A, Ivens A, Schneider A (2002) tRNAs in *Trypanosoma brucei*: genomic organization, expression, and mitochondrial import. Mol Cell Biol 22:3707–3717

Thomas BC, Li X, Gegenheimer P (2000) Chloroplast ribonuclease P does not utilize the ribozyme-type pre-tRNA cleavage mechanism. RNA 6:545–553

Turmel M, Lemieux C, Burger G, Lang BF, Otis C, Plante I, Gray MW (1999a) The complete mitochondrial DNA sequences of *Nephroselmis olivacea* and *Pedinomonas minor*. Two radically different evolutionary patterns within green algae. Plant Cell 11:1717–1730

Turmel M, Otis C, Lemieux C (1999b) The complete chloroplast DNA sequence of the green alga *Nephroselmis olivacea*: insights into the architecture of ancestral chloroplast genomes. Proc Natl Acad Sci U S A 96:10248–10253

Tyler BM, Tripathy S, Zhang X, Dehal P, Jiang RH, Aerts A, Arredondo FD, Baxter L, Bensasson D, Beynon JL et al (2006) *Phytophthora* genome sequences uncover evolutionary origins and mechanisms of pathogenesis. Science 313:1261–1266

Underbrink-Lyon K, Miller DL, Ross NA, Fukuhara H, Martin NC (1983) Characterization of a yeast mitochondrial locus necessary for tRNA biosynthesis. Deletion mapping and restriction mapping studies. Mol Gen Genet 191:512–518

Vioque A (1997) The RNase P RNA from cyanobacteria: short tandemly repeated repetitive (STRR) sequences are present within the RNase P RNA gene in heterocyst-forming cyanobacteria. Nucleic Acids Res 25:3471–3477

Walker SC, Engelke DR (2008) A protein-only RNase P in human mitochondria. Cell 135:412–414

Wang MJ, Davis NW, Gegenheimer P (1988) Novel mechanisms for maturation of chloroplast transfer RNA precursors. EMBO J 7:1567–1574

Wise C, Martin NC (1991a) Sequence analysis of *Saccharomyces exiguus* mitochondrial DNA reveals an RNase P RNA gene flanked by two tRNA genes. Nucleic Acids Res 19:4773

Wise CA, Martin NC (1991b) Dramatic size variation of yeast mitochondrial RNAs suggests that RNase P RNAs can be quite small. J Biol Chem 266:19154–19157

Yermovsky-Kammerer AE, Hajduk SL (1999) In vitro import of a nuclearly encoded tRNA into the mitochondrion of *Trypanosoma brucei*. Mol Cell Biol 19:6253–6259

Chapter 12
Human RNase P and Transcription

Nayef Jarrous, Robert Reiner, and Yana Dehtiar

12.1 Characterization of a Human RNase P Ribonucleoprotein

RNase P has been initially identified in Escherichia coli as the endoribonuclease that processes the 5′ leader sequence of precursor tRNA (Robertson et al. 1972). Based on this enzymatic activity, a nuclear RNase P was identified and partially purified from epithelioid carcinoma cells (Bothwell and Altman 1975; Koski et al. 1976). Later on, it has been shown through extensive biochemical purification procedures that nuclear RNase P from HeLa cells possesses a single RNA species, called H1 RNA (Bartkiewicz et al. 1989), and at least ten distinct protein subunits (see Eder et al. 1997; Jarrous and Altman 2001). These protein subunits, termed Rpp14, Rpp20, Rpp21 Rpp25, Rpp29 (also called Pop4), Rpp30, Rpp38, Rpp40, hPop1 and hPop5, were identified by phylogenetic sequence comparative analyses or mass spectrometry followed by molecular cDNA cloning and further biochemical analyses (Lygerou et al. 1996; Eder et al. 1997; Jarrous et al. 1998, 1999, 2001; van Eenennaam et al. 1999; Guerrier-Takada et al. 2002; van Eenennaam et al. 2001a, b) (Table 12.1).

RNase P is a ubiquitous and conserved ribonucleoprotein, even though it has been reported that the archaeon Nanoarchaeum equitans lacks RNase P activity (Randau et al. 2008) and 5′ end processing of tRNA could be based on protein enzyme in mitochondria (see Salavati et al. 2001; Holzmann et al. 2008). Archaeal and Eukaryal RNase P ribonucleoproteins are functionally, structurally and evolutionary related (Hartmann and Hartmann 2003). Four protein subunits of human RNase P are conserved in Archaea (Hall and Brown 2002; Hartmann and Hartmann 2003) (Table 12.1), while nine of them have homologs in yeast (see Chamberlain et al. 1998; Rosenblad et al. 2006) (Table 12.1). Archaeal Rpp21, Rpp29, Rpp30 and Pop5 have been shown to be part of an active RNase P ribonucleoprotein (Hall

N. Jarrous (✉), R. Reiner, and Y. Dehtiar
Department of Molecular Biology, The Hebrew University-Hadassah Medical School, Jerusalem 91120, Israel
e-mail: jarrous@md.huji.ac.il

F. Liu and S. Altman (eds.), *Ribonuclease P*, Protein Reviews 10,
DOI 10.1007/978-1-4419-1142-1_12, © Springer Science+Business Media, LLC 2010

Table 12.1 Subunits of human nuclear RNase P

Subunit[a]	Function related to tRNA processing or else	Chromatin binding[b]/ Pol I&III transcription[c]
Rpp14	Substrate binding, exonuclease[d]	+/+
Rpp20	ATPase[e]	+/+
Rpp21	H1 RNA binding, activity[f]	+/+
Rpp25	H1 RNA binding[g]	+/+
Rpp29	Substrate binding, activity[h]	+/+
Rpp30	H1 RNA binding, activity	+/+
Rpp38	H1 RNA binding	?/+
Rpp40		+/+
hPop1		+/+
hPop5	H1 RNA binding, activity	+/+
H1 RNA	Activity[f,h]	ND/+

[a]Jarrous and Altman 2001; Jarrous 2002
[b]Binding to chromatin of rDNA, tRNA or 5S rRNA genes. Reiner et al. 2006, 2008
[c]Required for Pol I or Pol III transcription in whole HeLa extracts and/or in cells. Reiner et al. 2006, 2008
[d]Jiang and Altman 2002
[e]Li and Altman 2001
[f]Reconstituted eukaryal or archaeal RNase P activity. Mann et al. 2003; Terada et al. 2006; Tsai et al. 2006
[g]Guerrier-Takada et al. 2002
[h]H1 RNA is active in vitro. See Kikovska et al. 2007
ND, not determined

and Brown 2002). Of note, the Alba protein, which is evolutionary related to the subunits Rpp20 and Rpp25 (Aravind et al. 2003), is not associated with purified Archaeal RNase P (Ellis et al. 2007; Hada et al. 2008). Hence, the integration of the two Alba-like proteins Rpp20 and Rpp25 into the RNase P complex in eukaryotes could signify a key event in the evolution of this ribonucleoprotein, and the acquirement of new abilities and functions through binding to chromatin in the nucleus (see below). These two proteins are also shared with the nucleolar ribonucleoprotein RNase MRP (Welting et al. 2006), which cleaves at the A_3 site in the internal transcribed spacer 1 of precursor rRNA (see Clayton 2001).

RNase P is gathered by an extensive network of RNA-protein and protein-protein interactions (Jiang and Altman 2001; Jiang et al. 2001). H1 RNA and its protein subunits are required for the efficient activity of RNase P in tRNA processing. A weak endonucleolytic activity of this ribonucleoprotein in tRNA processing could be reconstituted in vitro by the use of two recombinant protein subunits, Rpp21 and Rpp29, and H1 RNA, and under conditions of low ionic strength and neutral pH (Mann et al. 2003). Likewise, the archaeal counterparts of Rpp21 and Rpp29 are sufficient for reconstitution of the activity of an archaeal RNase P ribonucleoprotein (Tsai et al. 2006). H1 RNA has a conserved core structure (Frank et al. 2000; Marquez et al. 2006), binds to precursor tRNA in vitro (Mann et al. 2003) and exhibits a weak activity (~ 10^6-fold lower than that of its bacterial counterpart) under conditions of high ionic strength and pH 6 (Kikovska et al. 2007).

12.2 A Role for Human RNase P Ribonucleoprotein in Transcription by Pol III

While biochemical purification analyses show the existence of a large human nuclear RNase P in the cell (Eder et al. 1997; Jarrous and Altman 2001), cell biology studies could not demarcate a specific locale for this entity in the nucleus. Thus, in situ RNA hybridization (Jacobson et al. 1997) and indirect immunofluorescence analyses (Jarrous et al. 1999, 2001; van Eenennaam et al. 2001a, b) reveal that subunits of human RNase P are differentially localized in distinct subcellular compartments, such as the nucleoplasm, nucleolus, Cajal bodies, the perinucleolar compartment and cytoplasm. The molecular basis for the differential localization patterns of these subunits in the aforementioned intranuclear compartments is not clear. Nonetheless, it has been proposed that these protein subunits may have a possible differential sequestering role of families of transcribed SINE/Alu repeats in these specialized compartments (see Lunyak 2008). Protein subunits of human RNase P enter the nucleolus independently of their association with large complexes (van Eenennaam et al. 2001), and fluorescence recovery after photobleaching and fluorescence loss in photobleaching analyses show that Rpp29 rapidly cycles between the nucleoplasm and nucleolus (Chen and Huang 2001), an indication that this core component fulfills a dynamic function in the latter compartment. Localization of Rpp29 occurs in the dense fibrillar component (Jarrous et al. 1999), and it is sensitive to inhibition of transcription by actinomycin D (Jarrous et al. 2001). RNase P exhibits properties of a dynamic complex that is formed in a transcription-dependent manner (Jarrous et al. 2001; Jarrous 2002).

Coimmunoprecipitation studies demonstrate that monoclonal antibodies directed against specific protein subunits of human Pol III bring down an active RNase P from whole cell extracts (Reiner et al. 2006). Immunodepletion of RNase P from whole HeLa extracts using antibodies directed against its protein subunits, inhibits transcription of 5S rRNA and tRNA genes by Pol III. Moreover, inactivation of RNase P by targeting its protein subunits for destruction using small interfering RNAs or targeted cleavage of its H1 RNA moiety by RNase H digestion abolish transcription by Pol III. Figure 12.1 shows that transcription of distinct human small non-coding RNA genes, i.e., 5S rRNA, $tRNA^{Tyr}$, $tRNA_i^{Met}$ and 7SL RNA, is completely eliminated in whole extracts derived from HeLa cells efficiently transfected with siRNA against Rpp29 but not with control siRNA directed against luciferase. While RNase P may affect transcription of tRNA genes as processing factor, this is not the case for transcription of the 5S rRNA and 7SL RNA genes, whose transcripts are not recognized as substrates for this endoribonuclease. This latter conclusion is corroborated by nuclear run-on transcription assays that establish that inactivation of RNase P excessively reduces the level of nascent transcription of 5S rRNA genes by Pol III (Reiner et al. 2008).

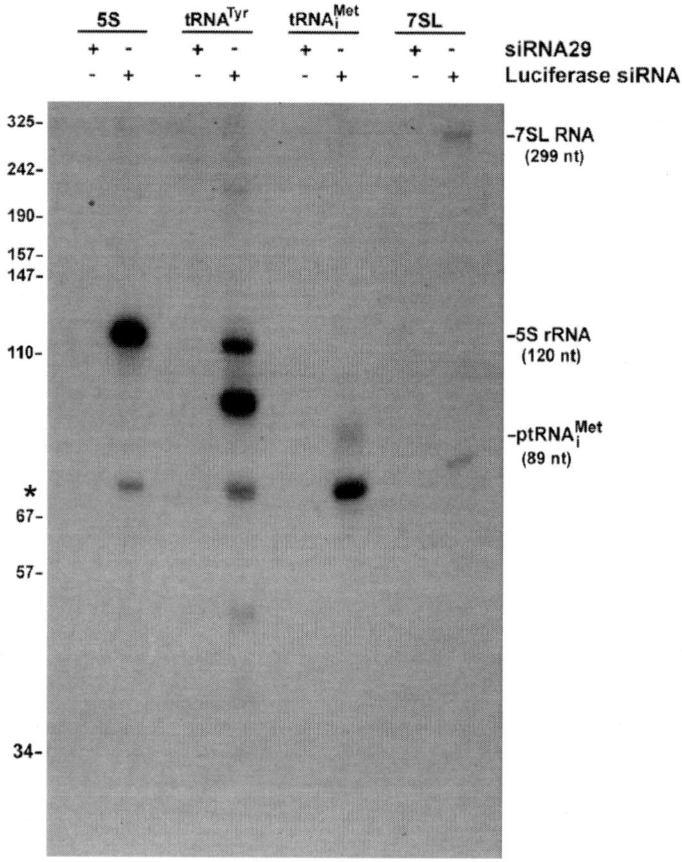

Fig. 12.1 Knockdown of a core component of human RNase P by RNA interference abolishes transcription of various Pol III genes. HeLa cells were transfected for 48 h with small interfering RNA against Rpp29 or with control siRNA against luciferase and then whole extracts were prepared and assayed for transcription of 5S rRNA, $tRNA^{Tyr}$, $tRNA_i^{Met}$ and 7SL RNA as described previously (Reiner et al. 2006). Positions of the labeled transcripts analyzed in 8% polyacrylamide/urea gel are indicated; the primary transcript of $tRNA^{Tyr}$ is 112 nt in length. Asterisk may point to labeled, endogenous $tRNA^{His}$. Positions of the ssDNA size marker are shown

12.3 A Novel Role for Human RNase P in rDNA Transcription by Pol I

The localization in nucleoli of protein subunits of human RNase P as described above, and the role of this ribonucleoprotein complex in Pol III transcription, raised the possibility that this factor is implicated in transcription of rRNA genes (rDNA) by Pol I. Functional analyses using RNA interference or external guide sequences that inactivate RNase P by knocking down its protein subunits (including its specific component Rpp21), or the use of RNase H to cleave its H1 RNA subunit, indicate that this ribonucleoprotein complex is required for efficient transcription of rDNA by Pol I (Reiner et al. 2008). Both transcription of rDNA and pre-rRNA synthesis precipitously decline as a result of inactivation of RNase P. Remarkably, biochemical purification analyses reveal that an active RNase P copurifies with components of Pol I and its associated transcription factors UBF and SL1 complex (Reiner et al. 2008). Moreover, coimmunoprecipitation experiments indicate that antibodies directed against Pol I, UBF or SL1 complex bring down an active RNase P. Chromatin immunoprecipitation analysis demonstrates that RNase P occupies the promoter and transcribed regions of rDNA (Reiner et al. 2008). However, RNase P does not capture the intergenic spacer region, an indication that this complex is not uniformly distributed on rDNA repeats.

12.4 Recruitment and Assembly of RNase P on Chromatin of Target Genes

Chromatin occupancy by human RNase P is dynamic in the sense that it is linked to the cell cycle. In synchronized cells, RNase P associates with chromatin of rRNA genes at G2 phase, in which transcription by Pol I is elevated (Grummt 2003), but it detaches from these loci in mitosis, which coincides with shut off of transcription (Reiner et al. 2006, 2008) (Fig. 12.2). Subunits of Pol I also detach from chromatin of the coding region of human rDNA (Reiner et al. 2008), a finding that is consistent with inhibition of rRNA transcription and exclusion of the Pol I subunit Rpa43 from the 35S rDNA gene region in anaphase (Clemente-Blanco et al. 2009). However, after exit from mitosis, RNase P is not recruited to chromatin of rDNA as a whole complex with its ten protein subunits (Fig. 12.2). Thus, the subunit Rpp20 and Rpp29 but not Rpp25 are promptly recruited to rDNA at early G1 phase (Reiner et al. 2008). Rpp25 rather appears on these genetic loci at late G1/S phase, which concurs with increased transcription by Pol I. This binding pattern indicates that the recruitment of the two former subunits is independent of that of the latter. Likewise, recruitment of RNase P to 5S rRNA and tRNA genes in the G1 phase proceeds through a dynamic and stepwise association process (Reiner et al. 2006, 2008). These findings rule out the prospect that RNase P preassembles as a whole particle with all its ten protein subunits before binding to its target genes. In support of this latter view, we found that knockdown of one protein subunit does not obstruct the recruitment of other subunits on target genes in cycling human cells, even though transcription by Pol I and Pol III is inhibited (Reiner et al. 2006, 2008). Thus, knockdown of Rpp25 by RNA interference does not lead to detachment of Rpp20 and Rpp29 from rDNA, while knockdown of Rpp29 does not affect chromatin occupancy of tRNA and 5S rRNA genes by Rpp21 (Reiner et al. 2006, 2008).

Recruitment and assembly of human RNase P in relation with the transcription cycle of rDNA and small non-coding RNA genes remains known. There are two scenarios for the assembly of preinitiation complexes on the promoter of rDNA:

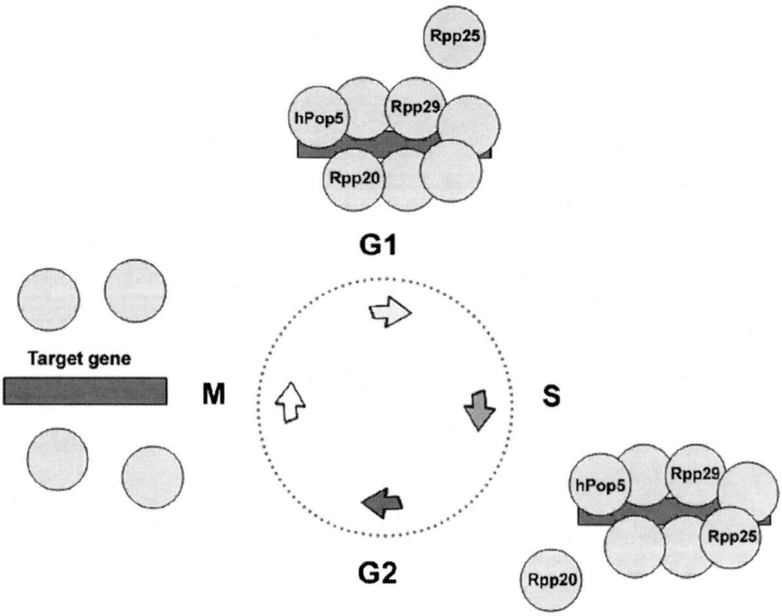

Fig. 12.2 A model for the recruitment of some protein subunits of human RNase P to chromatin of target genes during the cell cycle. In mitotic HeLa cells, protein subunits of RNase P disengage from chromatin of target genes, e.g., rDNA, tRNA or 5S rRNA, which concurs with cessation of transcription (*white arrow*). In mitosis, the steady state level of some subunits is down-regulated (by degradation or low expression) while that of others, such as Rpp20 and Rpp40, remains largely unchanged. Some subunits promptly reassociate with the target genes at early G1, while others, such as Rpp25, join in at late G1 or S phase, in which transcription increases (*gray arrows*). Chromatin occupancy by the Alba-like protein Rpp20 rather diminish at the S phase. At G2 phase, protein subunits are firmly bound to chromatin of efficiently transcribed genes (*dark gray arrow*). This model is based on data obtained in Reiner et al. 2006 and Reiner et al. 2008

One is recruitment of a preassembled Pol I holoenzyme and the other is a dynamic assembly of Pol I and its transcription initiation factors (see Grummt 2003; Prieto and McStay 2008). The second scenario is supported by the finding that subunits of Pol I enter the nucleolus as distinct entities rather than as part of a preassembled holoenzyme, and the assembly of Pol I proceeds in a sequential manner in each transcription cycle (Dundr et al. 2002). Moreover, the recruitment efficiency and retention of Pol I components at the promoter region of rDNA is used to control the transcriptional output of rRNA genes (Gorski et al. 2008). It remains unknown if RNase P assembly follows the sequential assembly pattern of Pol I on the rDNA promoter. Nonetheless, an active RNase P copurifies with Pol I and its transcription factors UBF and SL1 complex as determined by gel filtration chromatography and coimmunoprecipitation analyses (Reiner et al. 2008), an indication that these two molecular machines can be found assembled and associated in the nucleus.

12.5 What Is the Exact Role of RNase P in Transcription?

RNA interference reveals that inactivation of human RNase P in cells abolishes transcription of a variety of small non-coding RNA genes, an indication that this factor is implicated in general transcription by Pol III (Fig. 12.1; Reiner et al. 2006, 2008). Moreover, RNase P is required for transcription of Pol III genes, regardless of the type of their promoters (Fig. 12.1), and it acts on U6 snRNA promoter fused to a reporter gene in cells. Additionally, nuclear run-on transcription assays show that inactivation of RNase P considerably reduces the level of nascent transcription of 5S rRNA gene, while a more moderate decrease was measured for transcription of rDNA (Reiner et al. 2008). The level of nascent transcription reflects the number of polymerases that are engaged in active transcription. However, ChIP analysis demonstrates that inactivation of RNase P by RNA interference does not significantly alter chromatin occupancy of rDNA and 5S rRNA genes by RPB6 or RPB8, an indication that RNase P is not essential for the recruitment of these two core components (Reiner et al. 2008). Thereby, part of Pol I and Pol III appear to be loaded onto their target genes in cells with inactivated RNase P, but the transcription output is affected. Thus, a function for RNase P in the later assembly of Pol I and Pol III, initiation, elongation and termination is feasible. Moreover, it is unknown if RNase P (or its RNA) acts as a scaffold or a recruiting factor of the Pol I and Pol III transcription machineries. RNase P may also affect transcription in the nucleus by modifying chromatin. Histone methylation and acetylation associate with transcriptional activation (Hassan et al. 2001; Smith and Peterson 2005). RNase P may also alter chromatin structure and function, by acting as a new type of chromatin remodeling complex. Chromatin remodeling complexes are large protein complexes that share an SNF2-like related ATPase subunit (Lusser and Kadonaga 2003). Purified HeLa nuclear RNase P has ATPase activity, which is associated with its Rpp20 subunit (Li and Altman 2001), but this subunit does not show sequence similarity to the SNF2-like family of ATPases. Since Rpp20 does not exhibit RNA helicase activity, and RNase P does not require ATP for cleavage of precursor tRNA substrates, it is likely that Rpp20 and/or RNase P utilizes ATP for other functions (Li and Altman 2001), such as in chromatin binding and remodeling. Rpp20 belongs to the Alba-like superfamily of chromatin proteins (Aravind et al. 2003) and it dynamically binds to chromatin of active rDNA, 5S rRNA and tRNA genes (Reiner et al. 2006, 2008).

12.6 Does RNase P Affect Pol II Transcription?

It has been shown that inactivation of human RNase P, by the use of external guide sequences, or RNA interference that target one of its protein subunits, is associated with coordinate inhibition of the expression of other protein subunits and even unrelated proteins (Kovrigina et al. 2003; Cohen et al. 2003). Thus, knockdown of

Rpp38 leads to the coordinate inhibition of expression of Rpp21, Rpp25, Rpp29, Pop5, as well as lamins A and C (Kovrigina et al. 2003; Zhang and Altman 2004). The means of control of this phenomenon could be at the transcriptional level (Kovrigina et al. 2003). In principle, inactivation of RNase P can lead to deregulation of the expression of protein-coding genes by inhibition of transcription of Pol III transcripts that regulate Pol II (see Pagano et al. 2007; Mariner et al. 2008). Inhibition of RNase P may also alter chromatin structure, such as formation of heterochromatin (see Lunyak 2008), at genetic loci or barriers (tRNA genes)(Donze and Kamakaka 2001), which otherwise by their transcriptional activity and spatial organization, control the normal expression of nearby protein-coding genes. Alternatively, RNase P may act as specificity transcription factor for Pol II, thus regulating subsets of genes.

12.7 RNase P in Regulation of Expression of Noncoding RNA

RNase P is involved in regulation of non-coding RNA expression in yeast (Samanta et al. 2006, Yang and Altman 2007). Thus, global identification of non-coding RNA by modulating the subunit Rpp1, the homolog of Rpp30, of nuclear RNase P reveals that expression of numerous putative and verified non-coding RNA is affected in S. cerevisiae (Samanta et al. 2006). These non-coding RNA include antisense transcripts (e.g., HRA1–21), promoter-based transcripts (MAN1–50) and intergenic transcripts (TLN1–27). HRA1 RNA is a substrate for RNase P (Yang and Altman 2007). Those non-coding RNA that were located within the promoter region of protein-coding genes may possibly function as regulatory RNAs in transcriptional control of the downstream genes (Samanta et al. 2006). A more recent genome-wide search of an RNase P RNA mutant strain of *S. cerevisiae* broadens the role of this ribonucleoprotein in regulation of non-coding RNA to include the expression of a subset of intron-encoded box C/D snoRNAs (Coughlin et al. 2008). It is not known if the splicing-independent processing of these intron-encoded box C/D snoRNAs by RNase P occurs cotranscriptionally.

New findings unveil that long non-coding RNA are cleaved by mammalian RNase P. Thus, the metastasis-associated lung adenocarcinoma transcript 1, MALAT1, and the *MEN* β non-coding RNA, two nuclear-retained transcripts, are primarily processed by RNase P at a tRNA-like structure located upstream of the 3′ end cleavage/polyadenylation signals (Wilusz et al. 2008; Sunwoo et al. 2009). In the case of MALAT1, cleavage by RNase P generates the 3′ end of the abundant 6.7 kb nuclear-retained transcript with a short poly(A) tail-like structure and the 5′ end of tRNA-like transcript, which is further processed by other canonical tRNA processing machineries to produce the mature tRNA-like transcript that lacks aminoacylation (Wilusz et al. 2008). It remains unknown if RNase P acts cotranscriptionally in processing of these non-coding RNA by interacting with Pol II, as is the case with the 3′ end processing of non-coding snRNAs by the Integrator complex that interacts with the carboxy terminal domain of Pol II (see Baillat et al. 2005).

The release of tRNA molecules from polycistronic mRNA and tRNA transcripts by RNase P has been shown in Bacteria using biochemical and genetic means (see Alifano et al. 1994; Li and Altman 2003; Mohanty and Kushner 2007; Mohanty and Kushner 2008). A temperature sensitive rnpA mutation, A49, in the single C5 protein subunit of E. coli RNase P results in global alterations in the cleavage of many substrates, including pre-tRNAs, pre-4.5S RNA and polycistronic mRNAs (Li and Altman 2003). Thus, the cleavage of these distinct RNA substrates by RNase P does not require more than one protein subunit. This also suggests that the invention of numerous protein subunits for nuclear RNase P is consistent with the involvement of this ribonucleoprotein in fundamental biological tasks other than RNA processing, such as in transcription and gene regulation.

12.8 Prospects

Human RNase P is required for normal transcription by Pol I and Pol III, but the exact role of this ribonucleoprotein therein remains unknown. Current work in our laboratory focuses on the question if RNase P acts directly on Pol I and Pol III, and/or on chromatin to facilitate transcription in the nucleus. The use of naked DNA templates (without chromatin) in transcription assays (Reiner et al. 2006, 2008) suggests that RNase P does not need chromatin to act in vitro. However, the mode of action of RNase P in cells might be quite different.

The findings that nuclear RNase P has multiple tasks in fundamental cellular processes, i.e., transcription, tRNA processing and biogenesis of short and long ncRNA, raise the possibility that there could be various forms of this ribonucleoprotein (with common core and distinct auxiliary components) that fulfill separate tasks in specialized nuclear compartments. These complexes need to be characterized in terms of their subunit composition. Moreover, RNase P that is required for transcription of rDNA and 7SL RNA and 5S rRNA genes, whose transcripts are not recognized as substrates for this endoribonuclease, exhibits canonical tRNA processing activity (Reiner et al. 2006, 2008). Hence, it remains to be determined if this tRNA processing activity is required for transcription or if it merely reflects an intrinsic capability of H1 RNA that serves as a scaffold or recruiting factor for protein subunits of RNase P to bind to hundreds of target genes.

Acknowledgments This research is supported by the United States–Israel Binational Science Foundation (grant no. 2005/009) and the Israel Science Foundation (grant no. 673/06).

References

Alifano P, Rivellini F, Piscitelli C et al (1994) Ribonuclease E provides substrates for ribonuclease P-dependent processing of a polycistronic mRNA. Genes Dev 8:3021–3031

Aravind L, Iyer LM, Anantharaman V (2003) The two faces of Alba: the evolutionary connection between proteins participating in chromatin structure and RNA metabolism. Genome Biol 4:R64

Baillat D, Hakimi MA, Näär AM et al (2005) Integrator, a multiprotein mediator of small nuclear RNA processing, associates with the C-terminal repeat of RNA polymerase II. Cell 123:265–276

Bartkiewicz M, Gold H, Altman S (1989) Identification and characterization of an RNA molecule that copurifies with RNase P activity from HeLa cells. Genes Dev 3:488–499

Bothwell AL, Altman S (1975) Partial purification and properties of an endoribonuclease isolated from human KB cells. J Biol Chem 250:1451–1459

Chamberlain JR, Lee Y, Lane WS et al (1998) Purification and characterization of the nuclear RNase P holoenzyme complex reveals extensive subunit overlap with RNase MRP. Genes Dev 12:1678–1690

Chen D, Huang S (2001) Nucleolar components involved in ribosome biogenesis cycle between the nucleolus and nucleoplasm in interphase cells. J Cell Biol 153:169–176

Clayton DA (2001) A big development for a small RNA. Nature 410:29–31

Clemente-Blanco A, Mayán-Santos M, Schneider DA et al (2009) Cdc14 inhibits transcription by RNA polymerase I during anaphase. Nature 458:219–222

Cohen A, Reiner R, Jarrous N (2003) Alterations in the intracellular level of a protein subunit of human RNase P affect processing of tRNA precursors. Nucleic Acids Res 31:4836–4846

Coughlin DJ, Pleiss JA, Walker SC et al (2008) Genome-wide search for yeast RNase P substrates reveals role in maturation of intron-encoded box C/D small nucleolar RNAs. Proc Natl Acad Sci USA 105:12218–12223

Donze D, Kamakaka RT (2001) RNA polymerase III and RNA polymerase II promoter complexes are heterochromatin barriers in Saccharomyces cerevisiae. EMBO J 20:520–531

Dundr M, Hoffmann-Rohrer U, Hu Q et al (2002) A kinetic framework for a mammalian RNA polymerase in vivo. Science 298:1623–1626

Eder PS, Kekuda R, Stolc V et al (1997) Characterization of two scleroderma autoimmune antigens that copurify with human ribonuclease P. Proc Natl Acad Sci USA 94:1101–1106

Ellis JC, Barnes J, Brown JW (2007) Is Alba an RNase P subunit? RNA Biol 4:169–172

Frank DN, Adamidi C, Ehringer MA et al (2000) Phylogenetic-comparative analysis of the eukaryal ribonuclease P RNA. RNA 6:1895–1904

Gorski SA, Snyder SK, John S et al (2008) Modulation of RNA polymerase assembly dynamics in transcriptional regulation. Mol Cell 30:486–497

Grummt I (2003) Life on a planet of its own: regulation of RNA polymerase I transcription in the nucleolus. Genes Dev 17:1691–1702

Guerrier-Takada C, Eder PS, Gopalan V, Altman S (2002) Purification and characterization of Rpp25, an RNA-binding protein subunit of human ribonuclease P. RNA 8:290–295

Hada K, Nakashima T, Osawa T, Shimada H, Kakuta Y, Kimura M (2008) Crystal structure and functional analysis of an archaeal chromatin protein Alba from the hyperthermophilic archaeon Pyrococcus horikoshii OT3. Biosci Biotechnol Biochem 72:749–758

Hall TA, Brown JW (2002) Archaeal RNase P has multiple protein subunits homologous to eukaryotic nuclear RNase P proteins. RNA 8:296–306

Hartmann E, Hartmann RK (2003) The enigma of ribonuclease P evolution. Trends Genet 19:561–569

Hassan AH, Neely KE, Worman JL (2001) Histone acetyltransferase complexes stabilize swi/snf binding to promoter nucleosomes. Cell 104:817–827

Holzmann J, Frank P, Löffler E et al (2008) RNase P without RNA: identification and functional reconstitution of the human mitochondrial tRNA processing enzyme. Cell 135:462–474

Ishiguro A, Kassavetis GA, Geiduschek EP (2002) Essential roles of Bdp1, a subunit of RNA polymerase III initiation factor TFIIIB, in transcription and tRNA processing. Mol Cell Biol 22:3264–3275

Jacobson MR, Cao LG, Taneja K, Singer RH et al (1997) Nuclear domains of the RNA subunit of RNase P. J Cell Sci 110:829–837

Jarrous N, Eder PS, Guerrier-Takada C, Hoog C, Altman S (1998) Autoantigenic properties of some protein subunits of catalytically active complexes of human ribonuclease P. RNA 4:407–417

Jarrous N, and Altman S (2001) Human Ribonuclease P. Methods Enzyml 342:93–100

Jarrous N (2002) Human ribonuclease P: subunits, function, and intranuclear localization. RNA 8:1–7

Jarrous N, Reiner R (2007) Human RNase P: a tRNA-processing enzyme and transcription factor. Nucleic Acids Res 35:3519–3524

Jarrous N, Wolenski JS, Wesolowski D et al (1999) Localization in the nucleolus and coiled bodies of protein subunits of the ribonucleoprotein ribonuclease P. J Cell Biol 146:559–572

Jarrous N, Reiner R, Wesolowski D et al (2001) Function and subnuclear distribution of Rpp 21, a protein subunit of the human ribonucleoprotein ribonuclease P. RNA 7:1153–1164

Jiang T, Altman S (2001) Protein-protein interactions with subunits of human nuclear RNase P. Proc Natl Acad Sci USA 98:920–925

Jiang T, Altman S (2002) A protein subunit of human RNase P, Rpp 14, and its interacting partner, OIP2, have 3′→5′ exoribonuclease activity. Proc Natl Acad Sci USA 99:5295–5300

Jiang T, Guerrier-Takada C, Altman S (2001) Protein-RNA interactions in the subunits of human nuclear RNase P. RNA 7:937–941

Kikovska E, Svärd SG, Kirsebom LA (2007) Eukaryotic RNase P RNA mediates cleavage in the absence of protein. Proc Natl Acad Sci USA 104:2062–2067

Koski RA, Bothwell AL, Altman S (1976) Identification of a ribonuclease P-like activity from human KB cells. Cell 9:101–116

Kovrigina E, Wesolowski D, Altman S (2003) Coordinate inhibition of expression of several genes for protein subunits of human nuclear RNase P. Proc Natl Acad Sci USA 100:1598–1602

Li Y, Altman S (2001) A subunit of human nuclear RNase P has ATPase activity. Proc Natl Acad Sci USA 98:441–444

Li Y, Altman S (2003) A specific endoribonuclease, RNase P, affects gene expression of polycistronic operon mRNAs. Proc Natl Acad Sci USA 100:13213–13218

Lunyak VV (2008) Boundaries. Boundaries...Boundaries??? Curr Opin Cell Biol 20:281–287

Lusser A, Kadonaga JT (2003) Chromatin remodeling by ATP-dependent molecular machines. Bioessays 25:1192–1200

Lygerou Z, Pluk H, van Venrooij WJ et al (1996) hPop1: an autoantigenic protein subunit shared by the human RNase P and RNase MRP ribonucleoproteins. EMBO J 15:5936–5948

Mariner PD, Walters RD, Espinoza CA et al (2008) Human Alu RNA is a modular transacting repressor of mRNA transcription during heat shock. Mol Cell 29:499–509

Mann H, Ben-Asouli Y, Schein A, Moussa S. Jarrous N (2003) Eukaryotic RNase P: Role of RNA and protein subunits of a primordial catalytic ribonucleoprotein in RNA-based catalysis. Mol Cell 12:925–935

Marquez SM, Chen JL, Evans D et al (2006) Structure and function of eukaryotic ribonuclease P RNA. Mol Cell 24:445–456

Mohanty BK, Kushner SR (2007) Ribonuclease P processes polycistronic tRNA transcripts in Escherichia coli independent of ribonuclease E. Nucleic Acids Res 35:7614–7625

Mohanty BK, Kushner SR (2008) Rho-independent transcription terminators inhibit RNase P processing of the secG leuU and metT tRNA polycistronic transcripts in Escherichia coli. Nucleic Acids Res 36:364–375

Pagano A, Castelnuovo M, Tortelli F et al (2007) New small nuclear RNA gene-like transcriptional units as sources of regulatory transcripts. PLoS Genet 3:e1

Prieto JL, McStay B (2008) Pseudo-NORs: a novel model for studying nucleoli. Biochim Biophys Acta 1783:2116–2123

Randau L, Schröder I, Söll D (2008) Life without RNase P. Nature 453:120–123

Reiner R, Ben-Asouli Y, Krilovetzky I et al (2006) A role for the catalytic ribonucleoprotein RNase P in RNA polymerase III transcription. Genes Dev 20:1621–1635

Reiner R, Krasnov-Yoeli N, Dehtiar Y et al (2008) Function and assembly of a chromatin-associated RNase P that is required for efficient transcription by RNA polymerase I. PLoS One 3:e4072

Robertson HD, Altman S, Smith JD (1972) Purification and properties of a specific Escherichia coli ribonuclease which cleaves a tyrosine transfer ribonucleic acid precursor. J Biol Chem 247:5243–5251

Rosenblad MA, López MD, Piccinelli P et al (2006) Inventory and analysis of the protein subunits of the ribonucleases P and MRP provides further evidence of homology between the yeast and human enzymes. Nucleic Acids Res 34:5145–5156

Salavati R, Panigrahi AK, Stuart KD (2001) Mitochondrial ribonuclease P activity of Trypanosoma brucei. Mol Biochem Parasitol 115:109–117

Samanta MP, Tongprasit W, Sethi H, Chin CS, Stolc V (2006) Global identification of noncoding RNAs in Saccharomyces serevisiae by modulating an essential RNA processing pathway. Proc Natl Acad Sci USA 103:4192–4197

Smith CL, Peterson CL (2005) ATP-dependent chromatin remodeling. Curr Top Dev Biol 65:115–148

Sunwoo H, Dinger ME, Wilusz JE, Amaral PP, Mattick JS, Spector DL (2009) MEN ε/β nuclear retained non-coding RNAs are up-regulated upon muscle differentiation and are essential components of paraspeckles. Genome Res 19:347–359

Terada A, Honda T, Fukuhara H et al (2006) Characterization of the archaeal ribonuclease P proteins from Pyrococcus horikoshii OT3. J Biochem 140:293–298

Tsai HY, Pulukkunat DK, Woznick WK et al (2006) Functional reconstitution and characterization of Pyrococcus furiosus RNase P. Proc Natl Acad Sci USA 103:16147–16152

van Eenennaam H, Pruijn GJ, van Venrooij WJ (1999) hPop4: a new protein subunit of the human RNase MRP and RNase P ribonucleoprotein complexes. Nucleic Acids Res 27:2465–2472

van Eenennaam H, Lugtenberg D, Vogelzangs JH et al (2001a) hPop5, a protein subunit of the human RNase MRP and RNase P endoribonucleases. J Biol Chem 276:31635–31641

van Eenennaam H, van der Heijden A, Janssen RJ et al (2001b) Basic domains target protein subunits of the RNase MRP complex to the nucleolus independently of complex association. Mol Biol Cell 12:3680–3689

Welting TJ, Kikkert BJ, van Venrooij WJ et al (2006) Differential association of protein subunits with the human RNase MRP and RNase P complexes. RNA 12:1373–1382

Wilusz JE, Freier SM, Spector DL (2008) 3′ End Processing of a Long Nuclear-Retained Noncoding RNA Yields a tRNA-like Cytoplasmic RNA. Cell 135:919–932

Yang L, Altman S (2007). A noncoding RNA in Saccharomyces cerevisiae is an RNase P substrate. RNA 13:682–690

Zhang H, Altman S (2004) Inhibition of the expression of the human RNase P protein subunits Rpp 21, Rpp25, Rpp29 by external guide sequences (EGSs) and siRNA. J Mol Biol 342:1077–1083

Chapter 13
RNase P as a Drug Target

Dagmar K. Willkomm, Patrick Pfeffer, Klaus Reuter, Gerhard Klebe, and Roland K. Hartmann

13.1 Introduction

RNase P, particularly from bacteria, has elicited considerable interest as a drug target, for two main reasons: (1) The enzyme belongs to the small fraction (~7%) of gene products that are indispensable for bacterial viability (Schedl et al. 1974; Waugh and Pace 1990; Kobayashi et al. 2003; Gößringer et al. 2006; Wegscheid et al. 2006; Wegscheid and Hartmann 2006), and (2) the enzyme's architecture differs substantially in Bacteria and Eukarya: in Bacteria, the enzyme consists of an RNA subunit of almost protein-independent functionality; nevertheless, *in vivo* the RNA requires a small protein cofactor (~one-tenth of the mass of the RNA) for activity; in humans, the nuclear enzyme involves a structurally more simplified version of the RNA subunit that has become dependent on complexation with ten protein subunits (Jarrous and Altman 2001; Jarrous 2002), and the human mitochondrial enzyme has given up an RNA subunit, with its three protein subunits recruited from related and unrelated biochemical pathways (Holzmann et al. 2008). The modes how inhibitors could possibly interfere with the function of RNase P are diverse (see Fig. 13.1), partly owing to RNase P being composed of two classes of macromolecules.

Historically, the first inhibitor of bacterial RNase P described in the literature was puromycin (Vioque 1989; see Fig. 13.5a). Puromycin inhibited precursor tRNA (ptRNA) processing by *Escherichia coli* RNase P RNA (P RNA), binding to P RNA with an affinity (K_d ~ 1 mM) only one order of magnitude lower than the drug's interaction with the ribosome. Next, Kirsebom and coworkers analyzed several aminoglycosides in the ptRNA processing reaction catalyzed by *E. coli* P RNA (Mikkelsen et al. 1999). Among those tested, neomycin B inhibited the reaction most efficiently (50% inhibition of cleavage activity at 35 μM in the RNA-alone and 60 μM in the holoenzyme reaction), and suppression of Pb^{2+}-induced

D.K. Willkomm, P. Pfeffer, K. Reuter, G. Klebe, and R.K. Hartmann (✉)
Institut für Pharmazeutische Chemie, Philipps-Universität Marburg, Marbacher Weg 6, D-35037, Marburg, Germany
e-mail: roland.hartmann@staff.uni-marburg.de

F. Liu and S. Altman (eds.), *Ribonuclease P*, Protein Reviews 10,
DOI 10.1007/978-1-4419-1142-1_13,

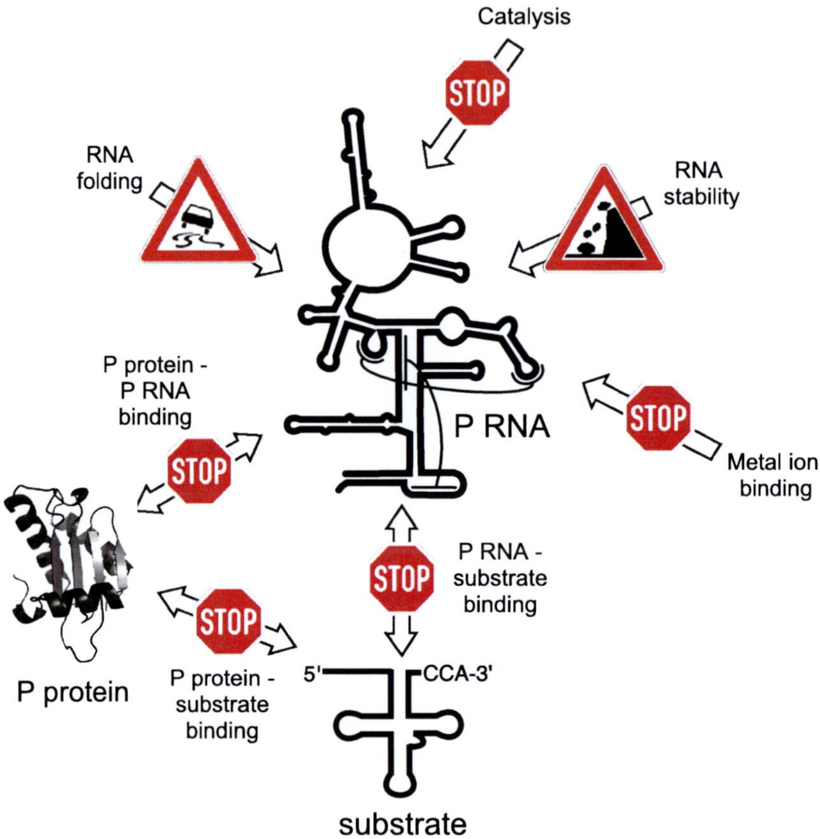

Fig. 13.1 Illustration of potential modes to interfere with bacterial RNase P function

cleavage of P RNA by neomycin B indicated that the aminoglycoside interferes with the binding of divalent metal ions to the RNA (Mikkelsen et al. 1999). Later, some or all of the six amino groups of neomycin B were converted to guanidinium groups or decorated with lysyl or arginyl groups to improve the inhibitory potential (Eubank et al. 2002; Kawamoto et al. 2008; Sect. 13.3). Apart from aminoglycosides such as neomycin B, paramomycin and kanamycin B, sisomycin and particularly 5-epi-sisomycin were observed to inhibit some RNAs at low micromolar or even submicromolar concentrations (Walter et al. 1999; Mikkelsen et al. 2001).

After the publication of the X-ray structure (Stams et al. 1998) of the *Bacillus subtilis* RNase P protein (P protein), two pharmaceutical companies became involved in exploring bacterial RNase P. Researchers at Smith Kline Beecham (now GlaxoSmithKline) reported an NMR structural analysis of the P protein from *Staphylococcus aureus* (Spitzfaden et al. 2000). This study substantiated conclusions drawn from the X-ray structure of the *B. subtilis* P protein (Stams et al. 1998) in conjunction with crosslinking results (Niranjanakumari et al. 1998), namely that the protein's central groove, the bottom of which is formed by its β-sheet, functions as a binding interface for single-stranded RNA as present in 5′-leader sequences of ptRNA substrates of RNase P (Spitzfaden et al. 2000). A second putative RNA binding epitope has been localized to helix $\alpha 2$, involving the most conserved region of bacterial P proteins and harbouring the so-called "RNR" motif; this region is thought to interact with P RNA (Stams et al. 1998; Niranjanakumari et al. 2007). Although these investigations laid the structural basis for drug discovery, no further respective efforts by GlaxoSmithKline have become public since then.

The second company that had bacterial RNase P on their agenda of drug targets was Message Pharmaceuticals. Screening compound libraries for inhibition of the *in vitro* assembled RNase P holoenzyme of *Neisseria gonorrhoeae* (US patent, Giordano et al. 2006; see also http://statusreports.atp.nist.gov/reports/95-01-0098.htm), this company introduced high-throughput screening into the field of RNase P: processing reactions, performed in microtiter plate format, were terminated by addition of stop buffer containing EDTA for Mg^{2+} chelation, as well as a DNA oligonucleotide with a 5′-terminal TAMRA fluorescent probe for hybridization to the complementary 10-nt long 5′-leader of the substrate, a precursor $tRNA^{Gln}$ from *Synechocystis* (Pascual and Vioque 1999). The extent of processing was then quantified by fluorescence polarization, exploiting the differences in molecular mobility of tRNA-linked and released 5′-leader. These enzyme activity assays identified the guanylhydrazones as potential inhibitors of bacterial RNase P (see Sect. 13.3). Some of these compounds selectively inhibited growth of certain bacteria, such as *S. aureus* or *Streptococcus pyogenes*. However, Message Pharmaceuticals suspended operations in early 2004.

Since these initial attempts, RNase P has been of continuous interest as a drug target, giving rise to several studies that follow up on the aminoglycosides, and explore other compounds as putative drugs. These studies are summarized in the following sections.

13.2 Antisense Inhibitors

The catalytic moiety of bacterial RNase P being a nucleic acid, inhibition of the enzyme by antisense oligonucleotide inhibitors provides a straightforward approach, once suitable target regions within the RNA have been identified. Four approaches have so far been pursued towards antisense inhibition of RNase P: (1) oligonucleotide-directed misfolding (Childs et al. 2003), (2) a rational approach addressing the L15/16 loop of type A P RNAs (Willkomm et al. 2003), (3) computer-based design according to rules established for targeting mRNAs (Willkomm et al. 2003), and (4) a SELEX strategy based on oligonucleotide-induced shifts in gel mobility of P RNA (own unpublished results).

Oligonucleotide-directed misfolding relies on the concept that antisense oligonucleotides present during transcription may direct an RNA into an inactive conformation by blocking contact sites relevant to the folding process. In the study by Childs et al. (2003), such antisense olignucleotides were added to an *in vitro* reaction transcribing *E. coli* P RNA in the presence of its protein cofactor. Among 32 12-mer DNAs complementary to consecutive regions of *E. coli* P RNA, one that was complementary to the L15/16 loop harboring the CCA binding site (Fig. 13.2a, nucleotides 289–300 of *E. coli* P RNA) was the most efficient inhibitor identified in this study, with an IC_{50} of 200 nM when added as a 2′-O-Me analog (IC_{50}, throughout text, is defined as the concentration of inhibitor at which product formation or cleavage activity is reduced by 50%). This inhibitor was shown to affect folding of J5/6, J3/4 and the L8 loop, according to structure probing with DEPC, and to be also effective on full-length renatured P RNA (IC_{50} of 3 μM). Considerable cotranscriptional interference of folding was also observed for oligonucleotides targeting nucleotides 61–72, 196–180, 181–192, 265–276, 283–294, and 295–306 of *E. coli* P RNA, with IC_{50} values of approx. 1 μM.

In an independent study, a rational approach (Willkomm et al. 2003) pinpointed the L15/16 loop of *E. coli* P RNA as a target region particularly amenable to antisense inhibitors. This region was initially chosen because it is exposed within the tertiary structure, forms Watson–Crick base pairs with the substrate, requiring it to be accessible for base pairing interactions, and because it is part of the catalytic core. Starting from an RNA hairpin construct that would allow an initial loop–loop interaction between oligonucleotide and P RNA, our successive optimization of the inhibitor (Willkomm et al. 2003; Gruegelsiepe et al. 2003) led to a single-stranded RNA 14-mer, fully complementary to nucleotides 291–304 of *E. coli* P RNA (Fig. 13.2a). This RNA oligonucleotide inhibited the RNA-alone reaction with an IC_{50} value of 2.2 nM, and showed a K_d of 0.7 nM for its target site, bound to P RNA over its entire length at the complementary site as predicted, and was sensitive to minor changes within the target sequence, thus qualifying for species-specific applications. Ribozyme inhibition was inferred to occur at four mechanistic levels: (1) direct blockage of base pairing of ptRNA 3′ NCCA ends to the P15 loop, (2) perturbation of the coordination of catalytically relevant Mg^{2+} [as a consequence of (1)] (Brännvall et al. 2003), (3) disruption of the P15 helix as part of the catalytic core, and (4) arresting the P RNA in an inactive conformation, in line with the study of Childs et al. (2003). Antisense inhibition via the nt 291–304 region of *E. coli* P RNA was also seen in reactions catalyzed by the holoenzyme reconstituted with the *E. coli* P protein (Willkomm et al. 2003).

The approach of computer-aided design of antisense inhibitors (Willkomm et al. 2003) employed rules deduced from designing antisense oligonucleotides that target mRNAs (Patzel et al. 1999; Kretschmer-Kazemi Far et al. 2001). After choosing potentially favourable target sites according to available secondary and tertiary structure models of *E. coli* P RNA, DNA antisense oligonucleotides of 18–20 nucleotides were designed to have their 3′ or 5′ end in the center of the putatively accessible target region. From *in vitro* RNA-alone reactions in the presence of these oligonucleotides, oligonucleotides targeting the P10-L11/12 region emerged as the most efficient inhibitors (Fig. 13.2a), with inhibition efficiency further increased when RNA variants of these DNA oligonucleotides were used.

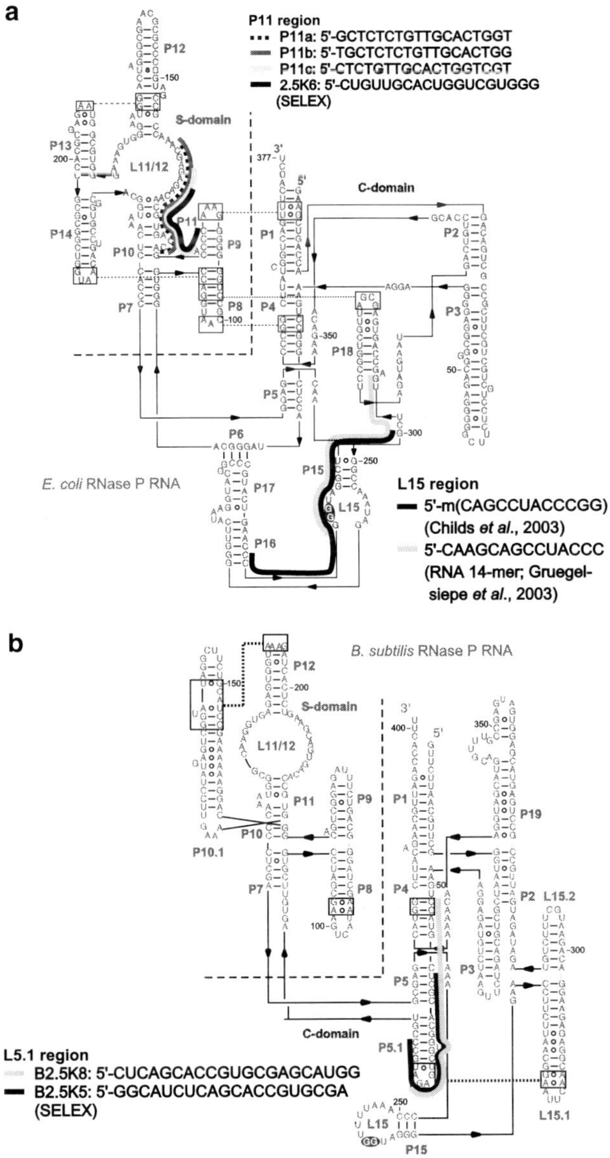

Fig. 13.2 Identified antisense inhibitors of bacterial P RNA from (**a**) *E. coli* and (**b**) *B. subtilis*. P RNA sequences complementary to the inhibitors are indicated by thick lines (*black, gray-shaded or dotted*) within the secondary structure (represented according to Massire et al. 1998), with the sequences of the inhibitors given adjacent to their target sites. Note that for the inhibitors identified by SELEX (unpublished results), only the part of the oligonucleotide sequence complementary to P RNA is given; during selection these core sequences were flanked by 18 and 28 nucleotides of unvaried linker sequence. The tandem G residues within L15, which interact with the ptRNA 3′-CCA, are highlighted

Whereas all previous approaches had been restricted to or were successful only for P RNAs of the structural type A, we recently collaborated with J. Kjems (University of Aarhus) to apply a SELEX strategy to both types of P RNA, A and B (unpublished results). For this purpose, a plasmid encoding the P RNA gene was fragmented to generate a T7 expression library in a multistep procedure involving DNA linker ligation, *Mme I* restriction endonuclease digestion for fragment size standardization and PCR (Jakobsen et al. 2004). T7 transcripts from this library included a central 20-nt stretch of varying sequence derived from the plasmid, flanked with 18 and 28 nt of unvaried linker sequence on the 5′- and 3′-side, respectively. Our selection was then based on excising bands of P RNA that were shifted in the presence of the T7 transcript library on native polyacrylamide gels at Mg^{2+} concentrations of 2.5 or 5 mM. Excised library transcripts bound to P RNA were amplified by RT-PCR via their 5′- and 3′-terminal linker sequences, followed by the next round of *in vitro* transcription and gel mobility shift assay. After 3–4 rounds, the enriched pool was sequenced. We thus identified an inhibitory 19-mer sequence complementary to nucleotides 115–133 of *E. coli* P RNA (Fig. 13.2a) – interestingly the same target region was identified in our previous computer-based screening. Applied to *B. subtilis* P RNA as a representative of the structural type B, most prominently two overlapping oligonucleotide binders emerged, complementary to the P5.1 stem-loop (nucleotides 52–72 and 58–77 of *B. subtilis* P RNA; Fig. 13.2b). Unique to type B RNAs, this structural element is known to form a long-range interaction with L15.1, according to the crystal structure of P RNA from *Bacillus stearothermophilus* (Kazantsev et al. 2005). Indeed, a 20-mer RNA complementary to nucleotides 58–77 (oligo B2.5K5, Fig. 13.2b), displaying an IC_{50} of ~40 nM in the RNA-alone reaction and of ~300 nM in the holoenzyme reaction, disrupted this interaction, and in addition, perturbed P RNA structure upstream of its binding site. Further, this had a long-range effect on L15, reducing the accessibility of G258/G259 involved in base pairing with ptRNA 3′-CCA ends (unpublished results).

An important issue when applying oligonucleotides as therapeutic agents *in vivo* is rapid degradation of DNA and RNA oligonucleotides. To circumvent this problem, “third generation” antisense agents have been devised that are highly resistant to nucleases (Wahlestedt et al. 2000; Good et al. 2001; Kurreck 2003): LNA (locked nucleic acids, with a methylene bridge between the 2′-oxygen and 4′-carbon atom that preorganizes and fixes the sugar pucker in A-type helical conformation) and PNA (peptide nucleic acid, with the nucleobases attached via methylene carbonyl linkages to an uncharged peptide backbone). PNA has been reported to be able to invade stable stem-loop structures not accessible to natural oligonucleotides (Egholm et al 1993). The thermal stability of LNA-RNA duplexes exceeds that of RNA–RNA helices (Singh and Wengel 1998), which entails the capacity of LNA oligonucleotides to invade stable RNA secondary structures (Nulf and Corey 2004).

To assess the effects of the chemical nature of the antisense agents on inhibition, we compared all-LNA (all residues LNA), all-DNA, all-PNA and all-RNA variants of the 14-mer inhibitor complementary to nucleotides 291–304 of *E. coli* P RNA, devised in our earlier study (Gruegelsiepe et al. 2006). IC_{50} values were similar for

RNA and LNA (around 3 nM), followed by PNA (13 nM) and DNA (25 nM), the fairly high IC_{50} for PNA possibly due to some aggregation of PNA oligomers. Helix stability was highest for the all-LNA 14-mer, at the cost of specificity, which was substantially lower for LNA than for PNA or RNA. The association rate k_{on} was highest for PNA ($22\times10^4 M^{-1} s^{-1}$) vs. $2\times10^4 M^{-1} s^{-1}$ for DNA, $7\times10^4 M^{-1} s^{-1}$ for RNA, and $9\times10^4 M^{-1} s^{-1}$ for LNA, at negligible dissociation rates for all except DNA. The PNA 14-mer also effectively inhibited the *E. coli* RNase P holoenzyme (Gruegelsiepe et al. 2006; Fig. S4 therein).

Finally, the major hurdle when applying antisense inhibitors to live bacteria is their uptake into the bacterial cell. Here, conjugates of PNA oligonucleotides with invasive peptides, derived from antimicrobial peptides of eukaryal innate immune systems, have been shown to enter Gram-negative and Gram-positive bacteria, and to specifically interact with the target RNA in the cell (Good et al. 2001; Nekhotiaeva et al. 2004). Following this concept, we appended a decamer peptide $(KFF)_3K$ to the 14-mer PNA targeting *E. coli* P RNA, using either 2-amino-ethoxy-2-ethoxy acetic acid (AEEA) or monoglycine (G) for linkage (Gruegelsiepe et al. 2006). At 10 μM concentration of PNA-peptide conjugate in the medium and 180 min incubation time, all cells of the two *E. coli* test strains (K12 and AS19) were killed when the conjugate was built with the G-linker (PNA-G-peptide); inhibition effects were weaker with the AEEA conjugates. Sequence specificity of the approach was demonstrated in two ways: (1) by using a scrambled version of the PNA-G-peptide, and (2) via growth rescue of PNA-G-peptide-treated *E. coli* cells by simultaneous expression of a plasmid-encoded *B. subtilis* P RNA, which was not targeted by the antisense inhibitor owing to sequence and structure variation in the L15 target region. Also, as deduced from RT-PCR experiments, the cellular levels of *E. coli* P RNA were reduced in *E. coli* AS19 cells expressing *B. subtilis* P RNA and treated with the PNA-G-peptide. Since such a reduction was not observed when the same bacteria were treated with the scrambled PNA-G-peptide, it can be concluded that binding of the antisense inhibitor to *E. coli* P RNA induced the RNA's degradation.

In conclusion, antisense inhibition appears to be a feasible approach for targeting P RNA. Accessible target regions have by now been identified in P RNAs of structural type A as well as B, and sequence specificity of inhibition has been demonstrated. Studies so far have shown that optimization of antisense inhibitors in terms of length and precise positioning is an important issue. While all types of antisense agents, DNA, RNA, LNA and PNA act as inhibitors, DNA is the least effective of all *in vitro*. For *in vivo* applications, to date, PNAs might be considered as the agent of choice due to their nuclease resistance, improved bacterial uptake and increased rates of association with target nucleic acids, the latter two features deriving from the neutral backbone of PNA. Yet LNA, owing to its enhancement of duplex stability, may open perspectives to reduce the length of antisense oligonucleotides in order to improve entry into bacteria. There is encouraging evidence that short LNA-oligomers (8- to 10-mers) maintain sufficient target selectivity without toxic effects (discussed in Gruegelsiepe et al. 2006). Lastly, linkage of the antisense agent to invasive peptides helps to overcome the uptake barrier of the bacterial cell envelope, where not only the type of invasive peptide, but also the nature of the linker appears to play an important role. Despite this progress brought about by the exploration of PNA, bacterial uptake of antisense inhibitors is still inefficient, a large hurdle to be taken before considering this approach for therapeutic purposes.

13.3 Aminoglycosides and Arginine Derivatives

Aminoglycosides bind to RNA molecules via (1) electrostatic interactions between their protonated amino groups and the RNA phosphate backbone, and (2) through shape complementarity combined with the capacity to form specific hydrogen bonds and van der Waals contacts with their RNA targets (Walter et al. 1999). The conformational flexibility of aminoglycosides further permits adaptation to different RNA target geometries. For example, neomycin-type aminoglycosides bind to the bacterial ribosomal A-site in a compact conformation, whereas a more extended conformer is present in the complex with TAR (Tor 2006). In case of the hammerhead ribozyme, a striking overlap in space between the charged amino groups of the aminoglycosides and the metal binding sites in the hammerhead was uncovered, suggesting that these compounds complement the negative electrostatic potential created by certain three-dimensional RNA folds (Hermann and Westhof 1998). Despite the considerable promiscuity of aminoglycosides, attributable to their highly charged nature and their conformational flexibility, they primarily target the ribosomal A-site. This has been explained by the A-site's encapsulating architecture and its function as a riboswitch of fundamental importance for the cell. Notably, the devastating effect of aminoglycosides on bacterial protein synthesis is achieved despite moderate target affinity, which nonetheless suffices because the aminoglycosides do not have to compete with endogenous high affinity ligands (Tor 2006). With their conformational adaptability and a core decorated with positve charges, the aminoglycosides are a natural paradigm of successful drug design against RNA. This has stimulated numerous attempts to develop aminoglycoside derivatives with diversified target specificity.

Aminoglycoside-arginine conjugates (AACs) were originally designed to disrupt interactions between the HIV regulatory proteins Tat and Rev, both of which contain arginine-rich RNA binding domains, and their respective RNA targets. The peptidomimetic AACs were found to be 2–4 orders of magnitude more effective than their parental aminoglycosides in disrupting the viral RNA-protein target complexes (Litovchick et al. 2001). Neomycin B (NeoB) inhibiting *E. coli* P RNA and holoenzyme in the micromolar concentration range (Mikkelsen et al. 1999) and the presence of an arginine-rich region, including the highly conserved "RNR" motif, in bacterial P proteins prompted Gopalan and coworkers in collaboration with the Lapidot group to explore AACs as inhibitors of bacterial RNase P. They initially analyzed two AACs, a hexa-arginine derivative of NeoB (termed NeoR, renamed as NeoR6 in the following, Fig. 13.3a) and a tri-arginine derivative of gentamycin

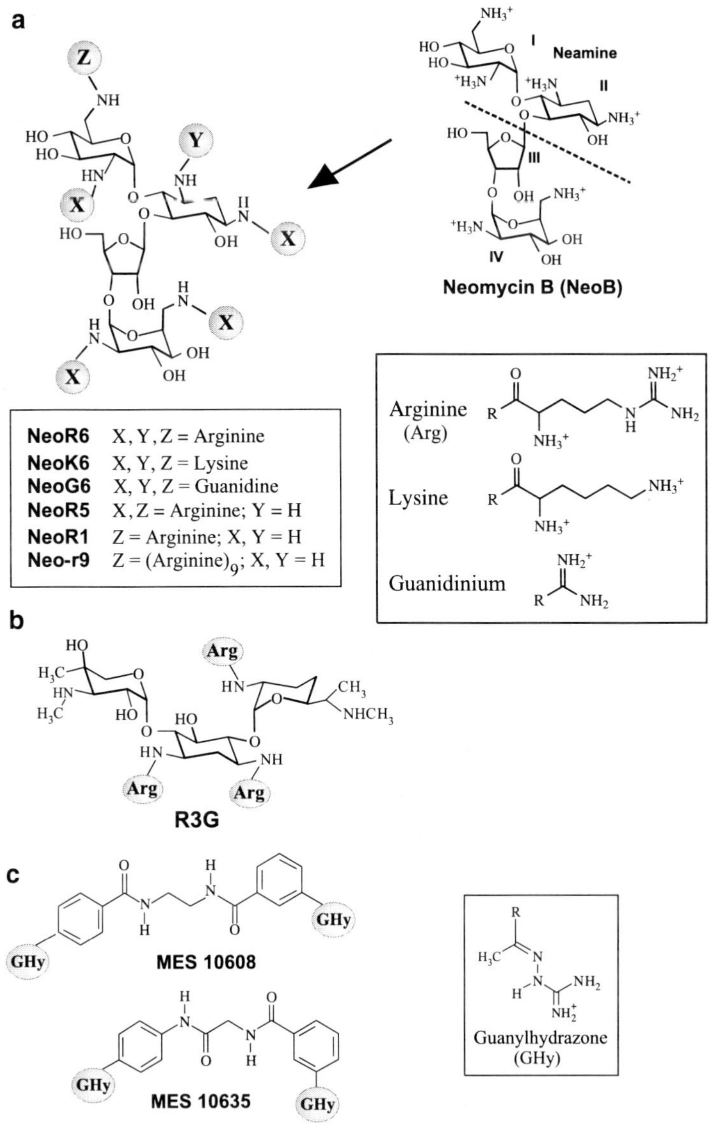

Fig. 13.3 Structures of (**a**) neomycin B and derivatives, (**b**) R3G, a tri-arginine derivative of gentamycin, and (**c**) two examples of guanylhydrazones, compounds MES 10608 and MES 10635. (**a**) In the neomycin B (NeoB) structure (top right), the individual rings are numbered as I–IV and the dashed line depicts the neamine part (rings I and II) of neomycin. For the aminoglycoside derivatives (left, specified in the box below), the arginine, lysine or guanidine moieties are attached at the positions highlighted by gray spheres and designated as X, Y and Z; in Neo-r9, the small "r" indicates D-arginine instead of L-arginine (indicated by "R"); Neam-r9 (see text), as Neo-r9 but lacking rings III and IV. The box on the right displays the molecular structure of the arginine, lysine and guanidinium groups; (**c**) The gray spheres marked as "GHy" represent the guanylhydrazone moieties depicted in the box on the right. For further details, see text

(R3G, Fig. 13.3b), for their inhibitory potential in ptRNA processing reactions catalyzed by several bacterial RNase P enzymes (Eubank et al. 2002). NeoR6 and R3G inhibited *E. coli* RNase P with IC_{50} values of about 0.1 and 0.3 μM, respectively; similar efficacies were seen with other bacterial holoenzymes, although some differences in sensitivity among the individual bacterial enzymes became evident. As a major outcome, NeoR6 was found to inhibit *E. coli* RNase P about 500-fold more effectively than the parental aminoglycoside neomycin B (Mikkelsen et al. 1999; Eubank et al. 2002). IC_{50} values for the processing reaction catalyzed by partially purified human nuclear RNase P were at least 10-fold higher than those for *E. coli* RNase P, demonstrating selectivity for bacterial enzymes. In addition, the authors carried out several controls to elucidate the mechanism of AAC-mediated inhibition. They found that: (1) both NeoR6 and R3G inhibit the RNA-alone in addition to the holoenzyme reaction, arguing against AAC-mediated displacement of the P protein from P RNA as the major mode of inhibition; this is in line with the observation that addition of NeoR6 or R3G to P RNA before or after assembly of the holoenzyme did not change inhibitory strength; the possibility of P protein displacement by AACs was definitely discarded in a follow-up study, where inhibition of *E. coli* RNase P by AACs could not be relieved by increasing the P protein concentration, and where patterns of P protein-mediated protection of P RNA from cleavage by RNases T1 and V1 remained unaffected by the presence of AACs (Kawamoto et al. 2008); (2) the presence of 1 μM of an 18-meric oligo(A) RNA (10-fold excess over the ptRNA concentration in the processing assay) failed to mitigate the inhibitory strength of NeoR and R3G, excluding the possibility that the compounds exerted their effects by substrate masking through nonspecific interactions with the single-stranded 5'-leader of ptRNA; (3) furthermore, the presence of 1 mM arginine remained without effect on *E. coli* RNase P activity, indicating that the structural framework of the aminoglycoside core that arranges the arginines is essential for AAC efficacy (Eubank et al. 2002). In conclusion, these findings concertedly suggest that the AACs target the bacterial P RNA and, like the parental NeoB (Mikkelsen et al. 1999), displace important metal ions.

In their follow-up investigation, Gopalan and coworkers compared RNase P inhibition by hexa-guanidinium (NeoG6) and -lysyl (NeoK6) conjugates of NeoB next to the penta-arginine derivative NeoR5 (Fig. 13.3a). IC_{50} values for inhibition of the *E. coli* RNase P holoenzyme were determined as ca 400, 6, 3 and 0.5 μM for NeoB, NeoG6, NeoK6 and NeoR5, respectively (Kawamoto et al. 2008). For further comparison, NeoR1 showed an IC_{50} of 4 μM in the same reaction (Berchanski and Lapidot 2008). Neither the presence of 1 mM arginine, lysine or guanidine, nor addition of poly(A) or elevated ptRNA concentrations affected the inhibitory efficacy of the tested AACs in the *E. coli* RNase P holoenzyme reaction. Inhibition by the most potent of the tested inhibitors, NeoR5, was largely attenuated at elevated Mg^{2+} concentrations: 1 μM NeoR5 reduced RNase P activity to ca 25% of the activity in its absence under conditions of 10 mM Mg^{2+}, but inhibition was essentially abrogated at 30 mM Mg^{2+}. Similar trends were seen with NeoG6 and NeoK6, substantiating the earlier findings that AACs compete with Mg^{2+} ions for binding to P RNA (Kawamoto et al. 2008). The 6 to 12 fold lower inhibitory potency of

NeoK6 and NeoG6 relative to NeoR5 permits to extract some important inferences regarding the nature of AAC-target RNA interactions: (1) the adverse effect caused by shortening of side chains to only the guanidinium functional groups in NeoG6 vs. NeoR5 illustrates the importance of side-chain flexibility, length and three-dimensional spacing of guanidinium groups; (2) NeoK6 is very similar to NeoR5 in terms of side-chain length and flexibility, but exposes terminal protonated amines instead of the planar, resonance-stabilized guanidinium moieties, suggesting the Arg side chains to be involved in H bonding and possibly π–π interactions in addition to electrostatic interactions (Kawamoto et al. 2008).

The selectivity of AACs was investigated by comparing their effect on reconstituted archaeal RNase P holoenzymes from *Methanothermobacter thermautotrophicus* (*Mth*), *Pyrococcus furiosus* (*Pfu*) and *Methanocaldococcus jannaschii* (*Mja*) RNase P; *Mth* and *Pfu* represent the P RNA type A architecture, similar to that of *E. coli*, whereas *Mja* is a prototype for the archaeal type M, whose more simplified RNA structure resembles that of eukaryotic P RNAs. NeoR5 inhibited the *Mth* and *Pfu* enzymes (reduction of *Mth* activity to ca 40% at 10 μM NeoR5), but even slightly stimulated the *Mja* enzyme (to 120%) at the same concentration. It was previously shown that mutations in the tRNA 3′-CCA binding loop L15 of *E. coli* P RNA, which weaken Mg^{2+} binding to L15, cause a threefold increase in the IC_{50} for NeoB (Mikkelsen et al. 1999). This finding pinpointed the L15 loop as a candidate target site for AACs. The absence of a bacterial-like L15 loop in the archaeal *Mja* RNA would explain the failure to inhibit this enzyme by AACs. Based on these considerations, a deletion variant of *E. coli* P RNA (ΔL15/P16/P17), lacking the L15/P16/P17 module, was constructed. However, the mutant holoenzyme was inhibited with the same efficiency by NeoR as the wild-type enzyme (Kawamoto et al. 2008). This finding suggested that either the L15 loop is not a major AAC target site or multiple AAC binding sites exist on bacterial P RNA, making inhibition insensitive to the loss of a single target site.

The Lapidot group (Berchanski and Lapidot 2008) further explored the structure-function relationships of AACs in comparison to a new set of aminoglycoside-polyarginine conjugates (APACs). Those included Neam-r9 and Neo-r9, neamine and NeoB derivatives with a D-arginine nonapeptide conjugated to ring I (Fig. 13.3a). The authors pursued a bioinformatic multistep docking approach to predict binding modes of AACs and APACs to *B. subtilis* P RNA, using the 3D model developed by the Westhof and Gopalan groups (Tsai et al. 2003) because of its good aggreement with the crystallographic data (Kazantsev et al. 2005; Torres-Larios et al. 2005) and since it avoids complications of the crystal structures owing to unresolved regions and crystal packing artifacts. The docking approach involved (1) energy minimization of AAC and APAC 3D structures, (2) geometric, geometric-electrostatic and geometric-hydrophobic Molfit scans, (3) defining the intersection from the three scans to predict putative binding sites, and (4) final refinement of predicted complexes by Discover3 (Berchanski and Lapidot 2008). Based on the docking results, three possible mechanisms of RNase P inhibition by AACs and APACs were suggested: competition with (1) the P protein and (2) ptRNA for binding to the P2/3/4-J19/4 region, which may entail displacement of Mg^{2+} ions from the P4 helix, and (3) interaction with the P15 region to displace catalytically important Mg^{2+} ions and to interfere with P protein and ptRNA binding to P RNA. The intermolecular energy was about tenfold lower for the analyzed APACs relative to the AACs, indicating stronger interactions of APACs than AACs with P RNA. Also, the preference for the aforementioned binding sites was less pronounced in the case of the AACs, and even more disperse binding was predicted for NeoB.

In this context, it is interesting to note that the apical guanylhydrazone moieties of the compounds investigated by Message Pharmaceuticals (see also Sect. 13.1) are chemically closely related to the arginine moieties of the AACs and APACs (see Fig. 13.3a and c). Thus, the mode of RNase P inhibition may be similar for the guanylhydrazones and the AACs/APACs.

In summary, the AACs and APACs (and possibly compounds with multiple guanylhydrazone moieties) appear to be promising inhibitors of bacterial RNase P, with up to several orders of magnitude higher affinities compared to their parental aminoglycoside core structures. Even selectivities have emerged: eukaryotic-like RNase P enzymes are much less sensitive to AACs than their bacterial counterparts. It will be intriguing to see how this class of compounds affects the second architectural type of bacterial RNase P, type B, as present in the *Firmicutes* such as *S. aureus*.

One potential strategy to increase the selectivity of AACs and APACs may be to constrain their conformation by covalently linking individual rings of the aminoglycoside core, as has been done for neomycin-type antibiotics (Tor 2006). Assuming that the affinity and selectivity of AACs and APACs can be further improved, one eagerly awaits studies on their uptake into bacterial and eukaryotic cells and the phenotypes caused, using assays such as the curing of HeLa cell cultures from *E. coli* infections (Good et al. 2001). This will help to solve critical questions, such as AAC and APAC selectivity at physiological Mg^{2+} concentrations (ca 1 mM free Mg^{2+}), and in a cellular context with many potential RNA targets being present simultaneously. Also, it will be interesting to see if toxic effects, such as the oto- and nephrotoxicity of neomycin B, are mitigated or exacerbated for these compounds.

13.4 Structure-Based Drug Design Using the Bacterial P Protein as Target

A virtual screening was performed using the crystal structure of the P protein from *B. subtilis* (Stams et al. 1998). Two crystal structures and one NMR structure of bacterial P proteins have revealed a conserved three-dimensional architecture despite low primary sequence conservation (Stams et al. 1998; Spitzfaden et al. 2000; Kazantsev et al. 2003). This is in line with the observation that the vast majority of bacterial P proteins are capable of functionally replacing the *B. subtilis* P protein *in vivo* (Gößringer and Hartmann 2007). Based on the protein's topology, the conserved central cleft known to interact with 5′-leader sequences of ptRNA substrates appeared to us as a promising binding region for drug-like molecules, because it is the most pronounced cavity of the molecule and supports an essential RNA-protein interaction. A potential drawback of this strategy may arise from the fact that small compound inhibitors will likely have to compete with the 5′-leaders of cellular ptRNAs for binding to the central cleft. Since structural information on drug-like molecules in complex with bacterial P proteins is not available, the drug-like subset of the ZINC database (Irwin and Shoichet 2005) was chosen as a good starting point to explore possible binding candidates. The docking program GOLD (Verdonk et al. 2003) was used to dock ~8 million compounds into the central cleft motif of the protein. Default parameters were set as suggested for docking calculations with GOLD. Subsequently, a rescoring of all computed binding geometries was performed using the scoring function DrugScore^{CSD} (Velec et al. 2005).

Figure 13.4a and b gives an overview of the protein and the binding site selected for the virtual screening study. Figure 13.4c shows the top scoring compound in its predicted geometry, with a corresponding two-dimensional structural interaction diagram depicted in Fig. 13.4e. Visual inspection of the top 15 scoring compounds revealed that most potential binders featured a piperazine-like core fragment (Fig. 13.4d) which was present in 12 of the predicted candidates. Overall, the binding geometries of the suggested molecules showed that they are similarly aligned within the central cleft of the protein, which indicates a good reliability of the computed results. Since all compounds from the ZINC database are commercially available, experimental testing of the suggested candidates is underway.

13.5 Inhibitors of RNase P from Eukaryotic Pathogens

Several antibiotics and other small ligands have been analyzed for their effect on RNase P. As a putative model system for eukaryotic pathogens, many of these screenings have been performed on partially purified nuclear RNase P from the slime mold *Dictyostelium discoideum*.

Among tested aminoglycosides, NeoB inhibited *D. discoideum* RNase P most effectively, an observation also made for bacterial RNase P (Mikkelsen et al. 1999). The aminoglycosides were found to act as classical noncompetitive inhibitors of *D. discoideum* RNase P, with K_i values (as defined by enzyme kinetics) of ca 140 µM, 730 µM, 1.1 mM, 1.4 mM and 1.9 mM for NeoB, gentamycin, tobramycin, paromomycin and kanamycin, respectively (Tekos et al. 2000, 2004). Furthermore, aminoglycosides competed with Mg^{2+} ions for the same binding sites on the enzyme, a feature that was also observed for nuclear RNase P from human epidermal keratinocytes (Tekos et al. 2003) as well as for bacterial RNase P (Mikkelsen et al. 1999).

The nucleoside analogs puromycin, blasticidin S and amicetin (Fig. 13.5a) inhibit the peptidyl transferase activity of both eukaryotic and prokaryotic ribosomes. For RNase P from *D. discoideum*, an IC_{50} of 5.5 mM for puromycin and 5 mM for amicetin was reported; at 10 mM blasticidin S, the maximum concentration

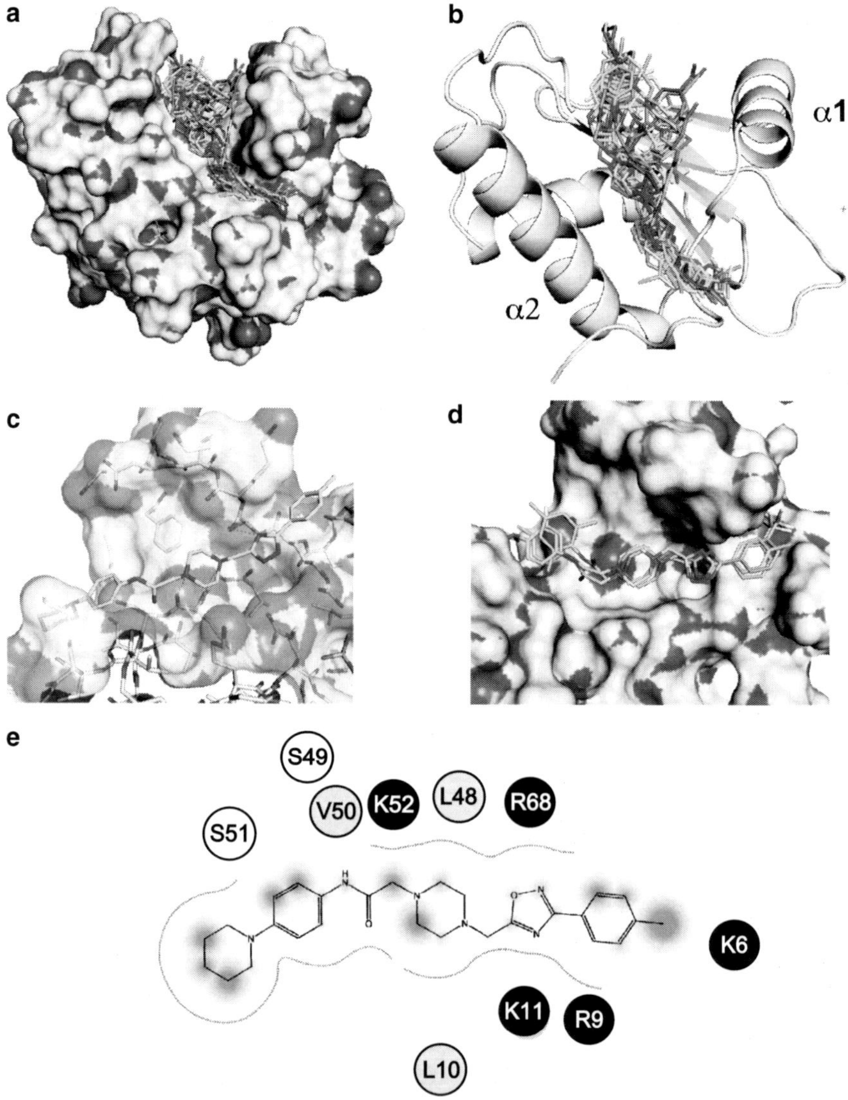

Fig. 13.4 Results of *in silico* screening for compounds binding to the central cleft of the *B. subtilis* P protein. The central cleft has been selected as the most promising binding site from a computational point of view. The figure shows the *B. subtilis* RNase P protein together with a set of docked compound geometries (**a**) in surface mode and (**b**) ribbon cartoon mode. The predicted geometries of (**c**) the top scoring compound and (**d**) the best three compounds (to illustrate the similar core fragment found in 12 of the 15 top candidates) are depicted. (**e**) Schematic overview of key residues of the P protein predicted to interact with the top scoring compound shown in panel C

that could be used because of low solubility, activity was reduced by 40% (Stathopoulos et al. 2000). More detailed kinetic analysis revealed that puromycin behaves as a competitive inhibitor with a K_i of 3.5 mM. This is consistent with the structural similarity of puromycin with the 3′-terminus of aminoacyl-tRNA, suggesting that puromycin interferes with substrate binding to RNase P. In contrast, blasticidin S and amicetin showed noncompetitive inhibition, with K_i values of 7.4 mM for blasticidin S and 2.8 mM for amicetin (Kalavrizioti et al. 2003). When present simultaneously, puromycin acted synergistically with blasticidin S as well as amicetin, suggesting different modes of action for puromycin versus blasticidin S and amicetin. In contrast, addition of blasticidin S to reactions containing amicetin weakened the inhibitory effect of amicetin (Stathopoulos et al. 2000), consistent with the two compounds having identical or overlapping binding sites on the enzyme.

Other peptidyl transferase inhibitors – chloramphenicol, spiramycin, lincomycin, acting on prokaryotic ribosomes, and anisomycin, acting on eukaryotic ribosomes – were tested in a similar way, but were without effect on RNase P activity in this test system (Stathopoulos et al. 2000).

Calcipotriol, a synthetic analog of vitamin D_3 (Fig. 13.5b) widely used in the treatment of psoriasis and other keratinization disorders, had a dose-dependent effect on RNase P from *D. discoideum*: calcipotriol acitvated the RNase P activity of *D. discoideum* cell lysates at concentrations below 50 μM, but was inhibitory at higher concentrations (IC_{50}=0.18 mM). Interestingly, neither the parent compounds cholesterol and 7-dehydrocholesterol, nor vitamin D_3 analogs very similar to calcipotriol, such as 1α-hydroxycholecalciferol and 1α,25-dihydroxycholecalciferol, displayed any effect on RNase P activity. The precise mode of action is unknown, but clearly independent of vitamin D-receptor mediated transcriptional control (Papadimou et al. 2000a). Similar to calcipotriol, anthralin (dithranol = 1,8-dihydroxy-anthrone; Fig. 13.5b), a derivative of chrysarobine also used for psoriasis treatment, inhibited RNase P in a cell-free *D. discoideum* system. Again, the mode of action remains unclear, and binding to allosteric inhibition sites as well as a direct effect of anthralin, or of the free radicals it is known to produce, on the RNA component of RNase P has been discussed. Additive inhibitory effects of calcipotriol and anthralin when present simultaneously point to different sites or modes of inhibition (Papadimou et al. 2000b).

Retinoids (Fig. 13.5c) are vitamin A analogs with functions ranging from the role in vision to multiple regulatory processes mediated by nuclear retinoid receptors which act as transcription factors. Their major therapeutic application is in a number of skin disorders. Analyzed for their effect on RNase P of *D. discoideum*, IC_{50} values of 80 and 500 μM and K_i values of 15 and 1500 μM were determined for all-*trans* retinoic acid and retinol, respectively. For the synthetic retinoids isotretinoin and acitretin, IC_{50} was 60 μM and 40 μM and K_i=20 and 8 μM, respectively, with dose response curves similar to those observed for the natural retinoids (Papadimou et al. 1998). Also some arotinoids, third generation retinoids with two aromatic rings, have been tested similarly. By far the most efficient inhibitor among these was Ro13-7410 (Fig. 13.5c) with a K_i of 45 μM, and a drastic increase in K_i

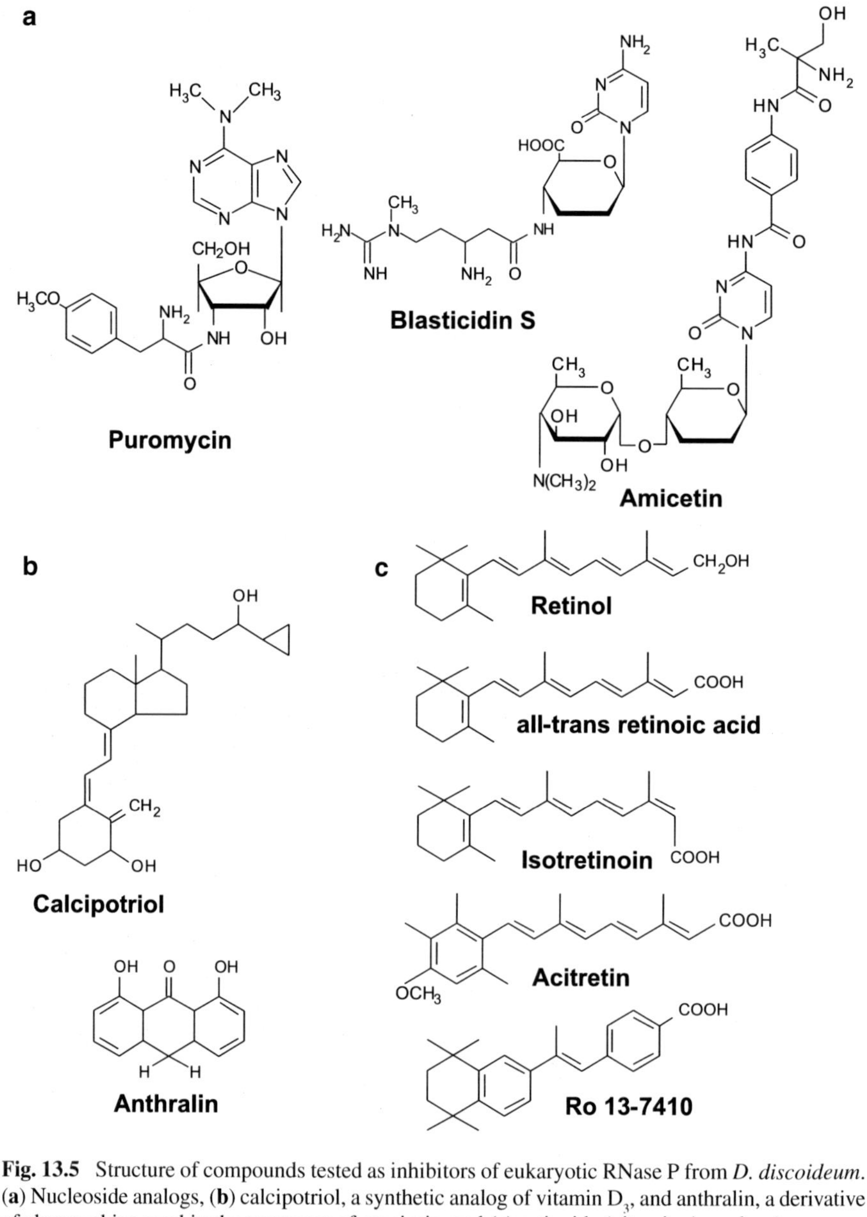

Fig. 13.5 Structure of compounds tested as inhibitors of eukaryotic RNase P from *D. discoideum*. (**a**) Nucleoside analogs, (**b**) calcipotriol, a synthetic analog of vitamin D_3, and anthralin, a derivative of chrysarobine used in the treatment of psoriasis, and (**c**) retinoids (vitamin A analogs)

when the terminal carboxyl group was removed or replaced (Papadimou et al. 2000c). Detailed kinetic analysis further revealed that all the above retinoids including Ro13-7410 acted as simple competitive inhibitors with only one retinoid molecule acting on each molecule of RNase P (Papadimou et al. 1998, 2000c). Some of the retinoids were also tested on partially purified nuclear RNase P from human keratinocytes, with K_i's mostly in the same range as for the *D. discoideum* enzyme (Papadimou et al. 2003).

In conclusion, a number of compounds currently used as therapeutic agents and known for their effect on ribosomes or on intracellular receptors, thus also affect tRNA processing by eukaryotic RNase P. On the one hand, this might contribute to the understanding of the drugs' modes of action, and on the other hand, this opens up perspectives for the design of drugs specifically targeted at RNase P from eukaryotic pathogens. However, such endeavors require the effect of such drugs on the eukaryotic host to be negligible compared to that on the pathogen, and at present it is not fully clear how close in terms of druggability RNase P from *D. discoideum* is to that of humans or eukaryotic pathogens. Its usefulness as a model system for either type of RNase P, therefore, requires further investigation. Finally, as with all data obtained from *in vitro* experiments, verification *in vivo* is crucial, and here the outcome might substantially differ from that *in vitro*, particularly in view of the low solubility and high K_i's of some of the compounds.

13.6 Other Small Ligand Effectors

13.6.1 Synthetic Inhibitors Which Act by Binding to the Substrate

A set of fully synthetic bis-benzimidazoles (Fig. 13.6a) have been tested for their effect on P RNA-mediated catalysis, based on the rationale that according to preliminary data one such compound bound to the T stem groove of *E. coli* $tRNA^{Phe}$ (Bichenkova et al. 1998), a region implicated in tRNA binding to RNase P. Indeed, several of these compounds inhibited ptRNA processing by *E. coli* P RNA at IC_{50} values between 5 and 21 μM (Hori et al. 2001). These inhibitors act at least partly on the substrate, as inferred from the fact that they perturbed the interaction of T- and D-loop with dissociation constants in the high nanomolar to low μM range. In line with a substrate-based mechanism, optimal inhibition required preincubation of the substrate with the compounds, whereas preincubation with P RNA had hardly any effect on the reaction (Hori et al. 2001).

Again motivated by evidence for binding of cationic porphyrins to tRNA at sites important for tRNA tertiary structure formation and possibly P RNA binding, the effect of several porphines and porphyrins on the cleavage reaction by *E. coli* P RNA was analyzed (Hori et al. 2005).

With K_i values of 0.8–4.1 μM, the porphyrins T4MPyP (meso-tetrakis(*N*-methylpyridyl)porphine), TMAP (meso-tetrakis[4-(trimethylammonio)phenyl] porphine) and PPIX (Protoporphyrin IX; Fig. 13.6b) are among the strongest small ligand inhibitors of the RNase P reaction described so far. Fluorescence data indicated that 1:1 complexes of these compounds form with *E. coli* P RNA. Since these compounds bind with similar affinity to P RNA (K_d's of around 50 nM) and ptRNAs or

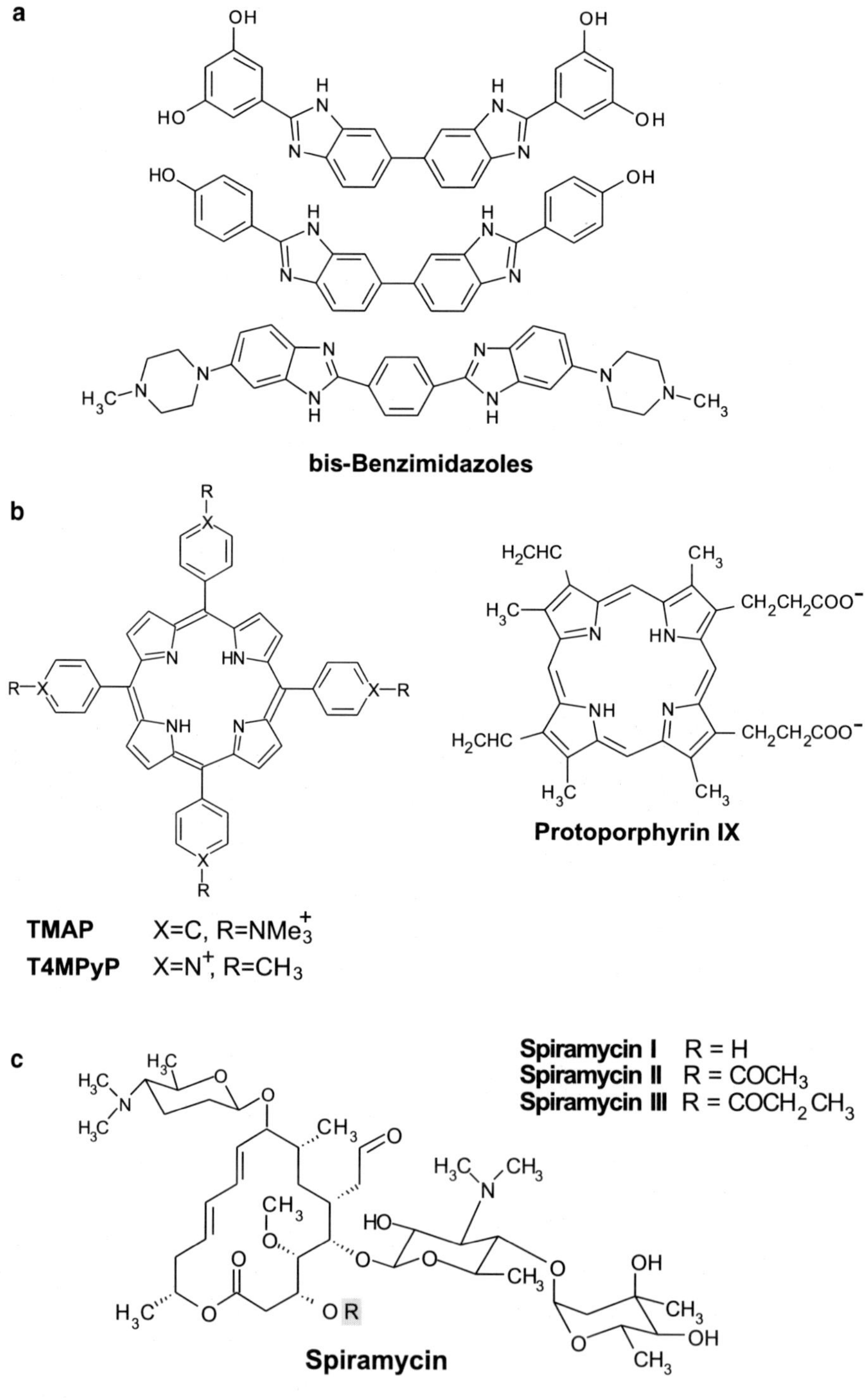

Fig. 13.6 Structure of (**a**) bis-benzimidazoles and (**b**) porphyrin derivatives, both inhibiting bacterial RNase P, and (**c**) the macrolide spiramycin that activates bacterial RNase P

tRNAs (K_d's of 0.1–1.2 μM), inhibition may be based on compound binding to P RNA, to substrate, or to both. Mechanistically, displacement of crucial metal ions, blockage of substrate-P RNA interaction and induction of conformational changes within RNA have been discussed (Hori et al. 2005).

13.6.2 Macrolides as Activators of Bacterial RNase P

Several macrolides tested at concentrations of up to 1 mM did not show any effect on RNase P of *D. discoideum* (see above). In contrast, the macrolides spiramycin (Fig. 13.6c), erythromycin, tylosin and roxithromycin affected the *E. coli* holoenzyme and P RNA-alone reaction in the low micromolar range, where they acted as dose-dependent activators (Toumpeki et al. 2008). Detailed analysis of the activation by spiramycin revealed a mixed-type activation mode with, at saturating concentrations of spiramycin, k_{cat}/K_s increased 18-fold in the holoenzyme reaction and 12-fold in the RNA-alone reaction. The activation was shown to be independent of pH in the range from 5 to 9, possibly indicative of hydrophobic interactions involved in binding of the macrolide to P RNA. Also, varying the Mg^{2+} concentration between 10 and 100 mM did not affect activation of the RNA-alone or holoenzyme reaction. There are two lines of evidence regarding the precise mode of spiramycin-mediated activation of RNase P: (1) Kinetic measurements demonstrate that spiramycin affects the catalytic step of the reaction in the RNA-alone as well as the holoenzyme reaction; (2) according to probing data, binding of spiramycin to *E. coli* P RNA leads to a structural rearrangement of the P10/11 region which is known to be involved in substrate binding, with A124, conserved in bacterial P RNAs, becoming more exposed. This may lead to increased affinity for the substrate and/or facilitated product release (Toumpeki et al. 2008).

13.7 Final Remarks

Various approaches and classes of potential inhibitors have been tested on RNase P as target and vivid research is going on. For some compounds, knowledge on the mode of inhibition has emerged, and further mechanistic insight is expected to be propelled by the recent advancements in our understanding of RNase P structure and function. However, up to now, therapeutic applications for any of these agents are far from being tangible. A fundamental question that arises in this context is whether natural inhibitors of bacterial RNase P exist, but have as yet escaped identification. Since RNase P is essential, differs substantially in bacteria versus eukarya, and has architectural properties that offer many routes to interfere with, one is tempted to entertain the suspicion that natural RNase P inhibitors exist in nature. On the other hand, the highly dynamic and flexible character of the enzyme, and particularly its RNA subunit (Kirsebom 2007), illustrated by the difficulties to obtain well-diffracting P RNA crystals (not to mention crystals of a holoenzyme or enzyme-substrate complex), may have prevented the evolution of RNase P-specific inhibitors. From this point of view, the bacterial P protein may be a more promising target than the RNA subunit. However, the bacterial P protein has characteristics of "intrinsically unstructured" proteins (Henkels et al. 2001) and binds via induced fit to the P RNA (Guo et al. 2006), suggesting that different conformers of the protein coexist in the cell, which might complicate its targetability. In conclusion, RNase P, although a promising drug target, remains a challenging one.

References

Berchanski A, Lapidot A (2008) Bacterial RNase P RNA is a drug target for aminoglycoside-arginine conjugates. Bioconjug Chem 19:1896–1906

Bichenkova EV, Sadat-Ebrahimi SE, Wilton AN et al (1998) Strong, specific, reversible binding ligands for transfer RNA: comparison by fluorescence and NMR spectroscopies with distamycin binding for a new structural class of ligand. Nucleosides Nucleotides Nucleic Acids 17:1651–1665

Brännvall M, Pettersson BM, Kirsebom LA (2003) Importance of the +73/294 interaction in *Escherichia coli* RNase P RNA substrate complexes for cleavage and metal ion coordination. J Mol Biol 325:697–709

Childs JL, Poole AW, Turner DH (2003) Inhibition of *Escherichia coli* RNase P by oligonucleotide directed misfolding of RNA. RNA 9:1437–1445

Egholm M, Buchardt O, Christensen L et al (1993) PNA hybridizes to complementary oligonucleotides obeying the Watson–Crick hydrogen-bonding rules. Nature 365:566–568

Eubank TD, Biswas R, Jovanovic M et al (2002) Inhibition of bacterial RNase P by aminoglycoside-arginine conjugates. FEBS Lett 511:107–112

Giordano T, Sturgess MA, Rao SJ (2006) Inhibitors of RNase P proteins as antibacterial compounds. Unites States Patent US 7,001,924 B2

Gößringer M, Hartmann RK (2007) Function of heterologous and truncated RNase P proteins in *Bacillus subtilis*. Mol Microbiol 66:801–813

Gößringer M, Kretschmer-Kazemi Far R, Hartmann RK (2006) Analysis of RNase P protein (*rnpA*) expression in *Bacillus subtilis* utilizing strains with suppressible *rnpA* expression. J Bacteriol 188:6816–6823

Good L, Awasthi SK, Dryselius R et al (2001) Bactericidal antisense effects of peptide-PNA conjugates. Nat Biotechnol 19:360–364

Gruegelsiepe H, Brandt O, Hartmann RK (2006) Antisense inhibition of RNase P: mechanistic aspects and application to live bacteria. J Biol Chem 281:30613–30620

Gruegelsiepe H, Willkomm DK, Goudinakis O et al (2003) Antisense inhibition of *Escherichia coli* RNase P RNA: mechanistic aspects. Chembiochem 4:1049–1056

Guo X, Campbell FE, Sun L et al (2006) RNA-dependent folding and stabilization of C5 protein during assembly of the *E. coli* RNase P holoenzyme. J Mol Biol 360:190–203

Henkels CH, Kurz JC, Fierke CA et al (2001) Linked folding and anion binding of the *Bacillus subtilis* ribonuclease P protein. Biochemistry 40:2777–2789

Hermann T, Westhof E (1998) Aminoglycoside binding to the hammerhead ribozyme: a general model for the interaction of cationic antibiotics with RNA. J Mol Biol 276:903–912

Holzmann J, Frank P, Loffler E et al (2008) RNase P without RNA: identification and functional reconstitution of the human mitochondrial tRNA processing enzyme. Cell 135:462–474

Hori Y, Bichenkova EV, Wilton AN et al (2001) Synthetic inhibitors of the processing of pretransfer RNA by the ribonuclease P ribozyme: enzyme inhibitors which act by binding to substrate. Biochemistry 40:603–608

Hori Y, Rogert MC, Tanaka T et al (2005) Porphyrins and porphines bind strongly and specifically to tRNA, precursor tRNA and to M1 RNA and inhibit the ribonuclease P ribozyme reaction. Biochim Biophys Acta 1730:47–55

Irwin JJ, Shoichet BK (2005) ZINC – a free database of commercially available compounds for virtual screening. J Chem Inf Model 45:177–182

Jakobsen MR, Damgaard CK, Andersen ES et al (2004) A genomic selection strategy to identify accessible and dimerization blocking targets in the 5′-UTR of HIV-1 RNA. Nucleic Acids Res 32:e67

Jarrous N (2002) Human ribonuclease P: subunits, function, and intranuclear localization. RNA 8:1–7

Jarrous N, Altman S (2001) Human ribonuclease P. Methods Enzymol 342:93–100

Kalavrizioti D, Vourekas A, Tekos A et al (2003) Kinetics of inhibition of ribonuclease P activity by peptidyltransferase inhibitors. Effect of antibiotics on RNase P. Mol Biol Rep 30:9–14

Kawamoto SA, Sudhahar CG, Hatfield CL et al (2008) Studies on the mechanism of inhibition of bacterial ribonuclease P by aminoglycoside derivatives. Nucleic Acids Res 36:697–704

Kazantsev AV, Krivenko AA, Harrington DJ et al (2003) High-resolution structure of RNase P protein from *Thermotoga maritima*. Proc Natl Acad Sci U S A 100:7497–7502

Kazantsev AV, Krivenko AA, Harrington DJ et al (2005) Crystal structure of a bacterial ribonuclease P RNA. Proc Natl Acad Sci U S A 102:13392–13397

Kirsebom LA (2007) RNase P RNA mediated cleavage: substrate recognition and catalysis. Biochimie 89:1183–1194

Kobayashi K, Ehrlich SD, Albertini A et al (2003) Essential *Bacillus subtilis* genes. Proc Natl Acad Sci U S A 100:4678–4683

Kretschmer-Kazemi Far R, Nedbal W, Sczakiel G (2001) Concepts to automate the theoretical design of effective antisense oligonucleotides. Bioinformatics 17:1058–1061

Kurreck J (2003) Antisense technologies. Improvement through novel chemical modifications. Eur J Biochem 270:1628–1644

Litovchick A, Lapidot A, Eisenstein M et al (2001) Neomycin B-arginine conjugate, a novel HIV-1 Tat antagonist: synthesis and anti-HIV activities. Biochemistry 40:15612–15623

Massire C, Jaeger L, Westhof E (1998) Derivation of the three-dimensional aruchitecture of bacterial ribonuclease P RNAs from comparaive sequence analysis. J Mol Biol 279:773–793

Mikkelsen NE, Brannvall M, Virtanen A et al (1999) Inhibition of RNase P RNA cleavage by aminoglycosides. Proc Natl Acad Sci U S A 96:6155–6160

Mikkelsen NE, Johansson K, Virtanen A et al (2001) Aminoglycoside binding displaces a divalent metal ion in a tRNA-neomycin B complex. Nat Struct Biol 8:510–514

Nekhotiaeva N, Awasthi SK, Nielsen PE et al (2004) Inhibition of *Staphylococcus aureus* gene expression and growth using antisense peptide nucleic acids. Mol Ther 10:652–659

Niranjanakumari S, Day-Storms JJ, Ahmed M et al (2007) Probing the architecture of the *B. subtilis* RNase P holoenzyme active site by cross-linking and affinity cleavage. RNA 13:521–535

Niranjanakumari S, Stams T, Crary SM et al (1998) Protein component of the ribozyme ribonuclease P alters substrate recognition by directly contacting precursor tRNA. Proc Natl Acad Sci U S A 95:15212–15217

Nulf CJ, Corey D (2004) Intracellular inhibition of hepatitis C virus (HCV) internal ribosomal entry site (IRES)-dependent translation by peptide nucleic acids (PNAs) and locked nucleic acids (LNAs). Nucleic Acids Res 32:3792–3798 Erratum in: Nucleic Acids Res 2004, 32:4954

Papadimou E, Georgiou S, Tsambaos D et al (1998) Inhibition of ribonuclease P activity by retinoids. J Biol Chem 273:24375–24378

Papadimou E, Monastirli A, Stathopoulos C et al (2000a) Modulation of ribonuclease P activity by calcipotriol. Eur J Biochem 267:1173–1177

Papadimou E, Monastirli A, Tsambaos D et al (2000b) Additive inhibitory effect of calcipotriol and anthralin on ribonuclease P activity. Biochem Pharmacol 60:91–94

Papadimou E, Monastirli A, Tsambaos D et al (2000c) Inhibitory effects of arotinoids on tRNA biogenesis. Skin Pharmacol Appl Skin Physiol 13:345–351

Papadimou E, Pavlidou D, Seraphin B et al (2003) Retinoids inhibit human epidermal keratinocyte RNase P activity. Biol Chem 384:457–462
Pascual A, Vioque A (1999) Substrate binding and catalysis by ribonuclease P from cyanobacteria and *Escherichia coli* are affected differently by the 3′ terminal CCA in tRNA precursors. Proc Natl Acad Sci U S A 96:6672–6677
Patzel V, Steidl U, Kronenwett R et al (1999) A theoretical approach to select effective antisense oligodeoxyribonucleotides at high statistical probability. Nucleic Acids Res 27:4328–4334
Schedl P, Primakoff P, Roberts J (1974) Processing of *E. coli* tRNA precursors. Brookhaven Symp Biol 26:53–76
Singh SK, Wengel J (1998) Universality of LNA-mediated high-affinity nucleic acid recognition. Chem Commun 12:1247–1248
Spitzfaden C, Nicholson N, Jones JJ et al (2000) The structure of ribonuclease P protein from *Staphylococcus aureus* reveals a unique binding site for single-stranded RNA. J Mol Biol 295:105–115
Stams T, Niranjanakumari S, Fierke CA (1998) Ribonuclease P protein structure: evolutionary origins in the translational apparatus. Science 280:752–755
Stathopoulos C, Tsagla A, Tekos A et al (2000) Effect of peptidyltransferase inhibitors on ribonuclease P activity from *Dictyostelium discoideum*. Effect of antibiotics on RNase P. Mol Biol Rep 27:107–111
Tekos A, Prodromaki E, Papadimou E et al (2003) Aminoglycosides suppress tRNA processing in human epidermal keratinocytes *in vitro*. Skin Pharmacol Appl Skin Physiol 16:252–258
Tekos A, Stathopoulos C, Tsambaos D et al (2004) RNase P: a promising molecular target for the development of new drugs. Curr Med Chem 11:2979–2989
Tekos A, Tsagla A, Stathopoulos C et al (2000) Inhibition of eukaryotic ribonuclease P activity by aminoglycosides: kinetic studies. FEBS Lett 485:71–75
Tor Y (2006) The ribosomal A-site as an inspiration for the design of RNA binders. Biochimie 88:1045–1051
Torres-Larios A, Swinger KK, Krasilnikov AS et al (2005) Crystal structure of the RNA component of bacterial ribonuclease P. Nature 437:584–587
Toumpeki C, Vourekas A, Kalavrizioti D et al (2008) Activation of bacterial ribonuclease P by macrolides. Biochemistry 47:4112–4118
Tsai HY, Masquida B, Biswas R et al (2003) Molecular modeling of the three-dimensional structure of the bacterial RNase P holoenzyme. J Mol Biol 325:661–675
Velec HF, Gohlke H, Klebe G (2005) DrugScore(CSD)-knowledge-based scoring function derived from small molecule crystal data with superior recognition rate of near-native ligand poses and better affinity prediction. J Med Chem 48:6296–6303
Verdonk ML, Cole JC, Hartshorn MJ et al (2003) Improved protein-ligand docking using GOLD. Proteins 52:609–623
Vioque A (1989) Protein synthesis inhibitors and catalytic RNA. Effect of puromycin on tRNA precursor processing by the RNA component of *Escherichia coli* RNase P. FEBS Lett 246:137–139
Wahlestedt C, Salmi P, Good L et al (2000) Potent and nontoxic antisense oligonucleotides containing locked nucleic acids. Proc Natl Acad Sci U S A 97:5633–5638
Walter F, Vicens Q, Westhof E (1999) Aminoglycoside-RNA interactions. Curr Opin Chem Biol 3:694–704
Waugh DS, Pace NR (1990) Complementation of an RNase P RNA (*rnpB*) gene deletion in *Escherichia coli* by homologous genes from distantly related eubacteria. J Bacteriol 172:6316–6322
Wegscheid B, Condon C, Hartmann RK (2006) Type A and B RNase P RNAs are interchangeable *in vivo* despite substantial biophysical differences. EMBO Rep 7:411–417
Wegscheid B, Hartmann RK (2006) The precursor tRNA 3′-CCA interaction with *Escherichia coli* RNase P RNA is essential for catalysis by RNase P *in vivo*. RNA 12:2135–2148
Willkomm DK, Gruegelsiepe H, Goudinakis O et al (2003) Evaluation of bacterial RNase P RNA as a drug target. Chembiochem 4:1041–1048

Chapter 14
Ribonuclease P as a Tool

Fenyong Liu

14.1 Introduction

Nucleic acid-based gene interference strategies, such as antisense oligonucleotides, ribozymes or DNAzymes, and RNA interference (RNAi), represent powerful research tools and promising therapeutic agents for human diseases (Dorsett and Tuschl 2004; Scherer and Rossi 2003). Each of these approaches has its own advantages and limitations in terms of targeting efficacy, sequence specificity, toxicity, and delivery efficiency in vivo. Antisense DNA and RNA have been extensively used in basic research and developed for clinical applications (Stein and Cheng 1993). RNA enzymes derived from hammerhead and hairpin ribozymes have been shown to be promising gene-targeting reagents to specifically cleave RNA sequences of choice (Sarver et al. 1990; Scherer and Rossi 2003; Yu et al. 1993). The RNAi approach is being widely used for studies of gene function and is being explored for therapeutic applications (Dorsett and Tuschl 2004; Hannon and Rossi 2004). This chapter summarizes the recent research progress of developing RNase P as a tool for basic research and for therapeutic application.

RNase P is a ribonucleoprotein complex responsible for the 5′ maturation of tRNAs (Evans et al. 2006; Gopalan and Altman 2007; Xiao et al. 2002). It catalyzes a hydrolysis reaction to remove a 5′ leader sequence from tRNA precursors (pre-tRNA) and several small RNAs. In *Escherichia coli*, RNase P consists of a catalytic RNA subunit (M1 RNA) of 377 nucleotides, and a protein subunit (C5 protein) of 119 amino acids (Frank and Pace 1998; Gopalan and Altman 2007). In vitro, M1 RNA can cleave its pre-tRNA substrate at high divalent ion concentration (e.g., 100 mM Mg^{2+}) in the absence of C5 protein (Guerrier-Takada et al. 1983). The addition of C5 protein dramatically increases the rate of cleavage by M1 RNA under low concentration of Mg^{2+} in vitro and is required for RNase P activity and cell viability in vivo. The human RNase P consists of a RNA subunit (H1 RNA),

F. Liu (✉)
Program in Comparative Biochemistry & Division of Infectious Diseases, School of Public Health, University of California, Room 326, 16 Barker Hall, Berkeley, CA, 94720, USA
e-mail: liu_fy@berkeley.edu

F. Liu and S. Altman (eds.), *Ribonuclease P*, Protein Reviews 10,
DOI 10.1007/978-1-4419-1142-1_14,

and at least ten protein subunits with molecular weights ranging from 14 to 115 kDa (Gopalan and Altman 2007).

14.2 Gene Targeting Strategy Based on RNase P

14.2.1 External Guide Sequence

Studies on substrate recognition by M1 RNA and RNase P have led to the development of a general strategy in which M1 RNA and RNase P can be used as gene targeting tools to cleave any specific mRNA sequences (Forster and Altman 1990; McClain et al. 1987; Yuan et al. 1992). The natural substrates for M1 RNA and RNase P include tRNA precursors as well as several other small RNAs such as the precursor to 4.5S RNA in *E. coli* (Fig. 14.1a) (Alifano et al. 1994; Bothwell et al. 1976; Komine et al. 1994). RNase P and M1 RNA recognize a common high-order structure of these substrates but not the primary sequence (Evans et al. 2006; Gopalan and Altman 2007; Xiao et al. 2002). Therefore, an mRNA can be hydrolyzed by M1 ribozyme and RNase P if a custom-designed sequence can be constructed to hybridize to the mRNA

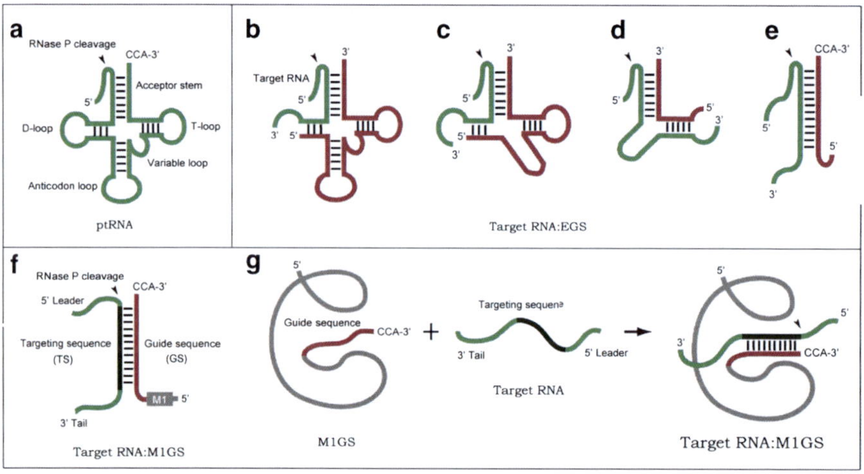

Fig. 14.1 (**a**, **b**) Representation of natural substrates (pre-tRNA (**a**). (**b**–**d**) A hybridized complex of a target RNA (e.g., mRNA) and an EGS that resembles a part of structure of a tRNA and can be cleaved by RNase P. (**c**) results from (**b**) by deleting the anticodon domain of the EGS, which is dispensable for EGS targeting activity, while (**e**) results from (**c**) by further deleting the D stem/loop and variable regions. (**d**) represents an alternative design based on (**c**) (Werner et al. 1998). Substrates in (**b**–**d**) can be cleaved by human RNase P and M1 ribozyme. In contrast, the stem structure in (**e**) can only serve as a substrate for M1 RNA and cannot be cleaved by human RNase P. (**f**) A complex formed between an M1GS ribozyme and a target mRNA substrate. (**g**) Representation of an M1GS RNA construct hybridizing a target RNA that contains a 5′ leader, targeting, and 3′ tail sequence. The *arrow* shows the site of the cleavage by RNase P and M1 RNA

to form a tRNA-like structure (Fig. 14.1b) (Forster and Altman 1990; Yuan et al. 1992). This custom-designed sequence is called an external guide sequence (EGS). An EGS directs RNase P to recognize the structure of the mRNA–EGS complex and cleave at the cleavage site (Fig. 14.1b–e) (Forster and Altman 1990; Yuan et al. 1992). The EGS may consist of two sequence elements: a sequence complementary to the targeted mRNA sequence, and a sequence which resembles a portion of the T-loop and stem, and the variable loop and stem of a tRNA. Further studies have indicated that the second element is essential for EGS function in reactions with human RNase P (Fig. 14.1c, d) but is dispensable in reactions with M1 RNA (Fig. 14.1e) (Forster and Altman 1990; Yuan et al. 1992). Moreover, in the reactions with human RNase P, the anticodon region has been shown to be dispensable for the EGS activity (Yuan and Altman 1994; Kawa et al. 1998) (Fig. 14.1c). An alternative design of shorter EGS constructs has also been reported, in which the targeted mRNA resembles the 5′ leader sequence, 5′ acceptor stem, the variable region, and the 5′ T-stem/loop of a tRNA, while the EGS of only 15–20 nucleotides looks like the 3′ acceptor stem and 3′ T-stem/loop regions (Fig. 14.1d) (Werner et al. 1998).

EGSs are antisense oligoribonucleotides that have been used to diminish gene expression in bacteria (Guerrier-Takada et al. 1995; Guerrier-Takada et al. 1997; McKinney et al. 2001) with the aid of either RNase P or M1 RNA. Studies have also shown that expression of EGSs in mammalian cell culture inhibits the gene expression and replication of herpes simplex virus 1 (HSV-1), human cytomegalovirus (HCMV), influenza virus, human immunodeficiency virus (HIV), and Kaposi's sarcoma-associated herpesvirus (KSHV) (Bai et al. 2008; Dunn et al. 2001; Hnatyszyn et al. 2001; Kawa et al. 1998; Plehn-Dujowich and Altman 1998; Zhu et al. 2004). In principle, any RNA could be targeted by a custom-designed EGS RNA for specific cleavage by RNase P.

14.2.2 M1GS RNA

Research on gene-targeting strategies based on M1 RNA and EGS has led to the development of a more efficient and easy-to-make agent, called M1GS RNA. M1GS RNA is simply constructed through insertion or 3′ extension of M1 RNA by a guide sequence, which base pairs with the target, and contains an unpaired 3′-NCCA end as present in natural *E. coli* tRNA substrates, to allow efficient cleavage by the tethered M1 moiety (Fig. 14.1f) (Frank et al. 1994; Liu and Altman 1995). This design is based on the idea that the guide sequence binds to its target mRNA and directs M1 RNA, which is in close proximity due to the covalent attachment to the guide sequence, to the site of cleavage (Fig. 14.1g). Subsequent studies have demonstrated that M1GS RNA is effective in blocking gene expression in bacteria, in cleaving various targets including oncogenic cellular and viral mRNAs in human cells, and in reducing viral infection and pathogenesis in animals (Bai et al. 2008; Cobaleda and Sanchez-Garcia 2000; Guerrier-Takada et al. 1995; Liu and Altman 1995; Trang et al. 2000a, b). Thus, M1GS ribozyme represents a promising tool for gene targeting application such as gene therapy.

14.3 In Vitro Characterization of RNase P-Mediated Targeting Approaches

14.3.1 Target Selection for EGS and M1GS Ribozyme

The EGS and M1GS ribozyme approaches have been used to target a wide range of RNAs including bacterial, viral, and cellular mRNAs. A limitation of these approaches is that the EGS and ribozyme may not get access to the target RNA if the potential target sites are buried within the organized and folded RNA structure. Therefore, a major prerequisite for M1GS design and EGS is to map the regions of the target mRNA that may be accessible to M1GS and EGS binding. Three different approaches have been used to determine the regions of a target mRNA that may be exposed to binding. First, computer-generated structure predictions are used to identify the targets that are likely to have accessible conformations (Zuker 2003). However, an accurate secondary structure prediction may not be possible using computer algorithms since it is almost impossible to simulate in vivo thermodynamic parameters for folding "in silico." Second, the secondary structure of the target mRNA can be probed with chemicals or enzymes that have different specificity in interacting with an RNA molecule (Brunel and Romby 2000). For example, the end-labeled mRNA sequence can be digested with RNases such as RNase T1, nuclease S1, and RNase V1 separately, and the cleavage products are resolved on a denaturing gel. Unlike RNase T1 and nuclease S1, which recognizes the single-stranded regions, RNase V1 only cleaves the regions of the RNA that are either base-paired or involved in tertiary interactions. The third approach is to map the accessible regions of a target RNA in cellular environment in which RNA folding is influenced by cellular proteins that may interact with the mRNA. Mapping of the accessible regions of an mRNA is possible using an in vivo mapping method based on nucleotide modifications by dimethyl sulphate (DMS) (Ares and Igel 1990; Liu and Altman 1995; Zaug and Cech 1995). When cells are incubated in medium containing DMS, the chemical compounds enter the cells and modify the nucleotides in the accessible regions of an mRNA. The regions of the mRNAs that are exposed and modified by DMS can then be determined by primer extension analysis using reverse transcriptase.

14.3.2 Designing Effective Guide Sequences and M1GS Ribozymes

The flanking sequence of the mRNA region to be targeted by RNase P or M1GS RNA needs to exhibit several sequence features in order to interact with an EGS and RNase P to achieve efficient cleavage. For M1GS targeting, the nucleotides 3′ and 5′ adjacent to the site of cleavage are a guanosine and a pyrimidine, respectively (Liu and Altman 1996). An additional sequence feature, in which a U is eight nucleotides downstream from this cleavage site, is needed for human RNase P targeting (Yuan et al. 1992; Yuan and Altman 1994). These sequence elements interact with the EGS to facilitate the formation of the mRNA–EGS complex into a tRNA-like structure. Furthermore, the interactions of these elements with RNase P are critical for recognition and cleavage by the enzyme (Yuan et al. 1992; Yuan and Altman 1994). In their studies of the unique design of EGS constructs, Werner et al. have also investigated the effects of different sequences of the EGSs on their targeting activity (Werner et al. 1998).

14.3.3 Potential Interactions Among RNase P/M1GS, Its Substrates, and Proteins

Studies have been carried out to understand how RNase P interacts with an mRNA:EGS complex molecule. These results suggested that those highly active EGSs that were selected in vitro, when in complex with the mRNA substrate, exhibited extensive tertiary interactions and an enhanced binding affinity with the target mRNA and in addition, resembled a portion of tRNA structure that optimally interacted with human RNase P, achieving highly efficient targeting activity (Yuan and Altman 1994; Zhou et al. 2002). These studies provide insight into the construction of highly active EGSs for gene targeting applications.

How a M1GS RNA interacts with a model mRNA substrate has also been investigated. The regions of the ribozyme that are in close contact with a model mRNA substrate were mapped by both UV crosslinking and nuclease footprint analysis (Kilani and Liu 1999; Trang et al. 1999). The results suggest that: (1) the cleavage site of the mRNA substrate is positioned at the same regions of the ribozyme which bind to the cleavage site of a ptRNA, (2) the target sequence (Fig. 14.1g) of a model mRNA substrate interacts with the same regions of the ribozyme that are in close proximity to the acceptor stem of a ptRNA, and (3) the 5′ leader and 3′ tail sequence of a model mRNA substrate (Fig. 14.1g) interact with some of the ribozyme regions (e.g., P12, P13, J11/14, and P14) which are not believed to interact with a ptRNA substrate (Kilani and Liu 1999; Trang et al. 1999). Identification of the binding sites of the ribozyme to different regions of the mRNA substrate serves as a starting point to investigate how a M1GS ribozyme recognizes its mRNA substrates and achieves sequence specificity. By mutating the binding site and manipulating the interactions between the binding site and the substrate, we may be able to construct ribozymes that achieve optimal substrate binding and cleavage efficiency.

The effects of the C5 protein on the interactions between a M1GS ribozyme and a model mRNA substrate has also been studied in order to understand how a M1GS RNA hydrolyzes a mRNA substrate in the presence of cellular proteins (Hsu et al. 2000). The C5 protein does not significantly affect the interaction of the ribozyme with the targeting sequence (Fig. 14.1g) of the substrate adjacent to the cleavage site (Hsu et al. 2000). However, differences in the interactions of the ribozyme to the 5′ leader and 3′ tail sequence (Fig. 14.1g) were found in the presence of the protein cofactor. It is reasonable to suggest that the differences in ribozyme–mRNA substrate interactions in the presence of C5 may account for the better substrate recognition and the increased cleavage efficiency observed in the presence of the protein cofactor. If this is the case, a ribozyme mutant, which, in the absence of C5, interacts with an mRNA substrate under low salt conditions in a similar way as the wild type ribozyme–C5 complexes, is expected to exhibit optimal catalytic activity to cleave an mRNA substrate under physiological conditions in the absence of the protein cofactor. Indeed, a ribozyme variant with base substitution mutations at nucleotide positions 224 and 225 of M1 RNA exhibits at least a tenfold higher efficiency to cleave a mRNA substrate in the absence of C5 (Kilani et al. 2000). Alternatively, ribozyme variants, which, upon binding a cellular protein, exhibit similar interactions to an mRNA substrate as those observed in the presence of C5, may function effectively in vivo.

Cellular extract fractions copurified with human RNase P have been shown to enhance M1GS activity (Altman and Kirsebom 1999; Liu and Altman 1995; Mann et al. 2003; Sharin et al. 2005). Specifically, human protein Rpp29, one of the protein subunits of human RNase P, has recently been identified to bind to M1 RNA as well as the tRNA substrate, and enhance its cleavage of a pre-tRNA substrate in vitro (Mann et al. 2003; Sharin et al. 2005). It is possible that other cellular proteins may also interact with M1GS RNA and its substrate, and enhance its cleavage of an mRNA substrate in cells. Further studies are needed to identify these proteins and investigate the mechanism of how these proteins modulate the activity of M1GS RNAs in cells. These studies will facilitate the construction of ribozymes that exhibit optimal substrate binding and high sequence specificity in vivo.

14.4 Characterization of RNase P-Mediated Approaches in Cultured Cells

14.4.1 Delivery and Expression of Ribozymes

The efficacies of M1GS ribozymes and EGSs have been investigated in vivo by either stably or transiently expressing them in cellular environment. Virus-based gene therapy vectors, including those derived from adenovirus, adeno-associated virus (AAV), and lentivirus as well as other retroviruses, can be used to deliver the sequences encoding ribozyme and EGS into numerous types of cells and tissues (Horster et al. 1999; Mautino 2002; Scherer and Rossi 2003). For example, retroviral (e.g., LXSN-based) vectors have been used for the stable expression of M1GS ribozyme and EGS in both human and murine cells, and M1GS and EGS were placed under the control of the U6 promoter and termination signal as the RNA transcripts synthesized from this promoter remain primarily in the nucleus where RNase P is localized (Bertrand et al. 1997; Kawa et al. 1998; Liu and Altman 1995; Reddy et al. 1987; Yuan et al. 1992). When the ribozyme needs to be delivered to target an mRNA in the cytoplasm, tRNA sequences may be linked with ribozymes and transcribed by native tRNA promoters in order to colocalize the target mRNA and ribozyme into the cytoplasm (Warashina et al. 2001). The utility of different promoters for ribozyme expression in vivo varies with the localization of target mRNA within the subcellular region or/and specific tissue type.

Ex vivo delivery of EGSs into human cells is gaining ground due to the selective advantage of this technique over the vector-based gene delivery approach. The EGSs can be chemically synthesized and the nucleoside residues could be modified appropriately so that the synthetic RNA molecules are not susceptible to cellular nuclease degradation. The oligonucleotides with 2′ hydroxyl modification and/or phosphorothioates are highly resistant to cellular endonucleases (Verma and Eckstein 1998). The delivery of such modified oligonucleotides can be accomplished by encapsulating them in liposomes or other biodegradable polymeric matrices (Jackson et al. 2002). The efficacy of this method in vivo is being explored for antisense, siRNA, ribozyme, and EGS molecules (Dunn et al. 2001; Ma et al. 1998, 2000; Scherer and Rossi 2003), and 2′-methyl modified EGS RNAs have been shown to be effective in blocking the expression of both viral and cellular genes (Ma et al. 2000; Zhu et al. 2004).

14.4.2 Kinetics of M1GS and EGS Action in Cells

Studies on hammerhead and hairpin ribozymes suggest that binding of the ribozyme to its target RNA appears to be rate-limiting in vivo (Lee et al. 1999; Sullenger and Cech 1993; zu Putlitz et al. 1999). To investigate the rate-limiting step of the M1GS and EGS-based approaches, M1GS and EGS RNAs were constructed to target the mRNA regions that were accessible to modification by DMS in cell culture and to ribozyme and EGS binding (Kilani et al. 2000; Yang et al. 2006; Zhou et al. 2002). Moreover, the ribozymes and EGSs were expressed primarily in the nuclei by using the promoter of small nuclear U6 RNA. This design would increase the probability for the constructed M1GS ribozymes and EGSs, to locate and bind to their target mRNA sequences in the nuclei. Under such conditions, it has been demonstrated that the effectiveness of the M1GS and EGS-based approaches in culture cells is dictated by the overall rate (k_{cat}/K_m) of the cleavage by the ribozyme, and EGS-directed cleavage by human RNase P (Kilani et al. 2000; Yang et al. 2006; Zhou et al. 2002). For example, R6 and R29, which are M1GS variants generated by an in vitro selection procedure, are at least 10- and 20-fold more active (k_{cat}/K_m) in cleaving HSV-1 TK mRNA in vitro than the ribozyme (i.e., M1-TK) derived from the wild type M1 sequence, respectively. A reduction of 95% and 99% in TK mRNA and protein expression was observed in cells expressing R6 and R29, respectively, while a reduction of 70% in TK expression was found in the M1-TK-expressing cells (Kilani et al. 2000). These results suggest that increasing the targeting efficiency of the ribozyme and EGS should lead to more effective inhibition of the target mRNA expression in cultured cells.

14.5 Engineering of EGS and M1GS RNAs by In Vitro Selection

Extensive research has been carried out to increase the targeting activity of EGSs and M1GS RNAs. Numerous novel M1GS and EGS RNAs that are highly active have been generated using in vitro selection processes. In vitro selection refers to a process where enrichment of molecules with desired properties is achieved through iterative cycles of isolation and amplification (Ellington and Szostak 1990; Joyce 1992; Tuerk and Gold 1990). The technology has been used for different purposes, from enhancing catalytic activity to enzyme engineering to isolating nucleic acids with novel functions (Berzal-Herranz et al. 1992; Joseph et al. 1993; Lorsch and Szostak 1994; Yuan and Altman 1994).

14.5.1 In Vitro Selection of M1GS RNAs

The in vitro selection technique was used to successfully isolate highly active M1GS RNA variants (Kilani et al. 2000) from a library of ribozymes contained random mutations in regions that are known to be conserved across all species of RNase P catalytic RNAs and are important for catalysis and substrate binding (Evans et al. 2006; Gopalan and Altman 2007; Xiao et al. 2002). The selection process was carried out in a series of steps that included: (1) annealing of M1GS RNA pool with a 5′ biotinylated substrate, (2) binding the complex to streptavidin-agarose column, (3) M1GS RNA-mediated cleavage of the substrate in the presence of divalent ion-containing buffer, (4) recovering M1 RNA with denaturing gel electrophoresis, (5) synthesis of cDNA copies of RNA molecules with reverse transcription-PCR (RT-PCR), followed by (6) in vitro transcription of the generated cDNAs with T7 RNA polymerase (Fig. 14.2) (Kilani et al. 2000). The sequences isolated after several rounds of selection were cloned and determined. The selected

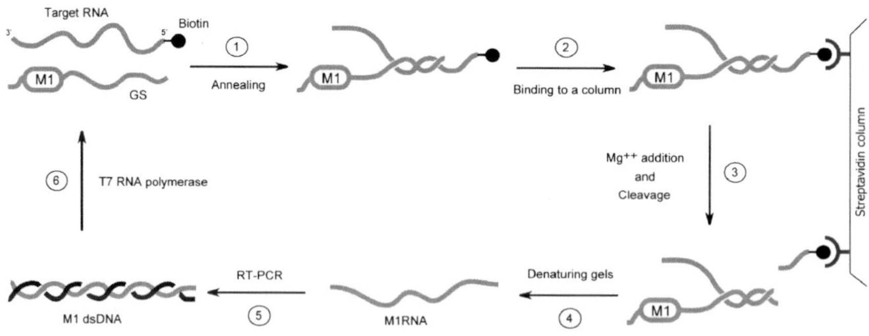

Fig. 14.2 Schematic representation of the in vitro evolution procedure for the generation of active M1GS RNA ribozyme variants that specifically cleave a target mRNA

RNase P ribozyme variants were shown to be highly efficient in cleaving target mRNA in vitro and effective in inhibiting the expression of target mRNA in cultured cells (Kilani et al. 2000; Trang et al. 2002; Zou et al. 2004).

14.5.2 In Vitro Selection of EGSs

RNase P-mediated cleavage of an mRNA complexed with an EGS that is derived from a natural tRNA is usually less efficient than that of a natural tRNA, possibly due to the nucleotide substitution at the mRNA sequence equivalent to the acceptor and D-stem regions (Figs. 14.1a–c and 14.3). These nucleotide substitutions may cause perturbations within interacting nucleotides and eventually disrupt some of the tertiary interactions important for maintaining the proper tRNA-like conformation that is essential for recognition by human RNase P. Restoration of these interactions or introducing additional sequence manipulation in other parts of the EGS sequence, such as those resembling the variable region, and T-stem and T-loop may increase the susceptibility of the mRNA–EGS complex to be cleaved by RNase P (Figs. 14.1a–c and14.3) (Yuan and Altman 1994; Zhou et al. 2002).

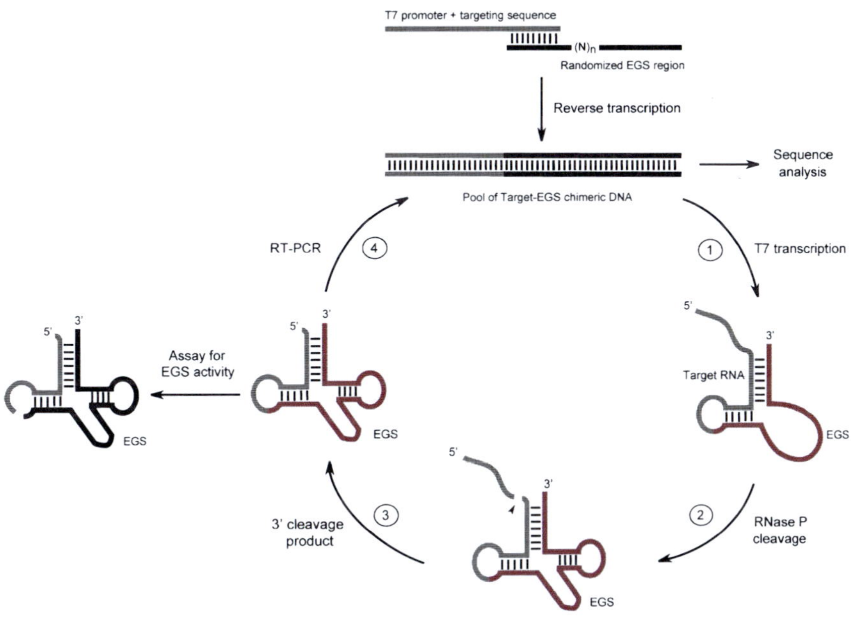

Fig. 14.3 Schematic representation of in vitro selection experiments for generation of active EGSs that target RNase P for cleavage of a target RNA. The *bold* and *red line* represents the EGS region and the randomized positions are each indicated as *N*. The site of cleavage by RNase P is marked with an *arrowhead*

To generate EGSs that are highly active in directing RNase P to cleave an mRNA, a pool of chimeric, covalently linked mRNA-EGS substrates that contain partially randomized sequences is constructed and selected, based on their ability to be a substrate for human RNase P (Fig. 14.3) (Yuan and Altman 1994; Zhou et al. 2002). The chimeric RNA, which integrates the mRNA sequence (e.g., tk46) and the EGS, contains a randomized sequence (e.g., 25 nucleotides) at the positions corresponding to the regions resembling the variable stem, T-stem and T-loop (Fig. 14.3), whereas the 5′ proximal sequence that spans the D-stem, D-loop, acceptor stem and the leader sequence, contains the mRNA sequence. In each round of selection, the RNA variants are digested with human RNase P and the 3′ cleavage products are isolated (Fig. 14.3). The cDNA molecules are synthesized from these RNA molecules by RT-PCR, and used as the templates for the synthesis of EGS RNA molecules for the next round of selection (Fig. 14.3). Novel and highly active EGSs have been generated and shown to be effective in inhibiting gene expression in cultured cells (Yang et al. 2006; Yuan and Altman 1994; Zhou et al. 2002).

14.6 Characterization of RNase P Targeting in Animals

To determine whether RNase P ribozyme is effective in vivo, a M1GS RNA was constructed to target the overlapping mRNA region coding for murine cytomegalovirus (MCMV) essential capsid proteins mAP and M80 (Bai et al. 2008). When the ribozyme-expressing constructs were delivered into MCMV-infected SCID mice via a modified "hydrodynamic transfection" procedure, expression of ribozymes was observed in the livers and spleens. Compared to the "control" animals that did not receive any M1GS constructs or received the "disabled" ribozyme construct, animals receiving the functional ribozyme construct exhibited a significant reduction of viral gene expression and infection (Bai et al. 2008). Viral titers in the spleens, livers, lungs, and salivary glands of the functional ribozyme-treated SCID mice were lower than those in the control animals. Moreover, survival of the infected animals significantly improved upon receiving the functional ribozyme construct. This study demonstrates the utility of M1GS ribozyme for inhibition of gene expression in animals (Bai et al. 2008). Further experiments have yet been reported to investigate the activity of EGSs to direct RNase P for gene targeting applications in animals.

14.7 Applications of RNase P as a Tool for Basic Research and for Therapy

14.7.1 Antibacterial Agents

Elegant studies by Altman and colleagues applied EGS technology to inhibit expression of essential genes of several bacteria, including *Salmonella*, *Yersinia*, and *Francisella tularensis*, and achieve an antimicrobial effect for *Salmonella* infection (Ko et al. 2008; McKinney et al. 2001, 2002, 2004; Xiao et al. 2008). More notably, in one of the studies, carefully designed EGSs were shown to be highly specific and were capable of species-specific targeting (McKinney et al. 2001). Sequences of mRNA differing between *E. coli* and *S. typhimurium*, but encoding identical proteins were both targeted. The expression of EGSs complementary to *E. coli* mRNAs resulted in reduction of the expression of the target mRNAs in *E. coli*, while no effects were shown against *S. typhimurium* mRNAs and vice versa. If EGS technology can be used in antimicrobial therapy, its ability to achieve species-specific inhibition of bacterial viability could become very useful in circumventing the current limitation of narrow spectrum antimicrobials in inhibiting commensal nonpathogenic bacteria. More recently, Altman and coworkers further used a novel selection scheme to identify highly active EGSs for directing RNase P to cleave target mRNAs and block gene expression in bacteria (Ko et al. 2008; Lundblad et al. 2008; Xiao et al. 2008). These results clearly demonstrate the feasibility of using EGS technology in antibacterial strategies.

14.7.2 Antiviral Agents

RNase P and M1 RNA have been extensively used for blocking infection of HIV (Hnatyszyn et al. 2001; Kraus et al. 2002), human influenza virus (Plehn-Dujowich and Altman 1998), and four herpesviruses including HCMV and MCMV (Bai et al. 2008; Trang et al. 2000b), HSV-1 (Kilani et al. 2000; Zhou et al. 2002; Zou et al. 2004), and KSHV (Zhu et al. 2004). When M1GS ribozyme was used to target the mRNA encoding ICP4 of HSV-1, the major transcriptional activator (Trang et al. 2000a), an 80% of ICP4 expression was suppressed, resulting in nearly 1,000-fold reduction in viral growth (Trang et al. 2000a). In the anti-HCMV studies, a M1GS ribozyme was used to target the overlapping region of two HCMV mRNAs, coding for the viral protease (PR) and assembly protein (AP), both of which are essential for viral replication (Kim et al. 2004a). Specific and potent inhibition of HCMV gene expression and replication was observed in cells expressing the ribozymes. Furthermore, when delivered in mice using a hydrodynamic transfection procedure, a M1GS ribozyme was expressed in the spleens and livers, and effectively blocked MCMV gene expression and infection in vivo (Bai et al. 2008).

To further enhance the efficiency of ribozymes, an in vitro selection procedure (Fig.14.2) was used to select for M1GS variants that efficiently cleaved a mRNA (Kilani et al. 2000; Kim et al. 2004b). Ribozyme variants that exhibited at least 20-fold higher cleavage efficiency (represented as k_{cat}/K_m) than the wild type M1 RNA were generated. These selected ribozymes were highly efficient in cleaving mRNAs of HSV-1 and HCMV and effective in blocking gene expression and replication of HSV-1 and HCMV in human cells (Kilani et al. 2000; Kim et al. 2004b; Yu et al. 2005; Zou et al. 2003; Zou et al. 2004).

EGS technology has also been applied to antiviral applications. For example, Altman and colleagues constructed EGSs that targeted the mRNAs coding for the polymerase and nucleocapsid protein of human influenza virus (Plehn-Dujowich and Altman 1998). This study further provided the direct evidence that targeting two different mRNAs simultaneously by the EGS technology inhibits viral growth more effectively than targeting of a single mRNA (Plehn-Dujowich and Altman 1998). In the anti-HIV studies with EGS RNAs, Hnatyzyn and coworkers have elegantly demonstrated that RNase P is highly effective in inhibiting HIV gene expression and replication (Hnatyszyn et al. 2001; Kraus et al. 2002). Expression of the constructed EGS RNAs in human heterogeneous T cell cultures upon HIV challenge was able to maintain CD4 levels, devoid of cytopathology, and did not produce significant level of HIV p24 through 30 days infection. Impressively, the cells that expressed the EGSs were resistant to HIV clinical isolates from clades A, B, C, and F (Hnatyszyn et al. 2001; Kraus et al. 2002). Thus, RNase P-associated EGSs may represent a new class of potential therapeutic agents for anti-HIV therapy.

EGS RNAs were also highly active in targeting RNase P to cleave the mRNAs of HSV-1 and HCMV in vitro, and block the gene expression and replication of these viruses in cultured cells (Kawa et al. 1998; Li et al. 2006). Furthermore, chemically modified EGS molecules were also constructed to target the mRNA encoding KSHV immediate-early transactivator Rta (Zhu et al. 2004). Exogenous administration of 2′-*O*-methyl-modified EGS to KSHV-infected human primary-effusion lymphoma cells significantly inhibited Rta expression and viral growth (Zhu et al. 2004).

In vitro selection has been utilized in EGS technology as well. EGS RNA molecules targeting a HSV-1 mRNA have been subjected to many rounds of selection and highly active EGS variants that exhibited up to 35 times higher activity than the natural tRNA substrate have been isolated (Zhou et al. 2002). The selected EGSs were highly effective in directing human RNase P-mediated inhibition of gene expression and replication of HSV-1 and HCMV in human cells (Yang et al. 2006; Zhou et al. 2002). Indeed, the selected EGSs were more effective than the EGS derived from a natural tRNA in down-regulating gene expression.

14.7.3 Anticancer Agents

In a series of elegant experiments, Sánchez-García and colleagues constructed M1GS ribozymes to specifically cleave chimeric RNA molecules originating from chromosomal abnormalities (Cobaleda and Sanchez-Garcia 2000). They used a well-characterized model of BCR and ABL genes where aberrant translocation results in BCR-ABL oncogenes that cause chronic myelogenous leukemia and acute lymphoblastic leukemias. M1GS RNAs that recognized the oncogenic mRNA at the fusion site appeared to be highly effective and specific in cleaving the target mRNA in vitro and blocking the effect of BCR-ABL function in cultured mammalian cells (Cobaleda and Sanchez-Garcia 2000).

Using the EGS technology, Ma, Stein, and colleagues have induced RNase P-mediated cleavage of the mRNA that encodes protein kinase C-α (PKC-α) (Ma et al. 2000). EGSs with 2′-*O*-methyl modification for enhanced stability were exogenously administered into T24 bladder carcinoma cells for specific down-regulation of PKC-α expression. No nonspecific cleavage, which is usually associated with RNase H-induced reactions, was found, providing direct evidence that RNase P-mediated cleavage induced by EGS is highly specific in targeting its mRNA (Ma et al. 2000). In addition, potent downregulation of antiapoptotic protein bcl-xL was observed, further suggesting a general applicability of the EGS technology for anticancer applications (Ma et al. 2000).

14.8 Advantage and Disadvantage of M1GS and RNase P-EGS Technology

14.8.1 The EGS Technology

Compared to other nucleic acid-based gene interference approaches, the EGS technology exhibits several unique and attractive features as a gene-targeting tool. First, the mechanism of the EGS technology for degradation of a specific mRNA is different from other RNA or DNA-based gene-targeting approaches. It uses the endogenous RNase P, which is one of the most ubiquitous, abundant, stable and efficient enzymes in all types of cells (Evans et al. 2006; Gopalan and Altman 2007; Kim and Liu 2007). This essential enzyme is highly expressed (5×10^4 copies per cell) and is responsible for the processing of all tRNA precursors that account for approximately 2% of total cellular RNA (Gopalan and Altman 2007). The action of RNase P with the EGS will result in irreversible cleavage of the target mRNA in a highly efficient catalytic fashion. Second, the sequence specificity of the EGS technology is governed by two different types of interactions between the EGS and the target mRNA: (1) the base-pairing interactions in which the sequence of 10–13 nucleotides in the EGS hybridizes with the target mRNA, and (2) the interactions between the target mRNA and the other part of the EGS sequence (equivalent to the T-stem and T-loop, and variable regions of a tRNA) which are required for folding of the RNase P-recognizable tertiary structure. Thus, the EGS-based technology is highly specific and does not generate nonspecific "irrelevant cleavage" that is observed in RNase H-mediated cleavage induced by conventional antisense phosphorothioate molecules (Ma et al. 2000; Yuan et al. 1992; Zhu et al. 2004). Third, EGSs exhibit little sign of cytotoxicity because cells expressing these molecules for more than 40 days appear to be normal (Kawa et al. 1998; Ma et al. 2000; Yuan et al. 1992; Zhu et al. 2004).

14.8.2 The M1GS RNA Approach

Compared to other nucleic acid-based gene targeting approaches including small ribozymes such as hammerhead and hairpin ribozymes, M1GS ribozyme possesses several unique features as a gene targeting tool. First, M1GS ribozyme can fold into a defined active conformation in the absence of its substrates, and catalyze the reaction upon substrate binding, potentially resulting in the most efficient cleavage of its target mRNA. Second, M1GS RNA is highly specific as this ribozyme can cleave one substrate over another, even though the two substrates share the first nine contiguous base pairs complementary to the guide sequence (Cobaleda and Sanchez-Garcia 2000). Furthermore, expression of M1GS RNAs over a period of 40 days does not lead to significant cytotoxicity in human cells. Third, M1GS can be designed to target and cleave any sequence. In contrast, hammerhead and hairpin ribozymes are limited by the requirement for the presence of specific nucleotide sequence (-GUX-) in the target mRNA in order for cleavage to occur (Doudna and Cech 2002; Scherer and Rossi 2003). Furthermore, a single point mutation in the required GUX sequence could render the ribozymes ineffective for target mRNA cleavage. The low sequence requirements at the cleavage site provide M1GS ribozyme strategies with the flexibility to be used against almost any target, including positionally fixed target sites such as the fusion junction of two chromosomes resulting in an oncogenic chimeric mRNA (Cobaleda and Sanchez-Garcia 2000).

In recent years, the use of the RNAi approach to degrade mRNA associated with human diseases has been the focus for nucleic acids-based gene interference studies (Dorsett and Tuschl 2004; Hannon and Rossi 2004). RNAi has the advantage of utilizing the cellular machinery in its process to knockdown mRNA, and can be effective in small concentration. However, the siRNA technology may "sequester or misguide" the cellular machinery which may have consequences on cell function not foreseeable at present. Recent studies have suggested that EGS RNAs are, at least, as effective as siRNA in down-regulating gene expression in cultured cells when they were constructed for the same targets (Zhang and Altman 2004). It will be interesting to compare the activity and effectiveness of M1GS RNA and RNAi approaches for knocking down gene expression in human cells.

As with any gene therapy design, stability and delivery of the agents remain a big concern. The delivery problem affects the siRNA technology to the same extent as the EGS and M1GS RNA technology. As an alternative to the viral vector approach, EGS and ribozymes can be delivered ex vivo by encapsulating them in liposomes or other biodegradable polymeric matrix (Ma et al. 2000; Werner et al. 1998; Zhu et al. 2004). For stability, the ribozymes and EGSs can be chemically synthesized with 2′ hydroxyl modification and/or phosphorothioates to resist cellular endonucleases (Verma and Eckstein 1998). In the case of M1GS, chemical synthesis of a functional active ribozyme is at present technically difficult and economically impractical due to its large size (~400 nucleotides). In contrast, EGSs, as short (15–60 nt) oligonucleotides, can be synthesized in vitro with specific chemical modification (Ma et al. 2000; Werner et al. 1998; Zhu et al. 2004). Thus, an EGS can be delivered directly (in naked form or with the aid of liposomes) to cells as well as delivered by expression vectors such as retroviral vectors.

14.9 Conclusion

Both the M1GS ribozyme and EGS technology have been shown to be efficient and specific in cleaving an mRNA sequence in vitro, and are effective in down-regulating the expression of both cellular and viral genes in cultured cells. Recent studies have also indicated that M1GS ribozyme is effective in blocking gene expression and infection of MCMV in animals. Furthermore, novel EGS sequences and ribozyme variants that are more active and effective in inhibiting gene expression can be generated using in vitro selection procedures. To develop EGS and M1GS RNA for therapeutic applications, the efficacy of the ribozymes and the EGSs needs to be further evaluated in animal models and ultimately, in human clinical trials. Several issues like delivery, stability, colocalization and a sustained effect by the EGSs and ribozymes on their targets may also need to be addressed in order to achieve successful therapeutic effects. Further studies on the biochemical mechanism of RNase P-EGS technology and M1GS RNA in vitro, and on their activity in cultured cells and animal models should facilitate the development of the EGSs and M1GS RNAs as novel classes of gene-targeting agents for both in vitro and in vivo applications.

Acknowledgments I apologize to the many authors whose publications were left unreferenced in this review due to space constraints. Special thanks go to Gerry Abenes, Kihoon Kim, Yong Bai, and Ed Yang for invaluable discussions on the manuscript and for preparations of the illustrations, and Sarah Viducich for editorial assistance. I also acknowledge the National Institutes of Health for their longstanding and generous support of our research on RNase P ribozyme and the EGS technology (AI041927 and DE014842).

References

Alifano P, Rivellini F, Piscitelli C, Arraiano CM, Bruni CB, Carlomagno MS (1994) Ribonuclease E provides substrates for ribonuclease P-dependent processing of a polycistronic mRNA. Genes Dev 8:3021–3031

Altman S, Kirsebom LA (1999) Ribonuclease P. In: Gesteland RF, Cech TR, Atkins JF (eds) The RNA world. Cold Spring Harbor Press, Cold Spring Harbor, pp 351–380

Ares M Jr, Igel AH (1990) Lethal and temperature-sensitive mutations and their suppressors identify an essential structural element in U2 small nuclear RNA. Genes Dev 4:2132–2145

Bai Y, Trang P, Li H, Kim K, Zhou T, Liu F (2008) Effective inhibition in animals of viral pathogenesis by a ribozyme derived from RNase P catalytic RNA. Proc Natl Acad Sci U S A 105:10919–10924

Bertrand E, Castanotto D, Zhou C, Carbonnelle C, Lee NS, Good P, Chatterjee S, Grange T, Pictet R, Kohn D, Engelke D, Rossi JJ (1997) The expression cassette determines the functional activity of ribozymes in mammalian cells by controlling their intracellular localization. RNA 3:75–88

Berzal-Herranz A, Joseph S, Burke JM (1992) In vitro selection of active hairpin ribozymes by sequential RNA-catalyzed cleavage and ligation reactions. Genes Dev 6:129–134
Bothwell AL, Garber RL, Altman S (1976) Nucleotide sequence and in vitro processing of a precursor molecule to Escherichia coli 4.5 S RNA. J Biol Chem 251:7709–7716
Brunel C, Romby P (2000) Probing RNA structure and RNA-ligand complexes with chemical probes. Methods Enzymol 318:3–21
Cobaleda C, Sanchez-Garcia I (2000) In vitro inhibition by a site-specific catalytic RNA subunit of RNase P designed against the BCR-ABL oncogenic products: a novel approach for cancer treatment. Blood 95:731–737
Dorsett Y, Tuschl T (2004) siRNAs: applications in functional genomics and potential as therapeutics. Nat Rev Drug Discov 3:318–329
Doudna JA, Cech TR (2002) The chemical repertoire of natural ribozymes. Nature 418:222–228
Dunn W, Trang P, Khan U, Zhu J, Liu F (2001) RNase P-mediated inhibition of cytomegalovirus protease expression and viral DNA encapsidation by oligonucleotide external guide sequences. Proc Natl Acad Sci U S A 98:14831–14836
Ellington AD, Szostak JW (1990) In vitro selection of RNA molecules that bind specific ligands. Nature 346:818–822
Evans D, Marquez SM, Pace NR (2006) RNase P: interface of the RNA and protein worlds. Trends Biochem Sci 31:333–341
Forster AC, Altman S (1990) External guide sequences for an RNA enzyme. Science 249:783–786
Frank DN, Harris M, Pace NR (1994) Rational design of self-cleaving pre-tRNA-ribonuclease P RNA conjugates. Biochemistry 33:10800–10808
Frank DN, Pace NR (1998) Ribonuclease P: unity and diversity in a tRNA processing ribozyme. Annu Rev Biochem 67:153–180
Gopalan V, Altman S (2007) Ribonuclease P: structure and catalysis. In: Gesteland RF, Cech TR, Atkins JF (eds) The RNA world. Cold Spring Harbor Laboratory Press, Cold Spring Harbor, New York
Guerrier-Takada C, Gardiner K, Marsh T, Pace N, Altman S (1983) The RNA moiety of ribonuclease P is the catalytic subunit of the enzyme. Cell 35:849–857
Guerrier-Takada C, Li Y, Altman S (1995) Artificial regulation of gene expression in Escherichia coli by RNase P. Proc Natl Acad Sci U S A 92:11115–11119
Guerrier-Takada C, Salavati R, Altman S (1997) Phenotypic conversion of drug-resistant bacteria to drug sensitivity. Proc Natl Acad Sci U S A 94:8468–8472
Hannon GJ, Rossi JJ (2004) Unlocking the potential of the human genome with RNA interference. Nature 431:371–378
Hnatyszyn H, Spruill G, Young A, Seivright R, Kraus G (2001) Long-term RNase P-mediated inhibition of HIV-1 replication and pathogenesis. Gene Ther 8:1863–1871
Horster A, Teichmann B, Hormes R, Grimm D, Kleinschmidt J, Sczakiel G (1999) Recombinant AAV-2 harboring gfp-antisense/ribozyme fusion sequences monitor transduction, gene expression, and show anti-HIV-1 efficacy. Gene Ther 6:1231–1238
Hsu AW, Kilani AF, Liou K, Lee J, Liu F (2000) Differential effects of the protein cofactor on the interactions between an RNase P ribozyme and its target mRNA substrate. Nucleic Acids Res 28:3105–3116
Jackson JK, Liang LS, Hunter WL, Reynolds M, Sandberg JA, Springate C, Burt HM (2002) The encapsulation of ribozymes in biodegradable polymeric matrices. Int J Pharm 243:43–55
Joseph S, Berzal-Herranz A, Chowrira BM, Butcher SE, Burke JM (1993) Substrate selection rules for the hairpin ribozyme determined by in vitro selection, mutation, and analysis of mismatched substrates. Genes Dev 7:130–138
Joyce GF (1992) Directed molecular evolution. Sci Am 267:90–97
Kawa D, Wang J, Yuan Y, Liu F (1998) Inhibition of viral gene expression by human ribonuclease P. RNA 4:1397–1406
Kilani AF, Liu F (1999) UV cross-link mapping of the substrate-binding site of an RNase P ribozyme to a target mRNA sequence. RNA 5:1235–1247

Kilani AF, Trang P, Jo S, Hsu A, Kim J, Nepomuceno E, Liou K, Liu F (2000) RNase P ribozymes selected in vitro to cleave a viral mRNA effectively inhibit its expressionin cell culture. J Biol Chem 275:10611–10622

Kim K, Trang P, Umamoto S, Hai R, Liu F (2004a) RNase P ribozyme inhibits cytomegalovirus replication by blocking the expression of viral capsid proteins. Nucleic Acids Res 32:3427–3434

Kim K, Umamoto S, Trang P, Hai R, Liu F (2004b) Intracellular expression of engineered RNase P ribozymes effectively blocks gene expression and replication of human cytomegalovirus. RNA 10:438–447

Kim K, Liu F (2007) Inhibition of gene expression in human cells using RNase P-derived ribozymes and external guide sequences. Biochim Biophys Acta 1769:603–612

Ko JH, Izadjoo M, Altman S (2008) Inhibition of expression of virulence genes of Yersinia pestis in Escherichia coli by external guide sequences and RNase P. RNA 14:1656–1662

Komine Y, Kitabatake M, Yokogawa T, Nishikawa K, Inokuchi H (1994) A tRNA-like structure is present in 10Sa RNA, a small stable RNA from Escherichia coli. Proc Natl Acad Sci U S A 91:9223–9227

Kraus G, Geffin R, Spruill G, Young AK, Seivright R, Cardona D, Burzawa J, Hnatyszyn HJ (2002) Cross-clade inhibition of HIV-1 replication and cytopathology by using RNase P-associated external guide sequences. Proc Natl Acad Sci U S A 99:3406–3411

Lee NS, Bertrand E, Rossi J (1999) mRNA localization signals can enhance the intracellular effectiveness of hammerhead ribozymes. RNA 5:1200–1209

Li H, Trang P, Kim K, Zhou T, Umamoto S, Liu F (2006) Effective inhibition of human cytomegalovirus gene expression and growth by intracellular expression of external guide sequence RNA. RNA 12:63–72

Liu F, Altman S (1995) Inhibition of viral gene expression by the catalytic RNA subunit of RNase P from Escherichia coli. Genes Dev 9:471–480

Liu F, Altman S (1996) Requirements for cleavage by a modified RNase P of a small model substrate. Nucleic Acids Res 24:2690–2696

Lorsch JR, Szostak JW (1994) In vitro evolution of new ribozymes with polynucleotide kinase activity. Nature 371:31–36

Lundblad EW, Xiao G, Ko JH, Altman S (2008) Rapid selection of accessible and cleavable sites in RNA by Escherichia coli RNase P and random external guide sequences. Proc Natl Acad Sci U S A 105:2354–2357

Ma M, Benimetskaya L, Lebedeva I, Dignam J, Takle G, Stein CA (2000) Intracellular mRNA cleavage induced through activation of RNase P by nuclease-resistant external guide sequences. Nat Biotechnol 18:58–61

Ma MY, Jacob-Samuel B, Dignam JC, Pace U, Goldberg AR, George ST (1998) Nuclease-resistant external guide sequence-induced cleavage of target RNA by human ribonuclease P. Antisense Nucleic Acid Drug Dev 8:415–426

Mann H, Ben-Asouli Y, Schein A, Moussa S, Jarrous N (2003) Eukaryotic RNase P: role of RNA and protein subunits of a primordial catalytic ribonucleoprotein in RNA-based catalysis. Mol Cell 12:925–935

Mautino MR (2002) Lentiviral vectors for gene therapy of HIV-1 infection. Curr Gene Ther 2:23–43

McClain WH, Guerrier-Takada C, Altman S (1987) Model substrates for an RNA enzyme. Science 238:527–530

McKinney J, Guerrier-Takada C, Wesolowski D, Altman S (2001) Inhibition of Escherichia coli viability by external guide sequences complementary to two essential genes. Proc Natl Acad Sci U S A 98:6605–6610

McKinney J, Guerrier-Takada C, Galan J, Altman S (2002) Tightly regulated gene expression system in Salmonella enterica serovar Typhimurium. J Bacteriol 184:6056–6059

McKinney JS, Zhang H, Kubori T, Galan JE, Altman S (2004) Disruption of type III secretion in Salmonella enterica serovar Typhimurium by external guide sequences. Nucleic Acids Res 32:848–854

Plehn-Dujowich D, Altman S (1998) Effective inhibition of influenza virus production in cultured cells by external guide sequences and ribonuclease P. Proc Natl Acad Sci U S A 95:7327–7332

Reddy R, Henning D, Das G, Harless M, Wright D (1987) The capped U6 small nuclear RNA is transcribed by RNA polymerase III. J Biol Chem 262:75–81

Sarver N, Cantin EM, Chang PS, Zaia JA, Ladne PA, Stephens DA, Rossi JJ (1990) Ribozymes as potential anti-HIV-1 therapeutic agents. Science 247:1222–1225

Scherer LJ, Rossi JJ (2003) Approaches for the sequence-specific knockdown of mRNA. Nat Biotechnol 21:1457–1465

Sharin E, Schein A, Mann H, Ben-Asouli Y, Jarrous N (2005) RNase P: role of distinct protein cofactors in tRNA substrate recognition and RNA-based catalysis. Nucleic Acids Res 33:5120–5132

Stein CA, Cheng YC (1993) Antisense oligonucleotides as therapeutic agents – is the bullet really magical? Science 261:1004–1012

Sullenger BA, Cech TR (1993) Tethering ribozymes to a retroviral packaging signal for destruction of viral RNA. Science 262:1566–1569

Trang P, Hsu AW, Liu F (1999) Nuclease footprint analyses of the interactions between RNase P ribozyme and a model mRNA substrate. Nucleic Acids Res 27:4590–4597

Trang P, Kilani AF, Kim J, Liu F (2000a) A ribozyme derived from the catalytic subunit of RNase P from *Escherichia coli* is highly effective in inhibiting replication of herpes simplex virus 1. J Mol Biol 301:817–826

Trang P, Lee M, Nepomuceno E, Kim J, Zhu H, Liu F (2000b) Effective inhibition of human cytomegalovirus gene expression and replication by a ribozyme derived from the catalytic RNA subunit of RNase P from Escherichia coli. Proc Natl Acad Sci U S A 97:5812–5817

Trang P, Hsu A, Zhou T, Lee J, Kilani AF, Nepomuceno E, Liu F (2002) Engineered RNase P ribozymes inhibit gene expression and growth of cytomegalovirus by increasing rate of cleavage and substrate binding. J Mol Biol 315:573–586

Tuerk C, Gold L (1990) Systematic evolution of ligands by exponential enrichment: RNA ligands to bacteriophage T4 DNA polymerase. Science 249:505–510

Verma S, Eckstein F (1998) Modified oligonucleotides: synthesis and strategy for users. Annu Rev Biochem 67:99–134

Warashina M, Kuwabara T, Kato Y, Sano M, Taira K (2001) RNA-protein hybrid ribozymes that efficiently cleave any mRNA independently of the structure of the target RNA. Proc Natl Acad Sci U S A 98:5572–5577

Werner M, Rosa E, Nordstrom JL, Goldberg AR, George ST (1998) Short oligonucleotides as external guide sequences for site-specific cleavage of RNA molecules with human RNase P. RNA 4:847–855

Xiao G, Lundblad EW, Izadjoo M, Altman S (2008) Inhibition of expression in Escherichia coli of a virulence regulator MglB of Francisella tularensis using external guide sequence technology. PLoS ONE 3:e3719

Xiao S, Scott F, Fierke CA, Engelke DR (2002) Eukaryotic ribonuclease P: a plurality of ribonucleoprotein enzymes. Annu Rev Biochem 71:165–189

Yang YH, Li H, Zhou T, Kim K, Liu F (2006) Engineered external guide sequences are highly effective in inducing RNase P for inhibition of gene expression and replication of human cytomegalovirus. Nucleic Acids Res 34:575–583

Yu M, Ojwang J, Yamada O, Hampel A, Rapapport J, Looney D, Wong-Staal F (1993) A hairpin ribozyme inhibits expression of diverse strains of human immunodeficiency virus type 1. Proc Natl Acad Sci U S A 90:6340–6344

Yu X, Trang P, Shah S, Atanasov I, Kim YH, Bai Y, Zhou ZH, Liu F (2005) Dissecting human cytomegalovirus gene function and capsid maturation by ribozyme targeting and electron cryomicroscopy. Proc Natl Acad Sci U S A 102(20):7103–7108

Yuan Y, Hwang ES, Altman S (1992) Targeted cleavage of mRNA by human RNase P. Proc Natl Acad Sci U S A 89:8006–8010

Yuan Y, Altman S (1994) Selection of guide sequences that direct efficient cleavage of mRNA by human ribonuclease P. Science 263:1269–1273

Zaug AJ, Cech TR (1995) Analysis of the structure of Tetrahymena nuclear RNAs in vivo: telomerase RNA, the self splicing rRNA intron, and U2 snRNA. RNA 1:363–374

Zhang H, Altman S (2004) Inhibition of the expression of the human RNase P protein subunits Rpp 21, Rpp25, Rpp29 by external guide sequences (EGSs) and siRNA. J Mol Biol 342:1077–1083

Zhou T, Kim J, Kilani AF, Kim K, Dunn W, Jo S, Nepomuceno E, Liu F (2002) In vitro selection of external guide sequences for directing RNase P-mediated inhibition of viral gene expression. J Biol Chem 277:30112–30120

Zhu J, Trang P, Kim K, Zhou T, Deng H, Liu F (2004) Effective inhibition of Rta expression and lytic replication of Kaposi's sarcoma-associated herpesvirus by human RNase P. Proc Natl Acad Sci U S A 101:9073–9078

Zou H, Lee J, Umamoto S, Kilani AF, Kim J, Trang P, Zhou T, Liu F (2003) Engineered RNase P ribozymes are efficient in cleaving a human cytomegalovirus mRNA in vitro and are effective in inhibiting viral gene expression and growth in human cells. J Biol Chem 278:37265–37274

Zou H, Lee J, Kilani AF, Kim K, Trang P, Kim J, Liu F (2004) Engineered RNase P ribozymes increase their cleavage activities and efficacies in inhibiting viral gene expression in cells by enhancing the rate of cleavage and binding of the target mRNA. J Biol Chem 279:32063–32070

Zu Putlitz J, Yu Q, Burke JM, Wands JR (1999) Combinatorial screening and intracellular antiviral activity of hairpin ribozymes directed against hepatitis B virus. J Virol 73:5381–5387

Zuker M (2003) Mfold web server for nucleic acid folding and hybridization prediction. Nucleic Acids Res 31:3406–3415

Index